Lecture Notes in Economics and Mathematical Systems

581

[edited by]
Kurt Marti · Yuri Ermoliev
Marek Makowski · Georg Pflug

Coping with Uncertainty

Modeling and Policy Issues

With 56 Figures
and 23 Tables

  Springer

Prof. Dr. Kurt Marti
Federal Armed Forces University Munich
Aero Space Engineering and Technology
85577 Neubiberg/Munich, Germany
E-mail: kurt.marti@unibw-muenchen.de

Prof. Dr. Yuri Ermoliev
International Institute for Applied System Analysis
Schloßplatz 1
2361 Laxenburg, Austria
E-mail: ermoliev@iiasa.ac.at

Dr. Marek Makowski
International Institute for Applied System Analysis
Schloßplatz 1
2361 Laxenburg, Austria
E-mail: marek@iiasa.ac.at

Prof. Dr. Georg Pflug
University of Vienna
Institute for Statistics and Informatics
Universitätsstrasse 5
1010 Vienna, Austria
E-mail: georg.pflug@univie.ac.at

ISBN-10 3-540-35258-9   Springer Berlin Heidelberg New York
ISBN-13 978-3-540-35262-4   Springer Berlin Heidelberg New York

Springer is a part of Springer Science+Business Media
springeronline.com

© Springer-Verlag Berlin Heidelberg 2006
Printed in Germany

Typesetting: Camera ready by author
Cover: Erich Kirchner, Heidelberg
Production: LE-TEX, Jelonek, Schmidt & Vöckler GbR, Leipzig

SPIN 11776215   Printed on acid-free paper – 88/3100 – 5 4 3 2 1 0

Preface

Uncertainties and risks have always been, and will remain, present; both in everybody's life and in policy making. This presence is not always recognized because humans tend to perceive the world in a deterministic way, and in most cases succeed in *somehow* dealing with uncertainties. However, ignoring uncertainties often results in serious problems that could at least be mitigated if the uncertainties were properly handled. The importance of the proper treatment of uncertainties is growing because the consequences of inadequate treatments are more and more costly, both in social and environmental terms. This is caused by the quickly changing world where one of the dominating driving forces is efficiency, which has led to globalization, increased interdependencies amongst more and more diversified socio-economic, technological and environmental systems, a reduction in many safety (both technological and social) margins, a concentration of assets in risk prone areas, and other factors which progressively contribute to the increasing vulnerability of the societies.

These ongoing global changes create fundamentally new scientific problems which require new concepts, methods, and tools. A key issue concerns a vast variety of practically irreducible uncertainties, which challenge our traditional models and require new concepts and analytical tools. This uncertainty critically dominates a number of policy making issues, e.g., the climate change debates. In short, the dilemma is concerned with enormous costs vs massive uncertainties of potentially extreme and/or irreversible impacts.

Traditional scientific approaches usually rely on real observations and experiments. Yet no sufficient observations exist for new problems, and "pure" experiments and learning by doing may be very expensive, dangerous, or simply impossible. In addition, available historical observations are often contaminated by "experimentator", i.e., our actions, policies. The complexity of new problems does not allow to achieve enough certainty just by increasing the resolution of models or by bringing in more links. They require explicit treatment of uncertainties using "synthetic" information composed of available "hard" data from historical observations, results of possible experiments, and scientific facts as well as "soft" data from experts' opinions, scenarios, stakeholders, and public opinion. As a result of all of these factors, our assessment will always have poor estimates. Finally, the role of science for new problems will increasingly deviate from traditional "deterministic predictions" analysis to the design of robust strategies against involved uncertainties and risks.

Addressing the new challenges of the proper treatment of uncertainties was the main aim of the workshop *Coping with Uncertainty* held at the International Institute for Applied Systems Analysis (IIASA), Laxenburg, Austria, on December 13-16, 2004. The workshop provided researchers and practitioners from different areas with an interdisciplinary forum for discussing various ways of dealing with uncertainties in various areas, including environmental and social sciences, economics, policy-making, management, and engineering. Presentations were prepared for an interdisciplinary audience, and addressed: open problems, limitations of known approaches, novel methods and techniques, or lessons from the applications of various approaches. The workshop contributed to a better understanding between practitioners dealing with the management of uncertainty, and scientists working on either corresponding modeling approaches that can be applied for improving understanding or management of uncertainty. In particular, the workshop focused on the following issues:

- modeling different types of uncertainty (probabilistic and non-probabilistic),
- the formulation of appropriate deterministic substitute problems for different types of uncertainty,
- the robustness of optimal solutions with respect to uncertainties,
- relevant solution techniques and approximation methods,
- open problems in the adequate treatment of uncertainties,
- concrete applications in economics, finance, engineering, energy, population, air quality, climate change, ecology, forestry, and other environmental problems.

The workshop was organized jointly by:

- IIASA - International Institute for Applied Systems Analysis, Laxenburg, Austria;
- Federal Armed Forces University Munich, Germany;
- Department of Statistics and Decision Support, University of Vienna, Austria.

The scientific Program Committee included: Yuri Ermoliev, IIASA, Laxenburg (A); Sjur Flam, University of Bergen (N); Leen Hordijk, IIASA, Laxenburg (A); Peter Kall, University of Zürich (CH); Marek Makowski, IIASA, Laxenburg (A); Kurt Marti, Federal Armed Forces University Munich (D); Georg Pflug, University of Vienna (A); Gerhard I. Schuëller, University of Innsbruck (A); and Roger Wets, University of California (USA).

The organizers gratefully acknowledge the support of:

- BMBF - the Federal Ministry of Education and Research, Germany;
- GAMM - International Association of Applied Mathematics and Mechanics;
- IFIP - International Federation for Information Processing;
- ÖFG - Österreichische Forschungsgemeinschaft.

Their generous support enabled the participation of many researchers who otherwise could not have attended the Workshop.

This volume is composed of chapters based on selected presentations from the *Coping with Uncertainty* Workshop. The contributions are organized into five parts:

1. *Uncertainty and decisions* presents the methodology for finding robust decisions, introduces new approaches to the evaluation of extreme risks, and discusses the opportunities of applying structured modeling technology to modeling endogenous uncertainties.
2. *Modeling stochastic uncertainty* presents different approaches to modeling structural reliabilities and catastrophic events, and discusses probabilistic modeling of scenes with Bayesian networks.
3. *Non-probabilistic uncertainty* deals with spatial and social heterogeneity, and discusses the applicability of the downscaling methods in economy and land-use; it also presents different approaches to addressing uncertainties in control processes, and their applications in industry and medicine.
4. *Applications of stochastic optimization* discusses the uncertainty impacts on sustainable energy development and climate change, in energy system planning, and in project planning; it also considers algorithmic issues of nonlinear stochastic programming.
5. *Policy issues under uncertainty* deals with the role of learning in the treatment of endogenous risks in climate change policy making, and risk pricing in related projects; finally, it discusses the public willingness to accept the costs of averting uncertain dangers.

We express our gratitude to all referees, and we thank all authors for the timely delivery of the final version of their contributions to this volume. Furthermore, we thank Ms Elisabeth Lößl of the Federal Armed Forces University Munich for her support in the preparation of this volume. Finally, we thank Springer-Verlag for including the Proceedings in the Springer Lecture Notes Series "LNEMS".

Kurt Marti, Munich   July, 2006
Yuri Ermoliev, Laxenburg
Marek Makowski, Laxenburg
Georg Pflug, Vienna

Contents

Part I. Uncertainty and Decisions

Facets of Robust Decisions 3
Y. Ermoliev, L. Hordijk
1 Introduction 3
2 Concepts of Robustness 5
3 Decision Problems Under Uncertainty 7
4 Uncertainty Modeling 10
5 Robust Stochastic Optimization 14
6 Temporal, Spatial and Social Heterogeneities 18
7 STO Methods for Robust Solutions 20
8 Sensitivity of Robust Strategies 22
9 Concluding Remarks 25
References 26

Stress Testing via Contamination 29
J. Dupačová
1 Introduction 29
2 Motivation: Stochastic Dedicated Bond Portfolio Management 31
3 Contamination and Stress Testing 32
4 Contamination and Stress Testing for CVaR 39
5 Conclusions 44
References 45

**Structured Modeling for Coping with Uncertainty in Complex
Problems** 47
M. Makowski
1 Introduction 47
2 Context 49
3 Uncertainty and Risk 50
4 Structured Modeling 54
5 Conclusions 60
References 61

Part II. Modeling Stochastic Uncertainty

**Using Monte Carlo Simulation to Treat
Physical Uncertainties in Structural Reliability** 67
D. C. Charmpis, G. I. Schuëller
1 Introduction 67

2   Direct Monte Carlo Simulation 69
3   Variance Reduction Techniques 70
4   Computational Efficiency Issues.............................. 74
5   Applications.. 76
6   Closing Remarks 81
References 82

**Explicit Methods for the Computation of Structural
Reliabilities in Stochastic Plastic Analysis** 85
I. Kaymaz, K. Marti
1   Stochastic Plasticity Analysis................................. 86
2   First Order Reliability Method (FORM) 90
3   Response Surface Method 91
4   Examples 96
5   Conclusions 102
References 102

Statistical Analysis of Catastrophic Events 105
J. L. Teugels, B. Vandewalle
1   WMO-Release 695 105
2   Extreme Value Statistics 106
3   SWISS-RE Casualties Table 112
4   Concluding Remarks 116
References 117

Scene Interpretation Using Bayesian Network Fragments 119
P. Lueders
1   Introduction... 119
2   Representation 120
3   Scene Interpretation 125
4   Related Work 128
5   Conclusion 128
References 129

Part III. Non-Probabilistic Uncertainty

**General Equilibrium Models with Discrete Choices
in a Spatial Continuum** 133
M. Keyzer, Y. Ermoliev, V. Norkin
1   Introduction... 133
2   The Continuum of Agents:
    Distribution of Spatial and Social Characteristics 136
3   Producer Behavior 139
4   Consumer Behavior 142
5   Existence of a Competitive Equilibrium....................... 146

6   Spatial Welfare Optimum: a Dual Approach 147
7   Deterministic Versus Stochastic Welfare Tâtonnement 151
References 153

Sequential Downscaling Methods
for Estimation from Aggregate Data 155
G. Fischer, T. Ermolieva, Y. Ermoliev, H. Van Velthuizen
1   Introduction 155
2   Downscaling Problems: Motivating Examples 157
3   Sequential Downscaling Methods 160
4   Minimax Likelihood and Maximum Entropy 163
5   Practical Applications 165
6   Concluding Remarks 167
References 168

Optimal Control for a Class of Uncertain Systems 171
F.L.Chernousko
1   Introduction 171
2   Statement of the Problem 172
3   Transformations 173
4   Solution of the Problem 174
5   Linear-Quadratic Performance Index 177
6   Examples 177
7   Conclusions 182
References 183

Uncertainties in Medical Processes Control 185
A.G.Nakonechny, V.P.Marzeniuk
1   Introduction 185
2   Generalized Pathologic Process 186
3   Simplified Model 189
4   Uncertainties 190
5   Conclusions 192
References 192

Part IV. Applications of Stochastic Optimization

Impacts of Uncertainty and Increasing Returns on Sustainable
Energy Development and Climate Change:
A Stochastic Optimization Approach 195
A. Gritsevskyi, H.-H. Rogner
1   Introduction 195
2   Modeling Approach and a Motivating Example 197
3   Model Structure 201
4   Implementation 209

5   Concluding Observations 211
References 214

Stochasticity in Electric Energy Systems Planning 217
A. Ramos, S.Cerisola, Á. Baíllo, J. M. Latorre
1   Introduction 217
2   Uncertainty Impact 218
3   Estimation Methods 220
4   Decision Making Methods 222
5   Characteristic Models 224
6   Conclusions 238
References 238

Stochastic Programming Based PERT Modeling 241
A. Gouda, D. Monhor, T. Szántai
1   Introduction 241
2   The Stochastic Programming Model of PERT 242
3   Numerical Results 243
References 254

Towards Implementable Nonlinear Stochastic Programming . . 257
L. Sakalauskas
1   Introduction 257
2   Stochastic Differentiation and Monte-Carlo Estimators 259
3   Statistical Verification of the Optimality Hypothesis 262
4   Optimization by Monte-Carlo Estimators with
    Sample Size Regulation 265
5   Numerical Study of Algorithms 271
6   Discussion and Conclusions 277
References 278

Part V. Policy Issues Under Uncertainty

**Endogenous Risks and Learning
in Climate Change Decision Analysis** 283
B. O'Neill, Y. Ermoliev, T. Ermolieva
1   Introduction 283
2   Endogenous Climate Change Risk: A General Model 285
3   A Basic Model with Linear Cost Functions 288
4   Extensions to the Basic Model 292
5   A Dynamic Stabilization Problem 296
6   Concluding Remarks 298
References 299

Pricing Related Projects 301
S. D. Flam,, H. I. Gassmann
1 Introduction... 301
2 Pooling of Single-Stage Linear Problems Subject to Uncertainty ... 303
3 Pricing of Linear Investment Projects........................... 306
4 Project Portfolios and Core Solutions........................... 307
5 Stochastic Production Games................................. 311
6 Concluding Remarks 311
References 312

**Precaution: The Willingness to Accept Costs
to Avert Uncertain Danger** 315
C. Weiss
1 Introduction... 315
2 Scales of Subjective Uncertainty 316
3 The Precautionary Principle and the Willingness to Incur Costs ... 318
4 A Proposed Principle of Innovation and Adaptive Management 322
5 Conclusion: A Framework for Balanced Precaution 324

Part I

Uncertainty and Decisions

Facets of Robust Decisions

Y. Ermoliev and L. Hordijk

Institute for Applied Systems Analysis, Laxenburg, Austria

Abstract. The aim of this paper is to provide an overview of existing concepts of robustness and to identify promising directions for coping with uncertainty and risks surrounding on-going global changes. Unlike statistical robustness, general decision problems may have rather different facets of robustness. In particular, a key issue is the sensitivity of decisions with respect to low-probability catastrophic events. That is, robust decisions in the presence of catastrophic events are fundamentally different from decisions ignoring them. Specifically, proper treatment of extreme catastrophic events requires new sets of feasible decisions, adjusted to risk performance indicators, and new spatial, social and temporal dimensions. The discussion is deliberately kept at a level comprehensible to a broad audience through the use of simple examples that can be extended to rather general models. In fact, these examples often illustrate fragments of models that are being developed at IIASA.

Key words: Robustness, decisions, uncertainty, stochastic optimization, discounting, downscaling, catastrophe modeling, extreme events, simulation.

1 Introduction

An alarming global tendency is the increasing vulnerability of our society. A thorough scientific policy analysis of related socio-economic, technological and environmental global change processes raises new methodological problems that challenge traditional approaches and demonstrate the need for new methodological developments. A key issue is the vast variety of inherent, practically irreducible uncertainties and "unknown" risks that may suddenly affect large territories and communities [13], [14], [21]. Traditional approaches usually rely on real observations and experiments. Yet, there are no adequate observations for new problems, responses of involved processes may have long term delays, and learning-by-doing experiments may be very expensive, dangerous, or simply impossible.

Large-scale catastrophic impacts and the magnitudes of the uncertainties that surround them particularly dominate the climate-change policy debates [2],[7], [8], [30], [34], [35], [37], [49]. The exact evaluation of overall global climate changes and vulnerability requires not only a prediction of the climate system, but also an evaluation of endogenous socioeconomic, technological, and environmental processes and risks. The main issue is the lack of historical data on potential irreversible changes occurring on large spatial, temporal, and social scales. The inertia of the overall climate change system, delayed responses, and the possibility of abrupt catastrophic changes [2] restricts purely adaptive wait-and-see approaches. Moreover, extreme events of heavy consequences playing such a decisive role are, on average, evaluated as improbable events during a human lifetime. A 500-year disaster (e.g., an extreme flood that occurs on average once in 500 years) may, in fact, occur next year.

The evaluation of complex heterogeneous global-change processes on "average" can be dramatically misleading. However, it is impossible to research all the details connected with these processes in order to achieve evaluations required by the traditional models in economics, insurance, risk-management, and extreme value theory. For example, standard insurance theory essentially relies on the assumption of independent, frequent, low-consequence (conventional) risks, such as car accidents, for which decisions on premiums, claims estimates and the likelihood of insolvency can be calculated via rich historical data. Existing extremal value theory [11] deals primarily with the maximum of also independent variables quantifiable by a single number (e.g., money). Catastrophes are definitely not quantifiable events in this sense. They have different patterns, spatial and temporal dimensions and induce heterogeneity of losses and gains which exclude the use of average characteristics. Globally, an average resident may even benefit from some climate-change scenarios, while some regions may be simply wiped out.

Under inherent uncertainty and heterogeneity of global processes the role of global change models rests on the ability to guide comparative analysis of the feasible decisions. Although exact evaluations are impossible, the preference structure among decisions can be a stable basis for a relative ranking of alternatives in order to design robust policies. As we know, finding out which of two parcels is the heavier without having the exact measurements is easier than saying how much heavier that parcel is.

Sections 2 and 3 analyze the known concepts of robustness in statistics, deterministic control theory and classical optimization. Global change decision problems call for new approaches. Sections 3 and 4 show that, contrary to the standard expected utility maximization, stochastic optimization (STO) models allow in a natural manner to represent different endogenous uncertainties and risks, spatial and temporal dependencies, equity constraints and abrupt changes. The ability of STO models to incorporate both anticipative ex-ante and adaptive ex-post decisions induces risk aversion among ex-ante decisions that implicitly depends on input data and goals and that practically cannot be characterized by an exogenous standard utility function [18]. In particular, even in the simplest linear model (Example 4), the co-existence of ex-ante and ex-post decisions induces VaR and CVaR type risk measures. Section 4 also indicates the misleading character of average characteristics, e.g., hazard maps, which are often used in the analysis of spatial exposures and vulnerability. This emphasizes the importance of distributional aspects, and the use of quantiles instead of average values. Unfortunately, the straightforward application of quantiles destroy additivity and concavity (convexity) of models and it makes the applicability of standard decomposition schemes problematic (Example 2). Section 5 introduces concepts of STO robustness. In particular, it shows that models with quantiles can be equivalently substituted by specific STO models preserving concavity (convexity). Section 6 emphasizes the role played by downscaling and catastrophe modeling to represent spatial and temporal distributions and vulnerability. Section 7 outlines the main ideas behind STO methods, especially, fast adaptive Monte Carlo optimization procedures which can be incorporated into catastrophe models and vulnerability analysis in order to evaluate robust strategies. Section 8 discusses the sensitivity of robust strategies with respect to extreme events. It introduces the concept of a stopping time which allows to focus the analysis on the most distractive potential extreme events (random scenarios). Combined with the catastrophe modeling, this concept opens up new approaches to spatio-temporal discounting in the presence of extreme events. Section 9 provides concluding remarks.

2   Concepts of Robustness

2.1   Statistical Robustness

The term "robust" was introduced into statistics in 1953 by Box and acquired recognition only after the publication of a path-breaking paper by Huber [28] in 1964. As Huber admits, researchers had long been concerned with the sensitivity of standard estimation procedures to "bad" observations (outliers), and the word "robust" was loaded with many, sometimes inconsistent connotations, frequently for the simple reason of confering respectability on it. According to Huber ([28], pp. 5, 6), "... any statistical procedure ... should be robust in the sense that small deviations from the model assumptions should impair the performance only slightly ..." This concept of robustness, in fact, corresponds to standard mathematical ideas of continuity: when disturbances become small, the performance of the perturbed and initial models also deviate slightly.

2.2   Bayesian Robustness

The concept of robustness was also introduced into Bayesian statistics [29] primarily as the insensitivity of statistical decisions to the uncertainty of prior probability distribution. A Bayesian sampling model P is often parameterized by a vector θ of unknown parameters. Let ξ be the observable random variables from P with true unknown parameters $\theta = \theta^*$ that have to be recovered from observations of ξ. In contrast to classical statistical models, it is assumed that there is a prior (probability) distribution $\pi(\cdot)$ characterizing the degree of beliefs about true vector θ^*, which in the presence of new information is updated by the Bayesian rule. In this case a statistical decision (estimate) about the true parameter θ^* can be characterized by an expected distance (loss function) $EL(x, \theta) = \int L(x, \theta)\pi(d\theta)$ from x to admissible θ. The efficiency of x is calculated by the posterior expected distance

$$E(L(x, \theta|\xi) = \int L(x, \theta)\pi(d\theta|\xi), \pi(d\theta|\xi) = \frac{P(\xi|\theta)\pi(d\theta)}{\int P(\xi|\theta)\pi(d\theta)}, \qquad (1)$$

where ξ is a given sample of data from $P(\xi|\theta^*)$. Bayesian robustness is characterized by the range of posterior expected distance, as the prior $\pi(\cdot)$ varies over the elicited class $\mathcal{P}$. An alternative approach is to choose a hyper-prior on the class of $\mathcal{P}$ and the standard Bayesian model.

2.3   Non-Bayesian Minimax Robustness

A probabilistic minimax robustness [29] consists of choosing x with respect to a worst-case distribution: minimize $\max_{\pi \in P} \int d(x, \theta)\pi(d\theta|\xi)$. This type of minimax ranking of x does not correspond to the Bayesian ranking w.r.t. a single distribution in $\mathcal{P}$. The worst-case distribution $\pi \in \mathcal{P}$ depends on x and ξ, i.e., it is a random endogenous distribution. Besides the probabilistic robustness we can distinguish also a stochastic minimax robustness (Section 5.5).

2.4   Deterministic Control Theory

As statistical robustness is similar to the local stability of dynamic systems, the robustness in deterministic control theory [43] was introduced as an additional requirement on the stability of optimal trajectories. In other words, additional constraints were introduced in the form of a stability criterion.

2.5   Robust Deterministic Optimization

Optimization theory provides tools for analyzing and solving various decision making problems. A standard deterministic problem is formulated as the maximization (minimization) of a function $f_0(x, \omega)$ subject to constraints $f_i(x, \omega) \geq 0$, $i = 0, 1, ..., m$, where $x = (x_1, ..., x_n)$ is a vector of decisions and ω are fixed variables characterizing the structure of the model, including the input data. Functions $f_i(x, \omega)$, $i = 0, 1, ..., m$, are assumed to be exactly known and analytically tractable, and ω belongs to an explicitly given set Ω of admissible scenarios, $\omega \in \Omega$. Robustness is defined [4] as the maximization of $\min_{\omega \in \Omega} f_0(x, \omega)$ over solutions x that satisfy all admissible values of uncertainty $f_i(x, \omega) \geq 0$, $i = 1, ..., m$, $\omega \in \Omega$. The set Ω is often characterized by a finite number of scenarios or simple sets such as intervals or ellipsoidal uncertainty $\Omega = \{\alpha_l + \sum_k \delta_{lk}\omega_k : \sum_k \omega_k^2 \leq 1\}$. These sets, in a sense, attempt to substitute for normal probability distributions in a simple but inconsistent with statistical analysis manner, which can be misleading (Section 4). It is clear that this type of deterministic worst-case robustness leads to extremely conservative decisions.

3   Decision Problems Under Uncertainty

Statistical decision theory deals with situations in which the model of uncertainty and the optimal solution are defined by a sampling model with an unknown vector of "true parameters" θ^*. Vector θ^* defines the desirable optimal solution, its performance can be observed from the sampling model and the problem is to recover θ^* from these data. Potential estimates of θ^* define feasible solutions x of the statistical decision problem. It is essential that x does not affect the sampling model so that the optimality and robustness of solutions can be evaluated by posterior distance (1).

The general problems of decision making under uncertainty deal with fundamentally different situations. The model of uncertainty, feasible solutions, and performance of the optimal solution are not given and all of these have to be characterized from the context of the decision making situation, e.g., socioeconomic, technological, environmental, and risk considerations. As there is no information on true optimal performance, robustness cannot be also characterized by a distance from observable true optimal performance. Therefore, the general decision problems, as the following Sections illustrate, may have rather different facets of robustness.

3.1   Expected Utility Maximization

Standard policy analysis, as a rule, uses a utility (disutility) maximization (minimization) model for the evaluation of desirable decisions. In the presence of uncertainty, any related decision x results in multiple outcomes characterized by functions $g_1(x, \omega)$, ..., $g_K(x, \omega)$ such as costs, benefits, damages, and risks, as well as indicators of fairness, equity, and environmental impacts. They depend on x, $x \in R^n$ and uncertainty from a set of admissible scenarios Ω, $\omega \in \Omega$.

A given decision x for different scenarios ω may have rather contradictory outcomes. In 1738 the mathematician Daniel Bernoulli introduced the concept of expected utility maximization as a rule for choosing decisions under multiple outcomes. It is assumed that all outcomes $g_1(x, \omega)$, ..., $g_K(x, \omega)$ can be converted in a single index of preferability $q(x, \omega)$, say, a monetary payoff. The standard expected utility model suggests that the choice of decision x maximizing an expected utility function $U(x) = Eu(q(x, \omega)) = \int u(q(x, \omega))P(d\omega)$, where $u(\cdot)$ is a utility associated with an aggregate outcome $q(x, \omega)$. The shape of u defines attitudes to risks. This model presupposes that, in addition to the knowledge of Ω, one can rank the alternative scenarios ω according to weights - objective or subjective probability measure P. The use of a probability measure as a degree of belief was formalized by Ramsey (1926). Savage (1954) published a thorough treatment of expected utility maximization based on subjective probability as a degree of belief (see discussion in [24]). As a result of this work the use of probability measure became a standard approach for modeling uncertainty by using "hard" observations and soft public and expert opinions in a consistent way within a single model.

3.2   Stochastic Optimization (STO) Model

The shortcomings of the expected utility maximization model are well known. Generally speaking, it is practically impossible to find a utility function that enables the aggregation of various attributes in one preferability index, including attitudes to different risks, the distributional aspects of gains and losses, the rights of future generations, and responsibilities for environmental protection. It is natural that, for complex problems, nonsubstitutable indicators should exist that have to be controlled separately in the same way as indicators of, say, health (e.g., temperature and blood pressure). Moreover, it is often practically impossible to identify exactly subjective (and objective) probability as a degree of beliefs. Most people cannot clearly distinguish [45] between probability ranging roughly from 0.3 to 0.7. Decision analysis often has to rely [9] on imprecise statements, for example, that event e_1 is more probable than event e_2, or that the probability p_1, p_2 of event e_1 or of event e_2 is greater than 50 percent and less than 90 percent. These statements may be represented by inequalities such as $p_1 \geq p_2$, $0.5 \leq p_1 + p_2 \leq 0.9$. A number of models with imprecise probabilities have been suggested (see, e.g., [44]) and these models were later integrated into classical probability theory.

The expected utility model is a specific case of STO [6], [20], [33], [40], [46] models that use various performance indicators $f_i(x, \omega)$, $i = 1, ..., m$, one of which can be the expected utility (disutility). These indicators depend on outcomes $g_k(x, \omega)$, $k = 1, ..., K$, on x and ω, i.e., $f_i(x, \omega) := q_i(g_1, ..., g_k, x, \omega)$. A rather general STO problem is formulated as the maximization (minimization) of the expectation function $F_0(x) = Ef_0(x, \omega) = \int f_0(x, \omega)P(d\theta)$, subject to constraints $F_i(x) = Ef_i(x, \omega) = \int f_i(x, \omega)P(d\theta) \geq 0$, $i = 1, ..., m$. The choice of proper indicators $f_i(x, \omega)$ and outcomes $g_k(x, \omega)$, $k = 1, ..., K$, is essential for the robustness of x. Globally or regionally aggregated outcomes are less uncertain but they may not reveal potentially dramatic heterogeneities induced by global changes on individuals, governments, and the environment. For instance, an aggregate income or growth indicators may not reveal an alarming gap between poor and rich regions, which may cause future instabilities. By choosing appropriate functions $g_k(x, \omega)$ and $f_i(x, \omega)$, STO models allow in a natural and flexible way to represent various risks, spatial, social, and temporal heterogeneities, and the sequential resolution of uncertainty in time. Often, as in Example 1, $f_i(x, \omega)$ are analytically intractable, non-smooth, and even discontinuous functions [17], and probability measure P is unknown, or only partially known, and may depend on x (Section 5, 6), which is essential for modeling endogenous catastrophic risks, e.g., due to increasing returns leading to concentrations of values in risk prone areas. Moreover, decisions x can be composed of anticipative ex-ante and adaptive ex-post components, which allows to model dynamic decision making processes with flexible adaptive adjustments of decisions when new information is revealed. The main challenge confronted by STO theory is that it is practically impossible in general to evaluate exact values of $F_i(x)$, $i = 0, 1, ..., m$, see, e.g., Example 1. As "deterministic" is a degenerated case of "stochastic", STO methods allow to deal with problems which are not solved by standard deterministic methods.

Example 1. Safety constraints: pollution control problems. A common feature of most models used in designing pollution-control policies [1] is the use of transfer coefficients a_{ij} that link the amount of pollution x_j emitted by source j to the pollution concentrations $g_i(x, \omega)$ at the receptor location i as $g_i(x, \omega) = \sum_{j=1}^{n} a_{ij}x_j$, $i = 0, 1, ..., m$. The coefficients are often computed with Gaussian type diffusion equations. These equations are solved over all possible meteorological conditions, and the outputs are then weighted by the frequencies of meteorological inputs over a given time interval, yielding average transfer coefficients a_{ij}. Deterministic models ascertain cost-effective emission strategies x_j, $j = 1, ..., n$ subject to achieving exogenously specified environmental goals, such as ambient average standard b_i at receptors $i = 1, ..., m$. These models can be improved by the inclusion of safety constraints that account for the random nature of coefficients a_{ij} and ambient standards b_i to reduce impacts of extreme events:

$$F_i(x) = Prob[\sum_{j=1}^{n} a_{ij}x_j \leq b_i] \geq p_i, i = 1, ...m, \tag{2}$$

namely, the probability that the deposition level in each receptor (region, grid, or country) i will not exceed uncertain critical load (threshold) b_i at a given probability (acceptable safety level) p_i.

Ignorance of risks defined by constraints (2) may cause irreversible catastrophic events. Although an average daily concentration of a toxicant in a lake is far below a vital threshold, real concentrations may exceed this threshold for only a few minutes and yet be enough to kill off fish. Constraints of the type (2) are important for the regulation of stability in the insurance industry, known as the insolvency constraints. The safety regulation of nuclear reactors requires $p_i = 1 - 10^{-7}$, i.e., a major failure occurs on average only once in 10^7 years. Stochastic models do not, however, exclude the possibility that a disaster may occur next year.

Remark 1. The constraints (2) are known as chance constraints [6], [20], [33], [40]. They can be written in the form of the standard STO model with discontinuous functions: $f_j(x, \omega) = 1 - p_i$ if $\sum_{j=1}^{n} a_{ij}x_j - b_i \leq 0$ and $f_j(x, \omega) = -p_i$, otherwise. If $p_i = 1$, $i = 1, ..., m$, the constraints (2) are reduced to constraints of deterministic robustness (Section 2.5).

The main computational complexity confronted by STO methods is the lack of explicit analytical formulas for goal functions $F_i(x)$, $i = 0, 1, , m$. For example, consider constraints (2). If there is a finite number of possible scenarios $\omega = (a_{ij}, b_i, i = \overline{1, m}, j = \overline{1, n})$ reflecting, say, prevailing weather conditions, then $F_i(x)$ are discontinuous piecewise constant functions, i.e., gradients of $F_i(x)$ are 0 almost everywhere. Hence, the straightforward conventional optimization methods cannot be used. Yet, it is possible to transform them into well defined convex (concave) optimization problems ([14], [18], Sections 5.2, 5.3)

4   Uncertainty Modeling

As discussed in Section 3, traditional statistical decision theory deals with situations where the model of uncertainty and the performance of optimal solution are given by a sampling model. In general decision problems the uncertainty, decisions and interactions among them have to be characterized from the context of the decision making situation.

Any relevant decision in the presence of essential uncertainty leads to multiple outcomes with potentially positive and negative consequences. A trade-off between them has to be properly evaluated which represents a challenging counterintuitive task. This is often used as a reason to ignore uncertainty with a plea for simple models or for postponing decisions until full information is available. The purpose of this section is to provide important motivations for the appropriate treatment of uncertainty.

4.1   Adaptive Control

Adaptive feedback control is often suggested as a way of dealing with the "unforeseen surprises" (ignored uncertainties) of deterministic models. A feedback control strategy depends on the current state of the system; therefore, when the state is perturbed, the strategy proceeds the control from a new state. The main issue in this approach is the inherent uncertainty, the delayed responses of socio-economic and environmental systems, and irreversibilities. The real consequences of decisions may remain invisible for long periods of time; thus, purely adaptive deterministic approaches can be compared to driving a car in the mountains on a foggy day facing backwards.

4.2   Simple Models

As the assumption of deterministic models about exact input data is often unrealistic, a number of simple models of uncertainty have been used. Simple models that provide an impression of explicit treatment of uncertainty may, in fact, produce misleading or wrong conclusions. One of the most popular ideas is to model uncertainty by a finite number of scenarios or states of the world. All members (agents) of the society know these states and their probabilities, i.e., they know "what-and-when" happens and can thus easily design compensation schemes or securities to spread risks around the world. As Arrow admits [3], catastrophes do not exist in such models (see also discussion in [7], [14]). Moreover, any of these scenarios in reality has the probability of 0.

4.3   Mean-Variance Analysis

This analysis substitutes real distributions by normal probability distributions. The following example illustrates its main danger. As discussed in [27], trajectories of the average annual atmospheric CO_2 changes were obtained from various monitoring stations. Analysts suggested characterizing the variability of these trajectories by calculating the sample mean, the standard deviation, and associated 95 percent confidence interval, which, in fact, contains only 13 percent of the observable CO_2 changes. The reason for this is that the histogram of the changes has a multimodal character that is fundamentally different from the normal distribution defined by the calculated sample mean and standard deviation. Multimodal distributions are typically used for characterizing the beliefs (opinions) of different political parties or movements, and heterogeneities induced by catastrophic events (see Fig. 5, 7 in [14]). In finance, a distribution of portfolio returns can be multimodal due to the contribution of different assets and asset classes.

4.4   Using Average Values

Average income, growth, daily pollutant concentration, average losses, expected utility, or expected returns may have a rather misleading character.

The projected global mean temperature changes fall within the difference between the average temperature of cities and their surrounding rural areas. Therefore, global climate change impacts can be properly evaluated only in terms of local temperature variability and related extreme events, in particular, heat waves, floods, droughts, windstorms, diseases, and sea level rise. The proper treatment of indicators with nonnormal, especially multimodal distributions requires special attention. The mean value of a multimodal distribution can be even outside the support of a distribution (the set of admissible values). Therefore, the use of average values orients the analysis on inadmissible values. Still, this value can be reasonably interpreted in the case of frequent repetitive observations. Subjective multimodal probability distributions and rare extreme events call for the use of quantiles, e.g., the median. Unfortunately, this destroys the additive structure and concavity (convexity) of standard models, as (in contrast to the average value) $median \sum_l v_l \neq \sum_l median(v_l)$ for random variables v_l. As a result this makes the applicability of well-known decomposition schemes and optimization methods problematic. Sections 5.2, 5.3 indicate a promising approach for dealing with arising problems.

Example 2. Optimal control problems. Discrete-time optimal control can be viewed as a specific case of STO models. In this case, x is composed of state variables $z(t)$, and control variables $u(t)$, that is, $x = \{z(t), u(t), t = 0, 1, ..., T\}$, where T is a given time horizon. Functions $f_i(x, \omega)$ are additive: $f_i(x, \omega) = \sum_{t=1}^{T} g_i(z(t), u(t), \omega_t, t)$, where ω_t is a stochastic disturbance at time t. Therefore, the use of $median(f_i(x, \omega))$ destroys the additive structure of optimal control problems essentially utilized in the Pontriagin's Maximim Principle and Bellman's recursive equations.

4.5   Deterministic Versus Stochastic Optimization

Deterministic decision problems are formulated in two steps. First of all, statistical procedures are used to estimate average values $\overline{\omega}$ of input data ω. After this intermediate task is performed, the deterministic problem with goal functions $f_i(x, \overline{\omega})$, $i = 0, 1, ..., m$ is solved. The use of $\overline{\omega}$ for multimode distributions orients decision analysis even on inadmissible scenarios. As well as for nonlinear $f_i(x, \omega)$, $Ef_i(x, \omega) \neq f_i(x, \overline{\omega})$. For example, if ω is uniformly distributed on $[-1, 1]$, then $\overline{\omega} = 0$ and $E(\omega x)^2 > (\overline{\omega}x)^2 = 0$.

STO methods deal directly with the variability of $f_i(x, \omega)$ affected by the variability of ω and decisions x, i.e., they deal simultaneously with uncertainty and decision analysis. Some decisions x can considerably reduce the variability of indicators $f_i(x, \omega)$, despite significant variability of ω, e.g., decisions $x_1 = 0$, $x_2 = 0$ for function $\omega_1 x_1 + \omega_2 x_2$. Besides, uncertainties often cancel each other for some decisions, e.g., decisions $x_1 = 1$, $x_2 = 1$ in the case of negatively correlated ω_1, ω_2. The later is especially important for optimization of insurance portfolios under catastrophic risks [13], [14], [21]). As a result, STO models can significantly reduce requirements on data quality in contrast to disconnected from decisions standard uncertainty analysis (see also Section 4.6).

The use of average values often smoothes the problem, but this may lead to wrong conclusions. The following simple model with abrupt changes shows that the use of average characteristics converts this model to a smooth and even linear deterministic version. Combined with sensitivity analysis, the resulting linear deterministic model is not able to detect abrupt changes: it plays a misleading role and can easily provoke an environmental collapse.

Example 3. Abrupt changes. Global changes with possible dramatic interactions among humans, nature and technology call for nonsmooth models. Nonsmooth and discontinuous processes are typical for systems undergoing structural changes and developments. In risk management, the possibility of an abrupt change is, by its very nature, present in the problem. The concept of nonsmooth and abrupt change is emphasized in the study of environmental and anthropogenic systems by such notions as critical load, surprise, and time bomb phenomena [1], [8], [17]. There are a number of methodological challenges involved in the policy analysis of nonsmooth processes. Traditional local or marginal analysis cannot be used because continuous derivatives do not exist, i.e., a nonsmooth, even deterministic, system cannot be predicted (in contrast to classical smooth systems) outside an arbitrary small neighborhood of local points.

The concentration of a pollutant $r_t = r_0 - xt + \sum_{k=1}^{N(t)} e_k$, where $\{e_k\}$ is a sequence of emissions from extreme episodes in interval $[0, t]$, $N(t)$, $t \geq 0$, is a counting process for the number of episodes in $[0, t]$, x is a rate of emission reduction, and r_0 is an initial concentration. The rate x pushes r_t down, whereas the random flow of emissions pushes r_t up. The main problem is to reduce the probability of a catastrophe associated with crossing a vital threshold ρ by r_t, $r_t > \rho$. The random jumping process r_t is often substituted by the expected concentration $\overline{r_t} = r_0 + (\alpha \overline{e} - x)t$, where $\alpha \overline{e}$ is the emission rate with intensity α and the emission mean value $\overline{e}$. That is, complex random jumping process r_t, is simply replaced by a linear function that decreases in time for $x > \alpha \overline{e}$. Thus, deterministic model $\overline{r_t}$ suggests, that if x slightly exceeds the average emission rate $\alpha \overline{e}$, then $\overline{r_t}$ decreases in time, which is the wrong conclusion. The sensitivity analysis of the linear deterministic model $\overline{r_t}$ under different scenarios for α and $\overline{e}$ produces the same trivial conclusions that robust x has to slightly exceed $\alpha \overline{e}$.

4.6   Probabilistic and Stochastic Models

There are two fundamental approaches to modeling uncertainty in probability theory, namely, probabilistic and stochastic models. Probabilistic models attempt to characterize processes completely and explicitly in terms of probability distributions or some of their characteristics. If one can evaluate explicitly multidimentional integrals $F_i(x) = Ef_i(x, \omega) = \int_\Omega f_i(x, \omega)P(d\omega)$, then the STO problem is reduced to a standard deterministic optimization model. Even the simplest situations illustrate difficulties. Thus, for two random variables ω_1, ω_2 with known probability distribution functions, the evaluation of probability distribution $\omega_1 + \omega_2$ is already an analytically intractable (in general) task requiring the evaluation of an integral. In addition, the distribution of $f_i(x, \omega)$, say, $\omega_1 x_1 + \omega_2 x_2$ significantly depends on x, e.g., compare $x_1 = 0$, $x_2 = 1$ and $x_1 = 1$, $x_2 = 0$. Exponential increase of computations occurs when one uses probability trees, transition probabilities, and variance-covariance matrices to represent the dynamics of uncertainties. The number of states of even the simplest discrete event systems (see, e.g., [17]) can be too large for complete explicit representations of them by matrices of transition probabilities. The computational "explosion" of probabilistic models, similar to the well-known "curse of dimensionality" of Bellmans equations, restricts their practical applicability for large scale global change problems.

Stochastic models deal directly with random variables $f_i(x, \omega)$ without the exact evaluation of $F_i(x)$. In combination with fast Monte Carlo simulations, some of the STO methods lead only to a linear increase of computations w.r.t. uncertain variables ω. In this case, goal functions are characterized by random laws (rules) and random processes (e.g., stochastic differential equations) rather than by transition probabilities, variance-covariance matrices, and partial differential equations. In fact, fast Monte Carlo procedures (Example 7) combine probabilistic and stochastic submodels, i.e., functions $F_i(x)$, in general, may have well defined analytical blocks.

5   Robust Stochastic Optimization

Although STO models allow to represent interdependencies among decisions, uncertainties and risks, yet inappropriate treatment of the variability of indicators $f_i(x, \omega)$ can be rather misleading.

5.1   Portfolio Selection

The Nobel prize laureate Markowitz [32] proposed the following mean-variance approach for designing robust portfolios of financial assets (and others, e.g., portfolios of technologies). Assume that $\overline{\omega_j}$ is the expected value of random returns ω_j from asset j, $j = 1, ..., n$, and x_j is a fraction of this asset in the portfolio, $\sum_{j=1}^{n} x_j = 1$, $x_j \geq 0$, $j = 1, ..., n$. The maximization of expected return $r(x) = \sum_{j=1}^{n} \overline{\omega} x_j$ from a portfolio $x = (x_1, ..., x_n)$ yields a trivial nonrobust solution: to invest all capital in the asset with the maximal expected return. The main idea [32] to achieve diversified robust portfolio is to consider a trade-off between expected returns and their variability characterized by the variance of returns $Var\rho(x, \omega)$, i.e., to maximize $r(x) - \mu Var\rho(x, \omega)$, $\rho(x, \omega) = \sum_{j=1}^{n} \omega_j x_j$, where μ is a trade-off (risk) parameter. Let us note that this approach requires that returns from portfolio $\sum_{j=1}^{n} \omega_j x_j$ have normal distribution.

Remark 2. The most important concerns in the case of more general portfolio selection problems are those related to the overestimation of real returns $\rho(x, \omega)$ by maximizing expected returns $r(x)$, i.e., when $\rho(x, \omega) < r(x)$. This calls for the maximization of a trade-off between expected returns and the risk of overestimation:
$r(x) + \mu E \min \{0, \rho(x, \omega) - r(x)\}$. It is easy to see that when the distribution of random returns $\rho(x, \omega)$ is normal, then the maximization of this function is equivalent to the maximization of the mean-variance criterion, as the absolute values of asymmetric risk function $E \min \{0, \rho(x, \omega) - r(x)\}$ are constant multiples of the standard deviation. Unfortunately, for nonlinear concave function $r(x)$ the straightforward application of the mean-variance approach leads to nonconcave optimization. The next section maneuvers this obstacle for rather general optimization problems.

5.2   Robust Utility Maximization

Consider the maximization of utility function $U(x) = Eu(q(x, \omega))$, (e.g., returns $r(x)$). If the distribution of random outcome $u(q(x, \omega))$ is not normal, for example, when the policy analysis involves the polarized beliefs of different communities, then the average value $U(x)$ may not belong to the set of attainable values $f(x, \omega)$ for $\omega \in \Omega$. Instead of $U(x)$ we can use a quantile $U_p(x)$ of $u(q(x, \omega))$ defined as maximal v such that $Prob[u(q(x, \omega)) \leq v] \leq p$, for $0 < p < 1$. The robust utility maximization problem can be formulated as the maximization of an adjusted to risk utility function $U_p(x) + \mu E \min \{0, u(q(x, \omega)) - U_p(x)\}$, which is not a concave function. As *Remark 2* indicates, for normal distributions and $p = 1/2$, this is equivalent to the mean-variance approach. Similar to Example 4, Section 5.4, one can conclude that the formulated problem is equivalent to the following concave STO optimization problem: maximize w.r.t. (x, z) function $\varphi(x, z) = z + \mu E \min \{0, \beta - z\}$, $\beta = u(q(x, \omega))$, $\mu = 1/p$.

Remark 3. This important fact can be seen from the following simple observations (see also Example 4): $\int_0^z (p - Prob[\beta \leq v]) dv = pz + E \min \{0, \beta - z\}$ for a random variable β with density. Let us also notice that for $\mu = 1/p$ we have $U_p(x) + \mu E \min \{0, u(q(x, u)) - U_p(x)\} = (1/p) \int_{u(q(x,\omega)) \leq U_p(x)} U(q(x, \omega)) dP = E[u(q(x, \omega)) | u(q(x, \omega)) \leq U(x)]$, i.e., the adjusted to risk utility function equals to the so-called expected shortfall (see, e.g., [11], [42]).

5.3 General STO Model

Similarly, a robust STO model can be written in the form: maximize w.r.t. (x, z) function $z_0 + \mu_0 E \min \{0, f_0(x, \omega) - z_0\}$ subject to $z_i + \mu_i E \min \{0, f_i(x, \omega) - z_i\} \geq 0$, $i = 1, ..., m$, where μ_i are weights. Components z_i^*, $i = 0, 1, ..., m$, of optimal solution (x^*, z^*) are quantiles of $f_i(x^*, \omega)$. The proof follows from the positivity of the Lagrange multipliers and *Remark 3*. Depending on the case, the robust model can also be formulated by using safety (Example 1) constraints $Prob[f_i(x, \omega) \geq 0] \geq p_i$ in combination, say, with constraints $E f_i(x, \omega) + \mu_i E \min \{0, f_i(x, \omega)\} \geq 0$, $i = 1, ..., m$ and other possible options [36].

5.4 Flexibility of Robust Strategies

The standard expected utility maximization model suggests two types of decisions in the response to uncertainty: either risk averse or risk prone decisions. These two options also dominate the climate change policy debates [34], [37], emphasizing either ex-ante anticipative emission reduction programs or ex-post adaptation to climate changes when full information becomes available. Clearly, a robust policy must include both options, i.e., the robust strategy must be flexible enough to allow for later adjustments of earlier decisions. The so-called (two-stage and multistage) recourse models of stochastic optimization [6], [20], [36], [46] incorporate both fundamental ideas of anticipation and adaptation within a single model and allow for a trade-off between long-term anticipatory strategies and related short-term adaptive adjustments. Therefore, the adaptive capacity can be properly designed ex-ante say, through emergency plans and insurance arrangements. The following example shows that the explicit incorporation of ex-ante and ex-post decisions induces risk aversion measures that cannot, in general, be imposed exogenously by a standard utility function.

Example 4. Mitigation versus adaptation: CVaR Risk measure.

A stylized static model of a climate stabilization problem [38] can be formulated as follows: let x denote an amount of emission reduction and let a random variable β denote an uncertain critical level of required emission reduction. Ex-ante emission reductions $x \geq 0$ with costs cx may underestimate β, $x < \beta$. It generates a linear total adaptation cost $az + dy$, where y is an ex-post adaptation, $y \leq z$ with cost dy; z is an ex-ante developed adaptive capacity with cost az.

To illustrate the main idea, let us assume that ex-post adaptive capacity is unlimited, $z = \infty$, and $c < d$. A two-stage stochastic optimization model is formulated as the minimization of expected total cost $cx + dEy$ subject to the constraint $x + y \geq \beta$. This problem is equivalent to the minimization of function $F(x) = cx + E \min \{dy | x + y \geq \beta\}$ or $F(x) = cx + dE \max \{0, \beta - x\}$, which is a simple stochastic minimax problem. Optimality conditions for these types of STO minimax problems show (see, e.g., [16], [17], [20], pp. 107, 416, [42] see also *Remark 3*) that the optimal ex-ante solution is the critical quantile $x^* = \beta_p$ satisfying the safety constraint $Prob[x \geq \beta] \geq p$ for $p = 1 - c/d$. This is a remarkable result: highly non-linear and even often discontinuous safety or chance constraint of type (2) is derived (justified) from an explicit introduction of ex-post second stage decisions y. Although the two stage model is linear in variables (x, y), the ex-post decisions y induce strong risk aversion among ex-ante decisions characterized by the critical quantile β_p.

Remark 4. If $c/d < 1$, then $x^* > 0$, i.e., it calls for coexistence of ex-ante and ex-post decisions. The optimal value $F(x^*) = dE\beta I(\beta > x^*)$, where $I(\cdot)$ is the indicator function. Again, according to Remark 3, this is the expected shortfall or Conditional Value-at-Risk (CVaR) risk measure [11], [42].

Remark 5. In more general two-stage models [14], [38], the risk aversion is not necessarily induced in the form of the critical quantile and CVaR risk measure. Despite this, the structure of robust policy remains the same. Only partial commitments are made ex-ante whereas other options are kept open until new information is revealed. In a sense, such flexible decisions incorporate both risk-averse and risk-prone components according to different "slices" of risks.

5.5   Uncertain Probability Distributions

Models of Section 3 assume that $P(d\omega)$ is known exactly. However, only some of its characteristics may be known. The elicited class $\mathcal{P}$ for admissible P is often given by constraints $\int \varphi_k(\omega) P(d\omega) \geq 0$, $k = \overline{1, K}$, $\int P(d\omega) = 1$; for example, constraints on joint moments $c_{s_1,...,s_l} \leq \int \omega_1^{s_1}...\omega_l^{s_l} P(d\omega) \leq C_{s_1,...,s_l}$, where $c_{s_1,...,s_l}$, $C_{s_1,...,s_l}$ are given constants. The robust STO problem can be formulated similar to Section 2.3 as a probabilistic maximin problem: maximize $F_0(x) = \min_{p \in \mathcal{P}} \int f_0(x, \omega) P(d\omega)$ subject to general constraints of Section 3.2. This probabilistic maximin approach was first initiated in STO in [12], [15], [50]. For specific sets $\mathcal{P}$, the solution of the inner minimization problem has a simple analytical form [29], [31]. For example, it is concentrated only in a finite number ([12], [15], [31], and Example 5) of admissible scenarios from Ω. Numerical methods for general problems were developed in [12], [15], [25], [26], [29].

Example 5. Robust stabilization and CVaR. The simple emission stabilization problem is defined (Example 4) by the minimization of $cx + dE \max\{0, \beta - x\} = z + d \int_z^\infty (\beta - x) P(d\beta)$. The robust stabilization problem with unknown P can be defined by minimization $cx + d \max_{p \in \mathcal{P}} \int_x^\infty (\beta - x) P(d\beta)$. To illustrate this possibility, suppose that β is a scalar random variable, $\Omega = [a, b]$, and an additional condition that defines the class $\mathcal{P}$ is $E\beta = \mu$. It is easy to see that the worst-case distribution is concentrated only in points a, b, with the probability mass $p(a) = \frac{b-\mu}{b-a}$, $p(b) = \frac{\mu-a}{b-a}$. Hence, the robust model is reduced to replacing the set of all admissible scenarios Ω by only two extreme scenarios a and b with probabilities $p(a)$, $p(b)$.

5.6   Stochastic Maximin Model

Probabilistic maximin robustness may not be sufficient to properly address the effects of extreme events. The classical extreme value theory [11] analyses so-called extreme value distributions, i.e., distributions of extreme values $\min_s \beta_s$ or, equivalently, $\max_s \beta_s$ of independent identically distributed (iid) random variables $\beta s(\omega)$, $s = 1, 2, ...$ Consider a more general problem. Let $\beta_1(\omega)$, $\beta_2(\omega)$,..., be a set of random variables affected by some decisions. Stochastic maximin models ([16]-[20]) attempt to maximize, in a sense, the extreme value distribution, e.g., its mean value $E \min_s \beta_s = \int \min_s \beta(\omega) P(d\omega)$. In contrast, the probabilistic maximin model deals with the maximization of the extreme mean value $\min_s E\beta_s$ or, equivalently, with the maximization of $\min_s \int \beta P_s(d\beta)$, where $P_s(\cdot)$ is the probability distribution of β_s. Therefore, this model evaluates impacts of extreme events (scenarios) $s = 1, 2, ...$, by extreme mean value, what may significantly underestimate them (Sections 4, 6).

A more general approach would be the combination of a probabilistic and a stochastic maximin model with $F_0(x) = \min_{p \in \mathcal{P}} E \min_{z \in Z} f_0(x, y, z, \xi)$, where ω is represented by variables y, z, ξ, $\omega = (y, z, \xi)$. Z is a set of variables z which are there to take into account potential extreme random scenarios, as in the extreme value theory [11]; the x variables are themselves decision variables; the y, $y \in Y$ variables correspond to uncertainty ranked by an objective or subjective probability measure P from $\mathcal{P}$; and ξ variables are ranked by a fixed probability measure as in the basic STO models. Thus in this model the worst case situation is evaluated with respect to the worst-case distribution for some uncertain variables y, whereas for other uncertain variables z it is evaluated from potential extreme random scenarios. In particular, this class of models includes purely stochastic maximin models with $F_0(x) = E \min_{y \in Y} f_0(x, y, \xi)$ as well as models with $F_0(x) = \min_{y \in Y} E f(x, y, \xi)$ combining the worst-case and the Bayesian approaches of Sections 2.2, 2.5 (see also discussion in [19], [20], pp. 105-106).

6   Temporal, Spatial and Social Heterogeneities

The significance of extreme events arguments in global climate changes has been summarized in [48] as follows: Impacts accrue ... not so much from slow fluctuations in the mean, but from the tails of the distributions, from extreme events. Catastrophes do not occur on average with average patterns. They occur as "spikes" in space and time. In other words, the distributional aspects, i.e., temporal and spatial distributions of values and risks are key issues to capture the main sources of vulnerability for designing robust policies.

6.1   Temporal Heterogeneity

Extreme events are usually characterized by their expected arrival time, for example, as a 1000-year flood, that is, an event that occurs on average once in every 1000 years. Accordingly, these events are often ignored as they are evaluated as improbable during a human lifetime. In fact, a 1000-year flood may occur next year. For example, floods across Central Europe in 2002 were classified as 1000-, 500-, 250-, and 100-year events. Another tendency is to evaluate potential extreme impacts by using so-called annualization, i.e., by spreading losses from a potential, say, 30-year catastrophe, equally over 30 years. In this case, roughly speaking, a potential 30-year crash of an airplane is evaluated as a sequence of independent annual crashes: one wheel in the first year, another wheel in the second year, and so on, until the final crash of the navigation system in the 30th year. The main conclusion from this type of deterministic analysis is that catastrophes do not exist. Section 8.1 introduces the notion of stopping time and related new approaches to discounting that allow for addressing the temporal variability of extreme events.

6.2   Spatial and Social Heterogeneity

A similar common tendency is the ignorance of real spatial patterns of catastrophes. A general approach is to use so-called hazard maps, i.e., maps showing catastrophe patterns that will never be observed as a result of a real episode, as a map is the average image of all possible patterns that may follow catastrophic events. Accordingly, social losses in affected regions are evaluated as the sum of individual losses computed on a location-by-location rather than pattern-by-pattern basis w.r.t. joint probability distributions. This highly underestimates the real impacts of catastrophes, as the following simple example shows.

Example 6. Social and individual losses. In a sense, this example shows that $100 \gg \overbrace{1 + 1 + \ldots + 1}^{100}$. Assume that each of 100 locations has an asset of the same type. An extreme event destroys all of them at once with probability 1/100. Consider also a situation without the extreme event, but with each asset still being destroyed independently with the same probability 1/100. From an individual point of view, these two situations are identical: an asset is destroyed with probability 1/100, i.e., individual losses are the same. Collective (social) losses are dramatically different. In the first case 100 assets are destroyed with probability 1/100, whereas in the second case 100 assets are destroyed only with probability 100^{-100}, which is practically 0. This example also illustrates the potential exponential growth of vulnerability from increasing network-interdependencies.

6.3   Downscaling, Upscaling and Catastrophe Modeling

So-called downscaling (see discussion in [5], [23]) and catastrophe modeling [47] are becoming increasingly important for estimating spatio-temporal vulnerability and catastrophic impacts. The aim of catastrophe models is to generate spatio-temporal patterns of potential catastrophic events and samples of mutually dependent losses. The designing of a catastrophe model is a multidisciplinary task requiring the joint efforts of environmentalists, physicists, economists, engineers and mathematicians. To characterize "unknown" catastrophic risks, that is, risks with the lack of historical data and large spatial and social impacts, one should at least characterize the random patterns of possible disasters, their geographical locations, and their timing. One should also design a map of values and characterize the vulnerabilities of buildings, constructions, infrastructure, and activities. Catastrophe models allow to derive histograms of mutually dependent losses for a single location, a particular hazard-prone zone, a country, or worldwide from fast Monte Carlo simulations rather than real observations [13], [14], [47].

The development of catastrophe models can be considered as a key risk management policy providing information for decision analysis in the absence of historical observations, in particular, on impacts of known events for new policies and potential extreme events that have never occurred in the past. This often raises new estimation problems. Traditional statistical methods are based on the ability to obtain observations from unknown true probability distributions, whereas new problems require information to be recovered from only partially observable or even unobservable variables. Rich data may exist on occurrences of natural disasters, incomes, or production values on global and national levels. Downscaling and upscaling methods in this case must - by using all available objective and subjective information - make plausible evaluations of local processes consistent with available global data, as well as, conversely, with global implications emerging from local data and tendencies.

7   STO Methods for Robust Solutions

7.1   Scenario Analysis

Outcomes of Monte Carlo simulations for a STO model are random sample functions $f_0(x, \omega)$, $f_1(x, \omega)$, ..., $f_m(x, \omega)$, that depend on the simulation run ω and a given vector of decisions x. Therefore, for a given x, outcomes vary at random from one simulation to another. The estimation of their mean values, variances, and other moments or histograms is time consuming in the presence of rare extreme events that require developments of specific fast Monte Carlo-type sampling procedures. Moreover, a change in policy variables x affects the probabilistic characteristics of outcomes and requires a new sequence of Monte Carlo simulations to estimate their new values. If functions $f_i(x, \omega)$, $i = 0, 1, ..., m$, have well defined analytical structure with respect to x for each simulated ω, then the following scenario analysis is often used. The Monte Carlo simulations generate scenarios ω^1, ω^2, ..., ω^N for each of which optimal solutions $x(\omega^1)$, $x(\omega^2)$, ..., $x(\omega^N)$ of the deterministic optimization model are calculated. Any of these solutions calculated for one scenario may not be feasible for other scenarios. The number of possible combinations of potential scenarios ω and decisions increases exponentially. Thus, with only 10 feasible decisions, for instance, levels of emission reductions in a given region, 10 regions and 10 possible scenarios for all of them, the number of "what-if" combinations is 10^{11}. The straightforward evaluation of these alternatives would require more than 100 years if a single evaluation takes only a second. Besides, the probability of each scenario ω^l, $l = 1, ..., K$, is in general, equal to 0. Therefore, the choice of final robust decisions is unclear and is not explicitly addressed.

7.2   Sample-Mean Approximations

STO models of Sections 3.2, 5 are able to explicitly characterize robustness by using proper indicators of different risks, flexible decisions and various equity and fairness constraints as goals of desirable policy. The main challenge is to design a search procedure that enables to find policy decisions specified by these goals. STO methods, in particular, adaptive Monte Carlo (AMC) optimization methods [14], avoid exact evaluations of all feasible alternatives. The problem confronted by STO methods is to estimate the maximum $F_0(x^*)$ of $F_0(x)$ subject to constraints $F_i(x) \geq 0$, $F_i(x) = Ef_i(x, \omega)$, $i = 1, ..., m$, by making use of only random outcomes from simulations $f_i(x, \omega)$, $i = 0, 1, ..., m$. Standard Monte Carlo methods can be regarded as estimating the value of multidimensional integrals $F_i(x) = \int f_i(x, \omega) P(d\omega)$,

$i = 0, 1, ..., m$, for fixed x. In particular, this can be done by using a sample mean $F_i^N(x) = 1/N \sum_{k=1}^N f_i(x, \omega^k)$. If functions $f_i(x, \omega)$ are analytically tractable w.r.t. x, then $F_i^N(x)$ can be used to find an approximate solution of the STO problem, assuming that $F_i^N(x)$ sufficiently approximates $F_i(x)$, $i = 1, ..., m$. Although in this case the original STO model is approximated by a deterministic optimization problem, its solution often requires new deterministic large-scale optimization methods (see, e.g., [6], [20], [33], [40], [46]), as well as the sample size N reduction techniques and fast Monte Carlo simulations. A principle complexity (Sections 5, 8) is that the measure P is often analytically intractable, that it may depend on x as in Section 5.5, and that samples are affected by current x and rare catastrophic events. In addition, the sample mean approximations $F_i^N(x)$ may destroy the concavity (convexity) of functions $F_i(x)$. For example, the expectation function ax^2, $a = p_1\omega_1 + p_2\omega_2 > 0$, $\omega_1 > 0$, $\omega_2 < 0$ is the convex function, but its sample mean approximation $(\frac{1}{N} \sum_{k=1}^N \omega^k)x^2$ may be the concave function even for rather large N in the case of a small probability p_1 and a large impact $\omega_1 > 0$. In these cases, in general, only AMC optimization is applicable. The asymptotic $(N \to \infty)$ rate of convergence for STO procedures is related to the properties of the sample mean estimates.

7.3   Adaptive Monte Carlo Optimization

An "Adaptive Monte Carlo" simulation [41] is a technique that makes online use of sampling information to sequentially improve the efficiency of the sampling itself. The notion "Adaptive Monte Carlo" optimization is used [14], [21] in a rather broad sense, where improvements of the sampling procedure with respect to the variability of estimates may be only a part of the improvements with respect to other goals of robust decisions.

Remark 6. A counterintuitive fact is that the estimation of a robust solution x^* and $F_0(x^*)$ starting from an initial solution x^0 often requires approximately the same (or an even smaller) number of simulations than the estimation of only $F_0(x^0)$ for fixed initial x^0. This is because of two forces. First of all, robust solutions x^* reduce risks and, hence, the variability of $F_0(x)$; therefore, movements toward $F_0(x^*)$ according to STO methods are themselves a variance-reducing process (see, e.g., numerical calculations in [21]). In contrast, $F_0(x^0)$ may have considerable variability due to the effects of extreme events; therefore, its estimation requires large samples. Secondly, the variance reductions can also be achieved by deliberate switches in the importance sampling.

Example 7. Environmental collapse. Let us illustrate the main idea of fast sample mean approximations and AMC optimization by a modification of Example 3. The concentration of a global pollutant at time t is calculated as $r_t = r_0 + \sum_{t=0}^t x_t e_t$, where x_t is the rate of global emission e_t reduction, $0 \leq x_t \leq 1$, and $e_0, e_1, ...$ are random dependent variables. At a random time moment τ, the critical threshold β for r_t is revealed and a collapse occurs when $r_t > \beta$. Assume that β is characterized by a probability distribution $B(z) = Prob[\beta < z]$ and $Prob[\tau = t] = p(1-p)^t$, $t = 0, 1, ...$, where probability p is characterized by a probability distribution in an interval $[p_*, p^*]$. The probability of a collapse $\Psi(x) = E \sum_{t=0}^{\infty} I(\beta < r_t)$, where $I(\beta < r_t) = 1$ or 0 if $\beta < r_t$ or $\beta \geq r_t$, respectively. Equivalently,

$$\Psi(x) = E \sum_{t=0}^{\infty} E[p(1-p)^t] B(r_0 + \sum_{t=0}^{t} x_t e_t) = E \sum_{t=0}^{\tau} EB(r_0 + \sum_{t=0}^{t} x_t e_t). \quad (3)$$

The probabilistic model is described by the analytically intractable function $\Psi(x)$. Moreover, an emission path e_0, e_0, ... is usually generated by solving a global energy/economy model, and e_t is a complex function of an emission reduction policy. The stochastic model in this example is described by the right hand side of (3) including the process r_t, the probability distribution for τ, and a stochastic generator of dependent emission path, (e.g., using global energy/economy model).

It is possible to use a straightforward Monte Carlo simulation to estimate $\Psi(x)$ for a fixed x. A simulation run s, $s = 1, 2, ...$ consists of sampling $p^s \in [p_*, p^*]$; $\tau = \tau_s$; a path e_t^s, $t = 0, 1, ..., \tau_s$ and β^s. The value $\Psi(x)$ is estimated as $\Psi^N(x) = \sum_{s=1}^{N} I(\beta^s < r^s)/N$. If p^s is a small probability then this straightforward approach requires large N. A stochastic model (3) allows much faster sample mean evaluations of $\Psi(x)$ and fast AMC optimization procedures [18], [21]. Conceptually, AMC optimization involves the following steps. An initial solution x^0 is fixed; p^0, τ^0, e_0^0, e_1^0, ..., $e_{\tau^0}^0$ are simulated. On this basis, by using known function $B(z)$, a so-called stochastic gradient is calculated allowing for adaptive adjustment of x^0 to x^1. For x^1, a new sample p^1, τ^1, e_0^1, e_1^1, ..., $e_{\tau^1}^1$ is calculated, and x^1 is adaptively adjusted in the same manner as x^2, and so on. It is important that evaluation of robust strategy in this manner proceeds with simulations $s = 1, 2, ...$ without intermediate evaluations of $\Psi^N(x)$. Details of this solution technique for rather general risk processes are discussed in [14], [18]. In parallel with adjustments of solutions x^s, the AMC optimization is able to change the sampling procedure [21] itself (importance sampling).

8   Sensitivity of Robust Strategies

Robust strategies for global changes require a proper focus on potential extreme events. As a result, the robust strategy with a small $\varepsilon > 0$ probability of extreme events can be significantly different from the policy that ignores these events by using $\varepsilon = 0$. Formally speaking, this is evident from Section 5.3, when $\varepsilon > 0$ results in shifts of ranges $f_i(x, \omega)$ to include potential catastrophic impacts (say, ranges of required emission reductions β in Example 4) that suddenly disappear for $\varepsilon = 0$. Informally speaking, the explicit introduction of extreme events with $\varepsilon > 0$ requires new sets of feasible decisions, new spatial, temporal, and social dimensions which suddenly disappear for $\varepsilon = 0$. Next Section shows that a key issue is the proper treatment of discounting and random time horizons of extreme events.

8.1   Extreme Events and Discounting

How can we justify strategies that may possibly turn into benefits over long and uncertain time horizons in the future? For example, how can we justify investment, say, in a flood defense system to cope with foreseen extreme 100-, 250-, 500- and 1000- year floods? A common approach is to discount future costs and benefits using a geometric (exponential) discount factors with the prevailing market interest rate as $V = \sum_{t=0}^{\infty} d_t V_t$, where $d_t = (1+r)^{-t}$, r is a discount rate, $V_t = E\nu_t$ for some random variables ν_t, $t = 0, 1, \dots$. An infinite deterministic stream of values V_t, $t = 0, 1, \dots$, can represent a cash-flow stream of a long-term investment activity. In economic growth models and integrated assessment models (see, e.g., [37]) the value V_t represents utility $U(x^t)$ of an infinitely living representative agent with consumptions x^t .

The infinite time horizon in V creates an illusion of truly long-term analysis. The choice of discount rate r as a prevailing interest rate within a time horizon of existing financial markets is well established. Uncertainties, especially related to extreme events, challenge the possibility of markets to offer proper rates. The following simple fact shows [22] that the standard discount factors obtained from markets orient policy analysis only on few decades, what precludes to properly address catastrophic impacts. It also indicates an important alternative approach to discounting.

Let $p = 1 - d$, $d = (1+r)^{-1}$, $q = 1 - p$, and let τ be a random variable with the geometric probability distribution $P[\tau = t] = pq^t$. It is easy to see that

$$\sum_{t=0}^{\infty} d_t V_t = E \sum_{t=0}^{\tau} \nu_t, \tag{4}$$

where $d_t = d^t$, $t = 0, 1, \dots$. This is also true for general discounting $d_t = (1+r_t)^{-t}$ with increasing positive r_t, where the stopping time τ is defined as $P[\tau \geq t] = d_t$.

That is, any discounted sum can be viewed as an expected value of the undiscounted sum within a random interval $[0, \tau]$. We can think of τ as a random "stopping time" associated with the first occurrence of an extreme stopping time (killing) event. The expected duration of τ, $E\tau = 1/p = 1 + 1/r \approx 1/r$ for small r. The same holds for the standard deviation $\sigma = \sqrt{q}/p$. Therefore, for the interest rate of 3.5 percent, $r \approx 0.035$, the expected duration is $E\tau \approx 30$ years, i.e., this rate orients the policy analysis on an expected 30-year time horizon. The bias in favor of the present in discounting with the rate of 3.5 percent is easily illustrated [39]. For a project with long-run benefits or costs, 1 Euro of benefits or costs in years 50, 100, and 200, has a present value respectively of 0.18, 0.003, and 0 Euros. Definitely, this rate has no correspondence with how society has to deal with a 200-year flood.

The equation (4) provides an alternative approach to discounting in the presence of catastrophic events: the use of the right hand side of (4) with τ defined by the arrival times of potential catastrophic events rather than by horizons of market interests. This also allows to address the variability of the value stream ν_0, ν_1,... by analyzing quantiles of random sum $\sum_{t=0}^{\tau} \nu_t$ (or $\sum_{t=0}^{\tau} V_t$ for deterministic flows V_0, V_1,...), e.g., as in Section 5. This approach was used in [13], [14], [21].

Example 8. Catastrophic risk management. The implications of (4) for long-term policy analysis are rather straightforward. It is realistic to assume [39] that typical cash-flow investment in a new nuclear plant has the following average time horizons: without a disaster, the first six years of the stream reflect the costs of constructions and commissioning, followed by 40-years of operating life when the plant is producing positive cash flows and, finally, a 70-year period of expenditure on decommissioning. The flat discount rate of 5 percent, according to (4), orients the analysis on a 20-year time horizon. It is clear that a lower discount rate places more weight on distant costs and benefits. For example, the explicit treatment of a potential 200-year disaster would require a discount rate of at least 0.5 percent instead of 5 percent. Similar examples are investments in mitigations to cope with climate change related extreme events. A rate of 3.5 percent, as is often used in integrated assessment models [37], [49] is definitely not appropriate.

Example 9. Time varying discounting. The use of undiscounted evaluation $V = E \sum_{t=0}^{\tau} \nu_t$ instead of $V = \sum_{t=0}^{\infty} d_t V_t$, as (4) shows, induces any standard exogenous discounting. This example shows that the induced discounting can be of a rather general form. Multipliers $E[p(1-p)^t]$ in (3) with random p can be viewed as time-varying discount factors. It is easy to see that the asymptotic of these multipliers are dominated by the least-probable extreme events. Indeed, assume that there is only a finite number of scenarios $p_1 < p_2 < ... < p_L$ ranked by probability weights v_1, v_2, ..., v_L. Then $E[p(1-p)^t] = (1-p_1)^t[v_1 p_1 + \sum_{s=2}^{L}(\frac{1-p_s}{1-p_1})^t] \sim v_1 p_1 (1-p_1)^t$.

Therefore, a given exogenous standard discount rate cannot match, in general, rather different sets of extreme events. This calls for the explicit introduction of stopping time τ and the use of random criterion $E \sum_{t=0}^{\tau} \nu_t$. As decisions affect the occurrence of extreme events (τ), this approach, in fact, is equivalent to using implicit spatio-temporal endogenous discounting. This approach allows also to treat distributional aspects by using quantiles of random sum $\sum_{t=0}^{\tau} v_t$, $V_t = Ev_t$ and STO methods (Section 5.2, 5.3).

8.2   Stopping Time and Stochastic Minimax

As Section 8.1 shows, the concept of stopping time allows to focus the analysis on the least-probable and the most destructive (killing) extreme events. There are strong connections [18] between the stopping time- and stochastic maximin type-problems defined in Section 5.5. These connections can be used for designing optimization methods [18] for stopping time problems.

The stopping time is often associated with the likelihood of some processes crossing "vital" thresholds. Consider a random process $R_t(x)$ and the threshold defined by a random β. Let us define the stopping time τ as the first time moment t when $R_t(x)$ is above β, that is, $\tau(x) = \max\{t \in [0, T] : R_s(x) \le \beta, 0 \le s \le t\}$. For example, climate change mitigations x deal with preventing the global temperature, say, R_t, from crossing its critical level β. In this case, the safety constraint can be defined by probability $Prob[\tau(x) \ge T]$, where T is a given horizon. Explicit analytical evaluation of this probability is practically impossible even for the simplest insurance risk processes [11]. This precludes the use of standard optimization methods. A promising idea is to use connections with stochastic minimax problems (see, e.g., [18]. Assume that r_t and β are one dimensional random variables, β is independent of r_t, $H(y) = Prob[\beta \ge y]$, and the performance indicators of the general STO problem depend on t, $f_i(t, x, \omega)$, $i = 0, 1, ..., m$. The robustness can be defined as in [14] by functions $Ef_i(t, x, \omega)$ at $t = \tau(x)$, $F_i(x) = Ef_i(\tau(x, \omega), x, \omega)$. Functions $F_i(x)$ can be written [18] as $F_i(x) = E \sum_{t=0}^{T} f_i(t, x, \omega) H(\max_{0 \le s \le t} R_s(x, \omega))$, i.e., a stopping time problem with implicit and, in general, discontinuous random function $\tau(x)$ is equivalently transformed into a stochastic minimax problem that can be solved by different methods [17].

9 Concluding Remarks

In the absence of sufficient information, models play a key role in comparative analysis of alternative solutions for designing robust policies. Any policy analysis focuses attention on situations where processes can be changed by decisions that should be selected in the best possible manner. In this paper we discussed various facets of robustness assuming that the policy analysis includes optimization models with given sets of goals and feasible decisions. In reality these sets are also uncertain and they can be specified through a dialogue of users with models, where optimization models create only some blocks of the overall decision support system. Advances in modeling and computational methods allow us to create a "laboratory world" [10], where we can test new policies never implemented in reality. This "learning-by-modeling" dialogue with models requires specific robust optimization methods which are able to maintain a consistency of outcomes under the changing environment of the "laboratory world" where goals and sets of feasible solutions are subject to modifications by users, new information and gained experience. In particular, the evaluation of robust policies often requires specific robust optimization methods that are able to correctly detect the effects of rare extreme events. A discussion of these is beyond the scope of this paper. At least, they require the development of specific fast Monte Carlo procedures (see, e.g., [18]). The use of quantiles, thresholds, and stopping times requires, in general, specific non-smooth stochastic optimization methods [13], [14], [17], [18]. Since the notion of robustness depends on the nature of decision problems, it is hopeless to provide a complete overview of all its feasible facets. Therefore, in this paper we have primarily focused on issues relevant to on-going modeling of global change processes at IIASA.

Acknowledgments. The authors are thankful to the participants of the IFIP/IIASA/GAMM workshop on Coping with Uncertainty (December 13-16, 2004, IIASA) and the anonymous referees for their critical and constructive comments which led us to important improvements of this paper.

References

1. Alcamo, J., Shaw, R., Hordijk, L. (Eds.) (1990): The RAINS Model of Acidification, Science and Strategies in Europe. Kluwer Academic Publishers, Dordrecht/Boston/London.
2. Alley, R.B., Marotzke, J., Nordhaus, W.D., Overpeck, J.T., Peteet, D.M., Pielke Jr., R.A., Pierrehumbert, R.T., Rhines, P.B., Stocker, T.F., Talley, L.D., Wallace, J.M. (2003): Abrupt Climate Change. Science **299**.
3. K.J. Arrow (1996): The theory of risk-bearing: small and great risks. Journal of Risk and Uncertainty **12**, 103-111.
4. Ben-Tal, A., Nemirovski, A. (1998): Robust Convex Optimization. Mathematics of Operation Research **23**, 769-805.
5. Bierkens, M.F.P., Finke, P.A., de Willigen, P. (2000): Upscaling and Downscaling Methods for Environmental Research. Kluwer, Dordrecht, The Netherlands.
6. Birge, J., Louveaux, F. (1997): Introduction to Stochastic Programming. Springer-Verlag, New York.
7. Chichilnisky, G., Heal, G. (1993): Global Environmental Risks. Journal of Economic Perspectives **7(4)**, 65-86.
8. Clark, W.C., Munn, R.E. (Eds.) (1985): Sustainable Development of the Biosphere. Cambridge University Press, Cambridge, UK.
9. Danielson, M., Ekenberg, L. (1998): A Framework for Analysing Decisions under Risk. European Journal of Operation Research **104/3**, 474-484.
10. Dantzig, G. (1979): The Role of Models in Determining Policy for Transition to a More Resilient Technological Society. IIASA distinguished lecture series **1**, 1979.
11. Embrecht, P., Klueppelberg, C., Mikosch, T. (2000): Modeling Extremal Events for Insurance and Finance. Applications of Mathematics, Stochastic Modeling and Applied Probability. Springer Verlag, Heidelberg.
12. Ermoliev, Y. (1970): On Some Stochastic Programming Problems. Kibernetika **1**, 3-9.
13. Ermolieva, T., Ermoliev, Y. (2005): Catastrophic Risk Management: Flood and Seismic Risks Case Studies. In Wallace, S.W. and Ziemba, W.T. (Eds.): Applications of Stochastic Programming, MPS-SIAM Series on Optimization, Philadelphia, PA, USA.

14. Ermoliev, Y., Ermolieva, T., MacDonald, G., and Norkin, V. (2000): Stochastic Optimization of Insurance Portfolios for Managing Exposure to Catastrophic Risks. Annals of Operations Research **99**, 207-225.

15. Ermoliev, Y.E., Gaivoronski, A., Nedeva, C. (1985): Stochastic Optimization Problems with Incomplete Information on Distribution Functions. SIAM J. Control and Optimization **23/5**, 697-708.

16. Ermoliev, Y., Leonardi, G. (1982): Some Proposals for Stochastic Facility Location Models. Mathematical Modeling **3**, 407-420.

17. Ermoliev, Y., Norkin, V. (1997): On Nonsmooth and Discontinuous Problems of Stochastic Systems Optimization. European Journal of Operation Research **101**, 230-244.

18. Ermoliev, Y., Norkin, V. (2004): Stochastic Optimization of Risk Functions. In: Marti, K., Ermoliev, Y., Pflug, G. (Eds.): Dynamic Stochastic Optimization. Springer Verlag, Berlin.

19. Ermoliev, Y., Nurminskiy, E. (1980): Stochastic Quasigradient Algorithms for Minimax Problems in Stochastic Programming. In: Dempster M.A.H. (Ed.): Stochastic Programming. Academic Press, London.

20. Ermoliev, Y., Wets, R. (Eds.) (1988): Numerical Techniques for Stochastic Optimization, Computational Mathematics, Springer Verlag, Berlin.

21. Ermolieva, T. (1997): The Design of Optimal Insurance Decisions in the Presence of Catastrophic Risks. IIASA Interim Report IR-97-068, Web: www.iiasa.ac.at.

22. Ermolieva, T., Ermoliev, Y., Fischer, G., Nilsson, S., Obersteiner, M. (2006): Extreme Events, Discounting, and Stochastic Optimization. Submitted to a journal (see also IIASA Interim Report IR-03-029, Web: www.iiasa.ac.at).

23. Fischer, G., Ermolieva, T., Van Veltuizen, H., Yermoliev, Y. (2004): On Sequential Downscaling Methods for Spatial Estimation of Production Values and Flows. Proceedings of the Conference on Data Assimilation and Recursive Estimation: Methodological Issues and Environmental Applications, Venice, Italy.

24. Fishburn, P. (1981): Subjective Expected Utility: A Review of Normative Theories. Theory and Decision **13**, 139-199.

25. Gaivoronski, A.A. (1986): Linearization Methods for Optimization of Functionals Which Depend on Probability Measures. Mathematical Progamming Study **28**, 157-181.

26. Golodnikov, A.N., Stoikova, L.S. (1978): Numerical Methods of Estimating Certain Functionals Characterizing Reliability. Cybernetics **2**, 73-77.

27. Hudz, H., M. Jonas, T. Ermolieva, R. Bun, Y. Ermoliev and S. Nilsson (2003): Verification times underlying the Kyoto Protocol: Consideration of risk. Background data for IR-02-066. International Institute for Applied Systems Analysis, Laxenburg, Austria. Available on the Internet: http://www.iiasa.ac.at/Research/FOR.

28. Huber, P.J. The Case of Choquet Capacities in Statistics. Bulletin of the International Statistical Institute **45**, 181-188.

29. Insua, D.-R., Ruggeri, F. (2000): Robust Bayesian Analysis. Springer Verlag, New York.

30. IPCC (2001). Climate Change 2001: The Scientific Basis. Technical Report. Intergovernmental Panel on Climate Change.

31. Kall, P., Ruszczynski, A., Frauendorfer, K. (1988): Approximation Techniques in Stochastic Programming. In: Ermoliev, Y., Wets, R. (Eds.): Numerical Techniques for Stochastic Optimization, Computational Mathematics, Springer Verlag, Berlin.

32. Markowitz, H.M. (1987): Mean Variance Analysis in Portfolio Choice and Capital Markets. Blackwell, Oxford.

33. Marti, K. (2005): Stochastic Optimization Methods. Springer, Berlin, Haidelberg.

34. Manne, A.S., Richels, R.G. (1995): The Greenhouse Debate: Economic Efficiency, Burden Sharing and Hedging Strategies. The Energy Journal **16/4**, 1-37.

35. Morgan, M.G., Kandlikar, M., Risbey, J., Dowlatabadi, H. (1999): Why Conventional Tools for Policy Analysis are Often Inadequate for Problems of Global Change: An Editorial Essay. Climatic Change **41/(3-4)**, 271-281.

36. Mulvey, J.M., Vanderbei, R.J., Zenios, S.A. (1995): Robust Optimization of Large Scale Systems. Operations Research **43**, 264-281.

37. Nordhaus, W.D., Boyer, J. (2001): Warming the World: Economic Models of Global Warming, MIT Press, Cambridge, Mass.

38. ONeill, B., Ermoliev, Y., Ermolieva, T. (2006): Endogenous Risks and Learning in Climate Change Decision Analysis. In: Marti, K., Ermoliev, Y., Pflug, G., Makovskii M. (Eds): Proceedings of the IFIP/IIASA/GAMM Workshop On Coping with Uncertainty, Springer-Verlag, Heidelberg.

39. OXERA (2002): A Social Time Preference Rate for Use in Long-term Discounting. OXERA Press, 1-74.

40. Prekopa, A. (1995): Stochastic Programming. Kluwer Academic Publishers, Dorbrecht, Netherlands.

41. Pugh, E.L. (1966): A gradient technique of adaptive Monte Carlo. SIAM Review **8/3**.

42. Rockafellar, T., Uryasev, S., Optimization of Conditional-Value-at-Risk, The Journal of Risk **2**, 21-41.

43. Reithmeier, E., Leitmann, G. (1996): Robust Constrained Control for Vibration Suppression of Mismatched Systems. Applied Mathematics and Computation **78**, 245-257.

44. Shafer, G. A. (1976): Mathematical Theory of Evidence. Princeton University Press.

45. Shapira, Z. (1995): Risk Taking: A M anagerial Perspective. Russel Sage Foundation.

46. Wallace, S.W., Ziemba, W.T. (2005): Applications of Stochastic Programming. MPS-SIAM Books Series on Optimization, **5**.

47. G.R. Walker (1997): Current Developments in Catastrophe Modeling, in: Financial Risk Management for Natural Catastrophes, eds. N.R. Britton and J. Oliver, Aon Group Australia Limited, Griffith University, Brisbane, 17-35.

48. Wigley, T.M.L. (1985): Impacts of Extreme Events. Nature **286**, 106-107.

49. Wright, E.L., Erickson, J.D. (2003): Incorporating Catastrophes into Integrated Assessment: Science, Impacts, and Adaptation. Climate Change **57**, 265-286.

50. Zackova, J. (1966): On Minimax Solutions of Stochastic Linear Programming. Casopis pro Pestovani Matematiky **91**, 430-433.

Stress Testing via Contamination

J. Dupačová

Charles University, Faculty of Mathematics and Physics, Dept. of Probability and Mathematical Statistics, Prague, Czech Republic

Abstract. When working with stochastic financial models, one exploits various simplifying assumptions concerning the model, its stochastic specification, parameter values, etc. In addition, approximations are used to get a solution in an efficient way. The obtained results, recommendations for the risk and portfolio manager, should be then carefully analyzed. This is done partly under the heading "stress testing", which is a term used in financial practice without any generally accepted definition. In this paper we suggest to exploit the contamination technique to give the "stress test" a more precise meaning. Using examples from portfolio and risk management we shall point out the directly applicable cases and will discuss also limitations of the proposed method.

Key words: Scenario-based stochastic programs, stress testing, contamination bounds, portfolio management, CVaR

AMS subject classification: 90C15, 90C31, 91B28

1 Introduction

In stochastic programming problems one aims at selection of the "best" decision or action which fulfills given "hard" constraints, say $x \in \mathcal{X}$, accepting that the outcome of this decision may be influenced by a realization of a random event ω. The realization of ω is not known at the time of decision making and to get the decision one uses the knowledge of the probability distribution P of ω. The random outcome of a decision $x \in \mathcal{X}$ is quantified as $f(x, \omega)$. Moreover, also "soft" constraints on x may be considered and their violation if ω occurs may be included into the random objective function f or treated separately in the form of probability constraints, such as $P\{g(x, \omega) \geq 0\} \geq p$ with p a given probability.

In the sequel we shall focus mainly on stochastic programs which may be written (after a suitable reformulation) in the following form: Given

$\mathcal{X} \neq \emptyset$, closed, $\mathcal{X} \subset R^n$,

$\omega \in \Omega \subset R^m$ random with probability distribution P known, independent of decision $x \in \mathcal{X}$,

$f : \mathcal{X} \times \Omega \to R^1$ measurable, with finite $E_P f(x, \omega) \, \forall x \in \mathcal{X}$

$$\text{minimize } F(x, P) := E_P f(x, \omega) = \int_\Omega f(x, \omega) dP(\omega) \text{ on the set } \mathcal{X}. \quad (1)$$

The optimal value of (1) will be denoted $\varphi(P)$, the set of its optimal solutions $\mathcal{X}^*(P)$.

With P known the main stumbling block for an algorithmic solution of such stochastic programs is the necessity to compute repeatedly at least the values of the multidimensional integrals in (1) of functions which themselves need not be defined explicitly. Various approximation schemes were designed. The prevailing approach is to solve a scenario-based form of (1) with P a discrete probability distribution which is carried by a finite number of points, say $\omega^1, \ldots, \omega^S$ with probabilities $p_1, \ldots, p_S$. The atoms of this discrete distribution are called *scenarios* and the scenario-based formulation of (1) reads:

$$\text{minimize} \sum_{s=1}^{S} p_s f(\boldsymbol{x}, \omega^s) \text{ on the set } \mathcal{X}. \tag{2}$$

There is an extensive evidence of successful applications of scenario-based stochastic programs in financial modeling, pricing and designing decision strategies, cf. [29], [30], and in other areas, cf. [28]. The origin of scenarios can be very diverse. They can be atoms of a known genuine discrete probability distribution, can be obtained in the course of a discretization or approximation scheme, by simulation or by a limited sample information. They can result from recognized regulations or from a preliminary analysis of the problem and probabilities of their occurrence may reflect an ad hoc belief or a subjective opinion of an expert; see Chapter II.5 in [14]. Under "scenario" one may also understand a single deterministic realization of all uncertainties and parameters up to the horizon; this setting covers not only a certain realization of ω or a choice of various input parameters but it may be also related to a specific macroeconomic or demographic situation.

Already the early applications of stochastic programming were aware of the fact that the obtained solution or policy can be influenced by the choice and an approximation of the probability distribution P. To analyze the results, the main tool has been sensitivity analysis via repeated runs of the optimization problem with a changed input, see e.g. [21].

Also possible simplifications of the model, e.g. using multiperiod two-stage program instead of a multistage one or relaxation of integrality constraints, misspecification of the approximated "true" probability distribution P or errors in estimating various input parameters may influence substantially the results; see e.g. [2], [3], [4], [19]. These are additional reasons for designing suitable validation techniques and tests. One may exploit parametric programming results, statistical methods, various sampling and simulation techniques, multimodeling, etc. The choice of the approach depends essentially on the structure of the solved problem, on the origin of scenarios and reflects sources of possible errors and misspecifications.

To validate results of financial applications, one uses mostly historical and empirical backtesting, stress testing and out-of-sample analysis. We suggest to complement these numerical techniques by the contamination approach which provides bounds to the errors. We shall explain the basic ideas of contamination technique and illustrate its application on a bond portfolio management problem and on CVaR criterion risk management. Finally, possible extensions and limitations of the proposed approach approach will be discussed.

2   Motivation: Stochastic Dedicated Bond Portfolio Management

Assuming known future short-term reinvestment interest rates i_t for period $(t, t+1)$, the dynamic dedicated bond portfolio model can be formulated as follows:

$$\text{minimize } \sum_{n=1}^{N} c_n x_n + y_0^+$$

subject to

$$\sum_{n=1}^{N} f_{nt}x_n + (1+i_{t-1})y_{t-1}^+ - y_t^+ = l_t,\ t = 1,\ldots,T,\ \boldsymbol{x} \geq 0,\ \boldsymbol{y}^+ \geq 0.$$

Here $\boldsymbol{x} = (x_1,\ldots,x_n)^\top$ is composition of the portfolio, $\boldsymbol{c} = (c_1,\ldots,c_n)^\top$ is the vector of acquisition prices and the T-vectors $\boldsymbol{f}_n$, $n = 1,\ldots,N$, $\boldsymbol{l}$ and $\boldsymbol{y}^+$ stand for the cash flows, liabilities and surpluses.

In reality, the future short-term reinvestment rates are hardly known. We assume instead that $\iota = (i_0,\ldots,i_{T-1})$ are random and that their probability distribution has been approximated by a discrete probability distribution P carried by a finite number of scenarios ι^s, $s = 1,\ldots,S$ with probabilities p_s. In addition, we allow for short-term shortfalls; this means that for some scenarios and time periods (except for the last one) nonzero discrepancies

$$y_t^{-s} = \left(l_t - \sum_n f_{nt}x_n - (1+i_{t-1}^s)y_{t-1}^{+s} + y_t^{+s} \right)^+$$

may occur. In such case, the investor borrows this amount and is obliged to repay it including the interest rate (higher than i_t^s for a positive spread δ between the short-term reinvestment and borrowing rates) in the next period. For each s, t we consider now the cash flow constraints which include scenario dependent surpluses y_t^{+s} and shortfalls y_t^{-s}. In addition, there is a penalty $\sum_s p_s \boldsymbol{q}^{s\top} \boldsymbol{y}^{-s}$ for borrowing included into the objective function. The resulting problem is

$$\text{minimize } \boldsymbol{c}^\top \boldsymbol{x} + y_0^+ + \sum_s p_s \boldsymbol{q}^{s\top} \boldsymbol{y}^{-s}$$

subject to

$$\sum_{n=1}^{N} f_{tn}x_n + (1+i_{t-1}^s)y_{t-1}^{+s} - y_t^{+s} - (1+i_{t-1}^s+\delta)y_{t-1}^{-s} + y_t^{-s} = l_t,\ \forall s,\ t = 1,\ldots,T-1$$

$$\sum_{n=1}^{T} f_{Tn}x_n + (1 + i^s_{T-1})y^{+s}_{T-1} - y^{+s}_T - (1 + i^s_{T-1} + \delta)y^{-s}_{T-1} = l_T \ \forall s$$

with $y^{+s}_0 = y^+_0$, $y^{-s}_0 = 0 \ \forall s$ and nonnegativity of all variables $\boldsymbol{x}, \boldsymbol{y}^{+s}, \boldsymbol{y}^{-s}, s = 1, \ldots, S$. Evidently, the optimal solution and the minimal cost depend on scenarios ι^s, on their probabilities and on spread δ.

This problem can be further generalized to accommodate random (scenario dependent) cash flows, liabilities and spread, to include trading possibilities and additional decision stages. To solve it, one has to generate sensible scenarios and provide other model parameters. To rewrite it in the form (2), with a fixed set of feasible first-stage decisions, we define the minimum cost for covering the discrepancies when the first-stage decision $\boldsymbol{x}, y^+_0$ is selected and scenario ι^s occurs:

$$u^s(\boldsymbol{x}, y^+_0) = \min \boldsymbol{q}^{s\top} \boldsymbol{y}^{-s}$$

subject to

$$\sum_{n=1}^{N} f_{tn}x_n + (1 + i^s_{t-1})y^{+s}_{t-1} - y^{+s}_t - (1 + i^s_{t-1} + \delta)y^{-s}_{t-1} + y^{-s}_t = l_t, \ 1 \le t \le T-1$$

$$\sum_{n=1}^{T} f_{Tn}x_n + (1 + i^s_{T-1})y^{+s}_{T-1} - y^{+s}_T - (1 + i^s_{T-1} + \delta)y^{-s}_{T-1} = l_T$$

with $y^{+s}_0 = y^+_0$, $y^{-s}_0 = 0$ and $y^{+s}_t \ge 0$, $y^{-s}_t \ge 0 \ \forall t$.

The full scenario-based problem reads now

$$\text{minimize } \boldsymbol{c}^\top \boldsymbol{x} + y^+_0 + \sum_s p_s u^s(\boldsymbol{x}, y^+_0) \tag{3}$$

with respect to $\boldsymbol{x} \ge 0$, $y^+_0 \ge 0$.

In the general case of a T-stage problem a sequence of decisions is built along each of considered data trajectories in such a way that decisions based on the same partial trajectory, on the same history, are identical (nonanticipativity) and the expected outcome (e.g., the expected gain or cost) of the decision process at time T is the best possible.

3 Contamination and Stress Testing

3.1 Basic Ideas

"Stress testing" is a term used in financial practice without any generally accepted definition. It appears in the context of quantification of losses or risks that may appear under special, mostly extremal circumstances. Such circumstances are frequently described by certain scenarios which may come from historical experience or may be judged possible in future given changes of macroeconomic, socioeconomic or political factors. The performance of the obtained optimal decision is then evaluated along these scenarios or the model is solved with an alternative input. We shall indicate now how it is possible to quantify such "stress testing" results.

Assume that the stochastic programming model for ALM, such as the stochastic dedicated bond portfolio management introduced in Section 2, was solved for a fixed set of scenarios ω^s, $s = 1, \ldots, S$, and that the influence of including other out-of-sample or stress scenarios should be considered. One could rewrite the program for the extended set of scenarios (and also constraints) and solve it. Another way is to think of this program put into the form

$$\min_{\boldsymbol{x} \in \mathcal{X}} \sum_s p_s u^s(\boldsymbol{x})$$

with a fixed set $\mathcal{X}$ of scenario-independent (first-stage) feasible solutions (the initial investments) and with performance measures u dependent on scenarios, compare with (3).

Denote by P the probability distribution concentrated on ω^s, $s = 1, \ldots, S$ with probabilities $p_s > 0$, $\sum_s p_s = 1$, by $\varphi(P)$ the optimal value of the problem and assume that the set of optimal solutions is nonempty and bounded; let $\boldsymbol{x}^*(P)$ be one of optimal solutions. Inclusion of additional scenarios means to consider another discrete probability distribution, say Q, carried by the out-of-sample or stress scenarios indexed by $\sigma = 1, \ldots, S'$, with probabilities $q_\sigma > 0$, $\sum_\sigma q_\sigma = 1$. Degenerated probability distribution Q carried only by one "stress" scenario is a special case. To quantify the consequences, one may construct the *contaminated distribution*

$$P_\lambda = (1 - \lambda)P + \lambda Q \qquad (4)$$

with a parameter $0 \leq \lambda \leq 1$. The contaminated probability distribution P_λ is carried by the pooled sample of the $S + S'$ scenarios that occur with probabilities $(1 - \lambda)p_1, \ldots, (1 - \lambda)p_S, \lambda q_1, \ldots, \lambda q_{S'}$.

The optimal value $\varphi(\lambda) = \varphi(P_\lambda)$ for the pooled sample is a finite concave function of λ on $[0, 1]$, it equals the initial value $\varphi(P)$ for $\lambda = 0$, and $\varphi(Q)$ for $\lambda = 1$. Moreover, under mild assumptions, see e.g. [8], one gets its continuity at $\lambda = 0^+$. An upper bound on its directional derivative at $\lambda = 0^+$ equals the difference between the value of the objective function $\sum_\sigma q_\sigma u^\sigma(\boldsymbol{x}^*(P))$ for the out-of-sample or stress scenarios evaluated at the optimal solution $\boldsymbol{x}^*(P)$ of the initial problem and $\varphi(P)$.

The bounds for the optimal value $\varphi(P_\lambda)$ of the problem based on the pooled sample follow from concavity of $\varphi(\lambda)$:

$$\varphi(P) + \lambda\varphi'(0^+) \geq \varphi(P_\lambda) \geq (1 - \lambda)\varphi(P) + \lambda\varphi(Q), \ 0 \leq \lambda \leq 1. \qquad (5)$$

Their final form results by substituting for $\varphi'(0^+)$:

$$(1 - \lambda)\varphi(P) + \lambda \sum_\sigma q_\sigma u^\sigma(\boldsymbol{x}^*(P)) \geq \varphi(P_\lambda) \geq (1 - \lambda)\varphi(P) + \lambda\varphi(Q) \qquad (6)$$

and is valid for all $\lambda \in [0, 1]$.

The additional numerical effort consists of

- Solving the problem

$$\min_{\boldsymbol{x} \in \mathcal{X}} \sum_{\sigma} q_\sigma u^\sigma(\boldsymbol{x}) \tag{7}$$

for the probability distribution Q carried by the out-of-sample, stress scenarios, the optimal decision is denoted $\boldsymbol{x}^*(Q)$.

In some papers stress testing is cut down to this procedure, i.e. to obtaining the optimal value $\varphi(Q)$ and comparing it with $\varphi(P)$. Such comparison may be a cause of misleading conclusions. Assume for example that $\varphi(P) = \varphi(Q)$. With exception of the constant contaminated objective function $\varphi(P_\lambda) = \varphi(P) \forall \lambda \in [0, 1]$, the concavity arguments imply that there exist values of λ for which $\varphi(P_\lambda) > \varphi(P)$.

- Evaluation and averaging the S' function values $u^\sigma(\boldsymbol{x}^*(P))$ for the new stress scenarios at the already obtained optimal solution.

This appears under the heading "stress testing" as well: one evaluates only the average performance of the obtained optimal solutions under the stress scenarios.

The assumption of discrete probability distributions P and/or Q is not important for derivation of contamination bounds. For example, for the general form (1), the average performance $\sum_\sigma q_\sigma u^\sigma(\boldsymbol{x}^*(P))$ of the optimal solution $\boldsymbol{x}^*(Q)$ in (6) is replaced by the expectation $\int_\Omega f(\boldsymbol{x}^*(P), \omega) dQ(\omega)$.

Provided that the set of optimal solutions of (7) is nonempty and bounded, similar bounds on the optimal value $\varphi(P_\lambda)$ may be also created by starting from the newly considered probability distribution Q and contaminating it by the initial one:

$$\lambda \varphi(Q) + (1 - \lambda) \sum_s p_s u^s(\boldsymbol{x}^*(Q)) \geq \varphi(P_\lambda) \geq (1 - \lambda)\varphi(P) + \lambda\varphi(Q) \tag{8}$$

for all $\lambda \in [0, 1]$. Together with the original bounds (6) one gets a tighter upper bound

$$\min\{(1 - \lambda)\varphi(P) + \lambda \sum_\sigma q_\sigma u^\sigma(\boldsymbol{x}^*(P)),\ \lambda\varphi(Q) + (1 - \lambda) \sum_s p_s u^s(\boldsymbol{x}^*(Q))\}$$

for $\varphi(P_\lambda)$. The cost is to evaluate also the performance of the optimal solution $\boldsymbol{x}^*(Q)$ of (7) along the initial S scenarios and averaging the results. See Figure 1 for illustration of these bounds.

Details can be found in [8], [10], for an application in ALM and in bond portfolio management see [12], [13], [15] and Chapter II.6 of [14].

Example 1. In the context of the stochastic bond portfolio management problem assume that the initial scenarios ι^s, $s = 1, \dots, S$ are equiprobable,

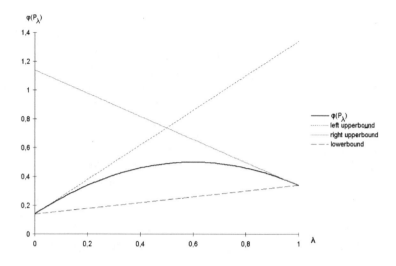

Fig. 1. Contamination bounds

i.e. $p_s = 1/S \, \forall s$ and that experts agreed on one additional interest rate scenario ι^* capturing an extremal event. This scenario is the only atom of the degenerated contaminating probability distribution Q and its probability $q = 1$. The contaminated distribution is carried then by the initial scenarios ι^s, $s = 1, \ldots, S$ and by the new scenario ι^*. Their probabilities are now $\frac{1-\lambda}{S}$ for $s = 1, \ldots, S$ and λ, respectively.

The best investment strategy $\boldsymbol{x}^*(Q)$ under contaminating scenario ι^* required in (8) can be found as an optimal solution of the corresponding deterministic program, which is a linear program in case of the linear utility function; its optimal value equals $\varphi(Q)$.

The next step reduces to evaluation of the performance of the initial optimal decision $\boldsymbol{x}^*(P)$ under the new scenario ι^*; the obtained value $u^*(\boldsymbol{x}^*(P))$ appears in (6) at the place of $\sum_\sigma q_\sigma u^\sigma(\boldsymbol{x}^*(P))$.

Probability λ assigns a weight to the view of experts and the bounds (6), (8) are valid for all $0 \leq \lambda \leq 1$. They indicate how much the weight λ, interpreted as the degree of confidence to the experts' view, affects the outcome of the investment decision.

The above results are then exploited to *quantify* the deviations in the performance of the obtained decision when new, extremal circumstances are taken into account, which is a true robustness result.

Also the impact of a modification of every single scenario according to the experts' views on the performance of the portfolio can be studied in a similar way. One uses the initial distribution P contaminated by Q which is carried now by equiprobable scenarios $\hat{\iota}^s = \iota^s + \delta^s$, $s = 1, \ldots, S$, and δ^s denotes the suggested (additive) modification of scenario ι^s. The contamination parameter λ relates again to the degree of confidence to the experts' view.

Contamination technique can be useful in postoptimality analysis with respect to inclusion of out-of-sample scenarios obtained by simulation or under disparate alternative input data, such as volatility curves in [12] or changed assumptions about behavior of insured in [15], for emphasizing the importance of a scenario by increasing its probability, in stress testing, and also in various stability studies, e.g. with respect to the assigned probabilities p_s. It is valid for multistage problems and may be also used to quantify changes due to inclusion of additional stages of the decision process, cf. [9].

Before the contamination technique can be applied the problem must be reformulated so that the set of feasible decisions is independent of P, see (3), continuity of the optimal value function $\varphi(\lambda) = \varphi(P_\lambda)$ at $\lambda = 0$ and existence of the directional derivative $\varphi'(0^+)$ must be proved and the form of the derivative which appears in the bounds must be derived. Solving the stochastic program of the same form for an alternative scenario-based probability distribution Q and evaluation the derivative means to apply a known procedure, usually for smaller optimization problems.

3.2   Comments and Extensions

Contamination technique was initiated in mathematical statistics as one of the tools for analysis of robustness of estimators with respect to deviations from the assumed probability distribution and/or its parameters. It goes back to von Mises and the concepts are briefly described e.g. in [27]. In stochastic programming it was developed first in [7] for stochastic programs written in the form (1)

$$\min_{\boldsymbol{x} \in \mathcal{X}} F(\boldsymbol{x}, P) := \int_\Omega f(\boldsymbol{x}, \omega) dP(\omega).$$

It helps to reduce the robustness analysis with respect to changes in P to a much simple analysis with respect to a scalar parameter λ : Possible changes in the probability distribution P are modeled using contaminated distributions (4) with $\lambda \in [0, 1]$ and Q another probability distribution under consideration. Due to this reduction, the results are directly applicable but they are less general than quantitative stability results with respect to arbitrary (but small) changes in P summarized in [25].

Being an expectation, the objective function in (1) is linear in P, hence

$$F(\boldsymbol{x}, \lambda) := \int_\Omega f(\boldsymbol{x}, \omega) dP_\lambda(\omega) = (1 - \lambda)F(\boldsymbol{x}, P) + \lambda F(\boldsymbol{x}, Q)$$

is linear in λ and its derivative with respect to λ equals $F(\boldsymbol{x}, Q) - F(\boldsymbol{x}, P)$.

We suppose that for all considered distributions, stochastic program (1) has an optimal solution. It is easy to prove that the optimal value function

$$\varphi(\lambda) := \min_{\boldsymbol{x} \in \mathcal{X}} F(\boldsymbol{x}, \lambda)$$

is concave for $\lambda \in [0, 1]$. This guarantees its continuity and existence of directional derivatives in the *open* interval $(0, 1)$, whereas continuity at the point $\lambda = 0$ is a property related with stability results for the stochastic program in question. In general, one needs a nonempty, bounded set of optimal solutions $\mathcal{X}^*(P)$ of the initial stochastic program (1). Various sets of assumptions are summarized in [8], the two most frequently used cases are listed below:

- Nonempty, compact $\mathcal{X}$ and $F(x, P)$, $F(x, Q)$ finite, continuous in x;
- Convex, closed $\mathcal{X}$, $F(x, Q)$ convex in x for all considered probability distributions (or $f(x, \omega)$ in (1) a convex function of x) and the set of optimal solutions $\mathcal{X}^*(P) \neq \emptyset$, bounded.

These assumptions together with stationarity of derivatives

$$\frac{dF(x, \lambda)}{d\lambda} = F(x, Q) - F(x, P)$$

are used to derive the form of the directional derivative

$$\varphi'(0^+) = \min_{x \in \mathcal{X}^*(P)} F(x, Q) - \varphi(0), \qquad (9)$$

which enters the upper bound for the concave function $\varphi(\lambda)$ in (5), cf. [8], [10] and references therein. If $x^*(P)$ is the *unique* optimal solution of (1),

$$\varphi'(0^+) = F(x^*(P), Q) - \varphi(0),$$

i.e., the local change of the optimal value function caused by a small change of P in direction $Q - P$ is asymptotically the same as that of the objective function at $x^*(P)$. If there are multiple optimal solutions of (1), each of them leads to an upper bound

$$\varphi'(0^+) \leq F(x(P), Q) - \varphi(0), \ x(P) \in \mathcal{X}^*(P).$$

Contamination bounds (6), and similarly also (8), can be then rewritten as

$$(1 - \lambda)\varphi(P) + \lambda F(x(P), Q) \geq \varphi(P_\lambda) \geq (1 - \lambda)\varphi(P) + \lambda\varphi(Q)$$

valid for an arbitrary $x(P) \in \mathcal{X}^*(P)$ and $\lambda \in [0, 1]$.

Concavity of the optimal value function $\varphi(\lambda)$ is important for constructing the above global bounds which hold true for all $\lambda \in [0, 1]$. It cannot be obtained, in general, when the set $\mathcal{X}$ depends on the probability distribution P. In such cases and under additional assumptions, only local stability results can be proved. On the other hand the results can be generalized to objective functions $F(x, P)$ convex in x and *concave* in P —the case appearing in the context of the *mean-variance objective function* and in *robust optimization* formulated in [22]; see [10], [11] for the related contamination results. To get these generalizations, it is again necessary to analyze persistence and stability properties of the parametrized problems $\min_{\boldsymbol{x} \in \mathcal{X}} F(\boldsymbol{x}, P_\lambda)$ and to derive the form of the directional derivative. Under the assumptions listed above, the optimal value function $\varphi(\lambda)$ remains concave on $[0, 1]$. Additional assumptions are needed to get the existence of the derivative

$$\varphi'(0^+) = \min_{\boldsymbol{x} \in \mathcal{X}^*(P)} \frac{d}{d\lambda} F(\boldsymbol{x}, P_\lambda)|_{\lambda=0^+}.$$

Example 2. Consider the mean-variance objective function

$$F(\boldsymbol{x}, P; \varrho) := -E_P r(\boldsymbol{x}, \omega) + \varrho \mathrm{var}_P r(\boldsymbol{x}, \omega) \tag{10}$$

with $r(\boldsymbol{x}, \omega)$ the random return of an investment $\boldsymbol{x} \in \mathcal{X}$ attained when the realization ω occurs; $\varrho > 0$ is a fixed parameter. By minimization of (10) for changing values of the parameter ρ one gets mean-variance efficient solutions and the points on the mean-variance frontier of the corresponding Markowitz model.

The variance of return for the contaminated probability distribution P_λ

$$\mathrm{var}_{P_\lambda} r(\boldsymbol{x}, \omega) = E_{P_\lambda} r^2(\boldsymbol{x}, \omega) - (E_{P_\lambda} r(\boldsymbol{x}, \omega))^2$$
$$= (1 - \lambda) E_P r^2(\boldsymbol{x}, \omega) + \lambda E_Q r^2(\boldsymbol{x}, \omega) - ((1 - \lambda) E_P r(\boldsymbol{x}, \omega) + \lambda E_Q r(\boldsymbol{x}, \omega))^2$$

is a concave function of λ for $\lambda \geq 0$. Its derivative

$$\frac{d\mathrm{var}_{P_\lambda} r(\boldsymbol{x}, \omega)}{d\lambda}|_{\lambda=0} = \mathrm{var}_Q r(\boldsymbol{x}, \omega) - \mathrm{var}_P r(\boldsymbol{x}, \omega) + (E_Q r(\boldsymbol{x}, \omega) - E_P r(\boldsymbol{x}, \omega))^2.$$

The objective function (10) for the contaminated probability distribution P_λ is

$$F(\boldsymbol{x}, P_\lambda; \varrho) = -(1 - \lambda) E_P r(\boldsymbol{x}, \omega) - \lambda E_Q r(\boldsymbol{x}, \omega) + \varrho \mathrm{var}_{P_\lambda} r(\boldsymbol{x}, \omega)$$

and its directional derivative

$$\frac{dF(\boldsymbol{x}, P_\lambda; \varrho)}{d\lambda}|_{\lambda=0^+} = F(\boldsymbol{x}, Q; \varrho) - F(\boldsymbol{x}, P; \varrho) + \varrho(E_Q r(\boldsymbol{x}, \omega) - E_P(r(\boldsymbol{x}, \omega))^2.$$

The optimal value function of the contaminated problem,

$$\varphi(\lambda; \varrho) := \min_{\boldsymbol{x} \in \mathcal{X}} F(\boldsymbol{x}, P_\lambda; \varrho)$$

is a concave function of λ and $\varphi(0; \varrho)$ coincides with the optimal value $\varphi(P; \varrho)$ of (10) on the set $\mathcal{X}$. Under similar conditions as for the expected value objective function in (1), its one-sided derivative exists,

$$\varphi'(0^+; \varrho) \leq F(\boldsymbol{x}(P), Q; \varrho) - \varphi(P; \varrho) + \varrho(E_Q r(\boldsymbol{x}(P), \omega) - E_P(r(\boldsymbol{x}(P), \omega))^2$$

and the contamination bounds of the type (6) follow.

Even without convexity with respect to x one may be able to prove the needed stability results, such as the *joint* continuity of $F(x, P)$, and apply Theorem 7 in [8] to get the existence and the form of the directional derivative. This was examined for two-stage stochastic integer programs, see e.g. [6].

We shall see in the next Section that an application of the above results to stability analysis and stress testing for the Conditional Value at Risk (CVaR) is straightforward.

There exist results for optimal solutions of contaminated stochastic programs and for the case that also constraints depend on P, but these results are not yet ready for a direct practical exploitation.

4   Contamination and Stress Testing for CVaR

4.1   Basic Formulas

Value at Risk (VaR) was introduced and recommended as a generally applicable risk measure to quantify, monitor and limit financial risks and to identify losses which occur with an acceptably small probability.

Denote

- $g(x, \omega)$ the loss if $x \in \mathcal{X}$ is selected and realization ω occurs,
- $P\{\omega : g(x, \omega) \leq k\} := G(x, P; k)$ the distribution function of the loss connected with a fixed decision x,
- $\alpha \in (0, 1)$ a selected confidence level.

Then the **Value at Risk** at the confidence level α is defined as

$$\mathrm{VaR}_\alpha(x, P) = \min\{k \in R : G(x, P; k) \geq \alpha\} \tag{11}$$

or

$$\mathrm{VaR}_\alpha^+(x, P) = \inf\{k \in R : G(x, P; k) > \alpha\}.$$

Hence, random losses greater than VaR occur with probability $1 - \alpha$. This interpretation is well understood in the financial practice.

It turns out, however, that there are various weak points of the recommended VaR methodology. To settle these problems new risk measures have been introduced. Here we shall discuss one of them—the Conditional Value at Risk.

The **Conditional Value at Risk**—CVaR$_\alpha$ is the mean of the α-tail distribution H_α of $g(x, \omega)$ defined as

$$
\begin{aligned}
H_\alpha(x, P; k) &= \qquad 0 \qquad\quad \text{for } k < \mathrm{VaR}_\alpha(x, P) \\
H_\alpha(x, P; k) &= \tfrac{G(x, P; k) - \alpha}{1 - \alpha} \ \ \text{for } k \geq \mathrm{VaR}_\alpha(x, P).
\end{aligned}
\tag{12}
$$

Assume that $E_P|g(\boldsymbol{x},\omega)| < \infty \, \forall \boldsymbol{x} \in \mathcal{X}$ and define

$$\Phi_\alpha(\boldsymbol{x},\psi,P) = \psi + \frac{1}{1-\alpha} E_P(g(\boldsymbol{x},\omega) - \psi)^+. \tag{13}$$

The fundamental minimization formula in [24] helps to evaluate CVaR and to analyze its stability including stress testing.

Theorem [24]. *As a function of ψ, $\Phi_\alpha(\boldsymbol{x},\psi,P)$ is finite and convex (hence continuous) with*

$$\min_\psi \Phi_\alpha(\boldsymbol{x},\psi,P) = \mathrm{CVaR}_\alpha(\boldsymbol{x},P) \tag{14}$$

and

$$\arg\min_\psi \Phi_\alpha(\boldsymbol{x},\psi,P) = [\mathrm{VaR}_\alpha(P,\boldsymbol{x}), \mathrm{VaR}_\alpha^+(\boldsymbol{x},P)]. \tag{15}$$

The auxiliary function $\Phi_\alpha(\boldsymbol{x},\psi,P)$ is linear in P and convex in ψ. To get persistence and stability properties with respect to P, it is enough to assume that the set (15) of optimal solutions of the simple stochastic program (14) is nonempty and bounded—a natural request concerning the quantiles of the probability distribution $G(\boldsymbol{x},P;\bullet)$.

There are various papers discussing properties of VaR, CVaR and relations between CVaR and VaR, see e.g. [5], [23]. We shall focus on the contamination-based stress testing for CVaR. The presence of probability constraints in definition of VaR requires that various distributional and structural properties are fulfilled, namely, for the unperturbed problem. These requirements rule out direct applications of contamination technique in case of empirical VaR whereas for normal distribution and parametric VaR one may exploit stability results valid for quadratic programs. Some related results on stress testing for VaR can be found e.g. in [16], [20].

4.2   Stress Testing for CVaR

Let P be a discrete probability distribution concentrated on $\omega^1,\dots,\omega^S$ with probabilities $p_s > 0$, $s = 1,\dots,S$ and $\boldsymbol{x}$ a fixed element of $\mathcal{X}$. Then the program (14) has the form

$$\min_\psi \psi + \frac{1}{1-\alpha} \sum_s p_s(g(\boldsymbol{x},\omega^s) - \psi)^+ \tag{16}$$

and can be further rewritten as

$$\min_{\psi, y_1,\dots,y_S} \{\psi + \frac{1}{1-\alpha} \sum_s p_s y_s : y_s \geq 0, \, y_s + \psi \geq g(\boldsymbol{x},\omega^s) \forall s\}.$$

Consider now a stress test of $\mathrm{CVaR}_\alpha(\boldsymbol{x},P)$, i.e., of the optimal value of (16). Let $\psi^* = \psi^*(\boldsymbol{x},P)$ be an optimal solution of (16) and ω^* be the stress scenario. We apply the contamination technique and proceed as explained in Example 1.

The $\text{CVaR}_\alpha(\boldsymbol{x}, Q)$ value for the degenerated probability distribution Q carried by the stress scenario ω^* equals $g(\boldsymbol{x}, \omega^*)$, the value $\Phi_\alpha(\boldsymbol{x}, \psi^*, Q) = \psi^* + \frac{1}{1-\alpha}(g(\boldsymbol{x}, \omega^*) - \psi^*)^+$. Hence, the bounds for the CVaR_α for the contaminated probability distribution P_λ carried by the initial scenarios ω^s, $s = 1, \ldots$, with probabilities $(1 - \lambda)p_s$ and by the stress scenario ω^* with probability λ have the form

$$(1 - \lambda)\text{CVaR}_\alpha(\boldsymbol{x}, P) + \lambda\Phi_\alpha(\boldsymbol{x}, \psi^*, Q) = \Phi_\alpha(\boldsymbol{x}, \psi^*, P_\lambda) \geq \qquad (17)$$

$$\geq \text{CVaR}_\alpha(\boldsymbol{x}, P_\lambda) \geq (1 - \lambda)\text{CVaR}_\alpha(\boldsymbol{x}, P) + \lambda\text{CVaR}_\alpha(\boldsymbol{x}, Q)$$

and are valid for all $\lambda \in [0, 1]$; compare with (6). The difference between the upper and lower bound equals

$$\lambda[\Phi_\alpha(\boldsymbol{x}, \psi^*, Q) - \text{CVaR}_\alpha(\boldsymbol{x}, Q)] = \lambda[\psi^* + \frac{1}{1 - \alpha}(g(\boldsymbol{x}, \omega^*) - \psi^*)^+ - g(\boldsymbol{x}, \omega^*)].$$

As the next step, let us discuss briefly **optimization problems with the $\text{CVaR}_\alpha(\boldsymbol{x}, P)$ objective function**

$$\text{minimize } \text{CVaR}_\alpha(\boldsymbol{x}, P)$$

on a closed, nonempty set $\mathcal{X} \in R^n$, cf. [1]. Using (14), the problem is

$$\min_{\boldsymbol{x}, \psi} \Phi_\alpha(\boldsymbol{x}, \psi, P), \; \boldsymbol{x} \in \mathcal{X}. \qquad (18)$$

For convex $\mathcal{X}$ and convex loss functions $g(\bullet, \omega)$ for all ω, $\Phi_\alpha(\boldsymbol{x}, \psi, P)$ is convex in $(\boldsymbol{x}, \psi)$ and standard stability results apply. Moreover, if P is the considered discrete probability distribution, $g(\bullet, \omega)$ a linear function of $\boldsymbol{x}$ and $\mathcal{X}$ convex polyhedral, we get a linear program

$$\min_{\psi, y_1, \ldots, y_S, \boldsymbol{x}} \{\psi + \frac{1}{1 - \alpha}\sum_s p_s y_s : y_s \geq 0, \; \boldsymbol{x}^\top \omega^s - \psi - y_s \leq 0 \, \forall s, \; \boldsymbol{x} \in \mathcal{X}\}. \quad (19)$$

Let $\psi^*(P)$, $\boldsymbol{x}^*(P)$ be an optimal solution of (18) and denote $\varphi_{C_\alpha}(P)$ the optimal value. To get contamination bounds for the optimal value of (18) with P contaminated by a stress probability distribution Q it is sufficient to assume a compact set of optimal solutions of (18). An evident instance is compact $\mathcal{X}$ and bounded interval (14). The bounds follow the usual pattern, compare with (6):

$$(1 - \lambda)\varphi_{C_\alpha}(P) + \lambda\Phi_\alpha(\boldsymbol{x}^*(P), \psi^*(P), Q) \geq \varphi_{C_\alpha}(P_\lambda) \geq (1 - \lambda)\varphi_{C_\alpha}(P) + \lambda\varphi_{C_\alpha}(Q).$$

To apply them one has to evaluate $\Phi_\alpha(\boldsymbol{x}^*(P), \psi^*(P), Q)$ and to solve (18) for the stress distribution Q. See Figure 2 for an example of contamination bounds obtained in the numerical example from [16].

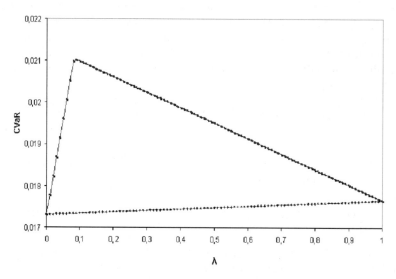

Fig. 2. Contamination bounds for CVaR

The values for $\lambda = 0$ and $\lambda = 1$ correspond to minimal CVaRs for distributions P and Q, respectively, both of them carried by different 5184 equiprobable scenarios. The optimal CVaR for the pooled sample of 10368 equiprobable scenarios lies in the interval $[0.0175, 0.0195]$ which corresponds to $\lambda = 1/2$. If the bounds are acceptably tight, the optimal CVaR for the pooled sample need not be computed.

4.3   Stress Testing for CVaR-Mean Return Efficient Problem

Similarly as for the Markovitz mean-variance problem, one considers two criteria – minimize $\mathrm{CVaR}_\alpha(\boldsymbol{x}, P)$ and maximize expected return $E_P r(\boldsymbol{x}, \omega)$ on a set $\mathcal{X}$. Two reformulations of this bi-criterial problem provide efficient solutions:

$$\min_{\boldsymbol{x} \in \mathcal{X}} \mathrm{CVaR}_\alpha(\boldsymbol{x}, P) - k E_P r(\boldsymbol{x}, \omega) \tag{20}$$

with $k \geq 0$ a parameter, compare with (10), or

$$\min \mathrm{CVaR}_\alpha(\boldsymbol{x}, P) \text{ s.t. } \boldsymbol{x} \in \mathcal{X}, E_P r(\boldsymbol{x}, P) \geq r \tag{21}$$

with parameter $r (\geq r_0)$.

Optimal solutions $\boldsymbol{x}_k^*(P)$, $\boldsymbol{x}_r^*(P)$ of (20) and (21) depend on the tradeof parameter values k and r, respectively.

The second reformulation is favored in the practice. Solving (21), one gets directly points $[\text{CVaR}_\alpha(\boldsymbol{x}_r^*(P), P), r]$ on the CVaR-mean efficient frontier in dependence on the specified value of parameter r.

Dependence of the set of feasible solutions of (21) on P means that in general, the optimal value for contaminated P_λ is *not concave* in λ. On the other hand, the set of feasible decisions of (20) is fixed, independent of the distribution, hence, contamination bounds for the optimal value function can be constructed as for CVaR evaluation or optimization. These, however, are not the bounds around the efficient frontier.

To trace out the **CVaR-mean return efficient frontier** one may solve (20) or (21) for many different values of k, r, respectively, or rely on parametric programming techniques. In the sequel we shall assume that $g(\boldsymbol{x}, \omega) = -r(\boldsymbol{x}, \omega) = \boldsymbol{x}^\top \omega$, $\mathcal{X}$ is a convex polyhedral set and P is a discrete probability distribution. Then both (20), (21) may be solved via *parametric linear programming* techniques, cf. [26].

Contamination of probability distribution P introduces an additional parameter λ into (20) and (21). As a consequence, nonlinearity with respect to k, λ appears in the objective function of (20) whereas both the objective function and the set of feasible solutions of (21) depend linearly on parameters.

Example 3. Assume in addition that $E_P \omega = E_Q \omega = \bar{\omega}$. Then the set of feasible decisions of (21) does not depend on λ, and the contamination bounds apply. Such assumption appears when scenarios are generated by the moment fitting method, see e.g. [18]. In this case, the nonlinear dependence of k and λ in the objective function of the contaminated program (20) disappears and contamination bounds for the CVaR-mean return problem can obtained by solving (21).

For solving the contaminated problem (21) one may apply the simplex based techniques of [17]. The problem is a linear parametric program with two independent parameters, λ in the objective function and r on the right-hand sides of constraints. Let us mention some favorable properties of such parametric programs related with their general form

$$\min \left\{ (\boldsymbol{c} + \lambda \hat{\boldsymbol{c}})^\top \boldsymbol{x} : \boldsymbol{A}\boldsymbol{x} = \boldsymbol{b} + r\hat{\boldsymbol{b}}, \boldsymbol{x} \geq 0 \right\} \tag{22}$$

with $(r, \lambda) \in \mathcal{A}$, a nonempty, closed two-dimensional interval, cf. Theorem 3.2 in [17].

In our CVaR-mean return problem , $\lambda \in [0, 1]$, $\boldsymbol{c}$ comes from P, $\hat{\boldsymbol{c}}$ from the "direction" $Q - P$ and $r \in [r_L, r_U]$ appears only in the mean return constraint $\bar{\omega}^\top \boldsymbol{x} \geq r$; we have obviously

$$r_L = \min\{\bar{\omega}^\top \boldsymbol{x}^*(P), \bar{\omega}^\top \boldsymbol{x}^*(Q)\}, \quad r_U = \max_{\boldsymbol{x} \in \mathcal{X}} \bar{\omega}^\top \boldsymbol{x}.$$

The assumption about $\mathcal{A} := [0, 1] \times [r_L, r_U]$ is fulfilled if $r_L < r_U$. In such case, existence of optimal solution of (22) is guaranteed for all $(r, \lambda) \in \mathcal{A}$ and for optimal solutions, the mean return constraint is active. Moreover,

- The optimal value function $\varphi(r, \lambda)$ is continuous on $\mathcal{A}$, convex in r, concave in λ.
- The two-dimensional interval $\mathcal{A}$ can be decomposed in a finite number of closed intervals, say, $\mathcal{A}^h_{(r,\lambda)}$ such that there exist optimal solution $\boldsymbol{x}^*_r(P_\lambda)$ of (22) which is linear on $\mathcal{A}^h_{(r,\lambda)}$ and the optimal value function is linear there in r and in λ. See Figure 3.

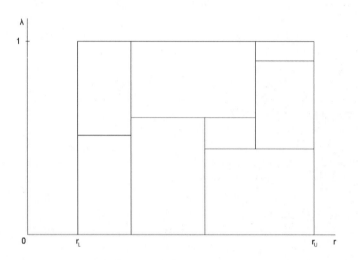

Fig. 3. Decomposition of set $\mathcal{A}$

The simplex-based algorithm detailed in Section 3.3 of [17] uses two columns for solution components and two rows for the criterium. The critical boundaries of intervals $\mathcal{A}^h_{(r,\lambda)}$ are obtained by discussion of feasibility and optimality conditions with respect to parameters r, λ.

These properties and Figure 3 indicate that for values of $\lambda \le \lambda_1$, $\lambda_1 > 0$ small enough, the efficient solutions of the contaminated problem are equal to optimal solutions of the noncontaminated problem (21), i.e., they do not depend on λ :

$$\boldsymbol{x}^*_r(P_\lambda) = \boldsymbol{x}^*_r(P) \, \forall r \in [r_l, r_U]$$

and $\varphi(r, \lambda)$, the optimal contaminated value $\mathrm{CVaR}_\alpha(\boldsymbol{x}^*_r(P_\lambda), P_\lambda)$ with a fixed mean return r, is linear in λ.

Hence, under assumptions of the example *small contamination of P does not influence composition of CVaR-mean return efficient portfolios.*

5 Conclusions

The contamination technique is presented as a tool suitable for postoptimality and sensitivity analysis of the optimal value with respect to various input perturbations. For scenario-based stochastic programs, it is easily applicable in out-of-sample analysis and stress testing for portfolio management models of the recourse and robust optimization type. This extends also to mean-variance and CVaR optimization whereas its application for CVaR-mean efficient portfolios is more involved.

Acknowledgments. This work was partly supported by research project MSM 0021620839 and by the Grant Agency of the Czech Republic (grants 201/05/2340 and 402/05/0115).

References

1. Andersson, F., Mausser, H., Rosen, D., Uryasev, S. (2001) Credit risk optimization with Conditional Value-at-Risk criterion. Math. Program. **B 89**, 273–291
2. Bertocchi, M., Dupačová, J., Moriggia, V. (2000) Sensitivity of bond portfolio's behavior with respect to random movements in yield curve: A simulation study. Ann. Oper. Res. **99**, 267–286
3. Bertocchi, M., Dupačová, J., Moriggia, V. (2005) Horizon and stages in applications of stochastic programming in finance. Ann. Oper. Res., to appear
4. Chopra, W. K., Ziemba, W. T. (1993) The effect of errors in means, variances and covariances on optimal portfolio choice. J. of Portfolio Management **19**, 6–11
5. Dempster, M. A. H. (ed.) (2002) Risk Management: Value at Risk and Beyond. Cambridge Univ. Press
6. Dobiáš, P. (2003) Contamination technique for stochastic integer programs. Bulletin of the Czech Econometric Society **18**, 65–80
7. Dupačová, J. (1986) Stability in stochastic programming with recourse – contaminated distributions. Math. Programing Study **27**, 133–144
8. Dupačová, J. (1990) Stability and sensitivity analysis in stochastic programming. Ann. Oper. Res. **27**, 115-142
9. Dupačová, J. (1995) Postoptimality for multistage stochastic linear programs. Ann. Oper. Res. **56**, 65–78
10. Dupačová, J. (1996) Scenario based stochastic programs: Resistance with respect to sample. Ann. Oper. Res. **64**, 21–38
11. Dupačová, J. (1998) Reflections on robust optimization. In: Marti, K., Kall, P. (eds.) (1998) Stochastic Programming Methods and Technical Applications. LNEMS **437**, Springer, Berlin, 111–127
12. Dupačová, J., Bertocchi, M. (2001) From data to model and back to data: A bond portfolio management problem. European J. Oper. Res. **134**, 33–50
13. Dupačová, J., Bertocchi, M. and Moriggia, V. (1998) Postoptimality for scenario based financial models with an application to bond portfolio management. In: Ziemba, W. T., Mulvey, J. (eds.) (1998) World Wide Asset and Liability Modeling. Cambridge Univ. Press, 263–285
14. Dupačová, J., Hurt, J., Štěpán, J. (2002) Stochastic Modeling in Economics and Finance, Part II. Kluwer Acad. Publ.. Dordrecht
15. Dupačová, J., Polívka, J. (2004) Asset-liability management for Czech pension funds using stochastic programming. SPEPS 2004–01 (downloadable from http://dochost.rz.hu-berlin.de/speps)

16. Dupačová, J., Polívka, J. (2005) Stress testing for VaR and CVaR. SPEPS 2005–01 (downloadable from http://dochost.rz.hu-berlin.de/speps)
17. Guddat, J., Guerra Vasquez, F., Tammer, K., Wendler, K. (1986) Multi-objective and Stochastic Optimization Based on Parametric Optimization. Akademie-Verlag, Berlin
18. Høyland, K., Wallace, S. W. (2001) Generating scenario trees for multistage decision problems. Manag. Sci. **47**, 295–307
19. Kaut, M., Vladimirou, H., Wallace, S. W., Zenios, S. A. (2003) Stability analysis of a portfolio management model based on the conditional value-at-risk measure. Submitted
20. Kupiec, P. (2002): Stress testing in a Value at Risk framework. In: Dempster, M. A. H. (ed.) Risk Management: Value at Risk and Beyond. Cambridge Univ. Press, 76–99
21. Kusy, M. I., Ziemba, W. T. (1986) A bank asset and liability management model. Oper. Res. **34**, 356–376
22. Mulvey, J. M., Vanderbei, R. J., Zenios, S. A. (1995) Robust optimization of large scale systems. Oper. Res. **43**, 264–281
23. Pflug, G. Ch. (2001) Some remarks on the Value-at-Risk and the Conditional Value-at-Risk. In: Uryasev, S. (ed.) Probabilistic Constrained Optimization, Methodology and Applications. Kluwer Acad. Publ., Dordrecht, 272–281
24. Rockafellar, R. T., Uryasev, S. (2001) Conditional value-at-risk for general loss distributions. J. of Banking and Finance **26**, 1443–1471
25. Römisch, W. (2003) Stability of stochastic programming problems. Chapter 8 in: Ruszczynski, A., Shapiro, A. (eds.) Handbook on Stochastic Programming. Elsevier, Amsterdam, 483–554
26. Ruszczyński, A., Vanderbei, R. J. (2003) Frontiers of stochastically nondominated portfolios. Econometrica **71**, 1287–1297
27. Serfling, R. J. (1980) Approximation Theorems of Mathematical Statistics. Wiley, New York
28. Wallace, S. W., Ziemba, W. T. (eds.) (2005) Applications of Stochastic Programming. MPS-SIAM Mathematical Series on Optimization, SIAM and MPS, Philadelphia
29. Ziemba, W. T. (2004) The Stochastic Programming Approach to Asset and Liability Management. AIMR, Charlotteville, Virginia
30. Ziemba, W. T., Mulvey, J. (eds.) (1998) World Wide Asset and Liability Modeling. Cambridge Univ. Press

Structured Modeling for Coping with Uncertainty in Complex Problems

M. Makowski

International Institute for Applied System Analysis, A-2361 Laxenburg, Austria

Abstract. Uncertainty is a key issue in many public debates and policy making, including climate change, pension systems, and integrated management of catastrophic risks. Rational treatment of uncertainty in many such situations requires new methods not only for the appropriate handling of endogenous uncertainties but also for modeling complex problems.

The paper first outlines the key issues related to uncertainties and risks, including some pitfalls of using traditional methods in situations when they are inappropriate. Then, new methods of modeling endogenous uncertainties and catastrophic risks are summarized. Next, structured modeling technology developed for handling the whole modeling process of model-based support for solving complex problems is discussed.

The development of the presented methods has been motivated by actual policy-making issues, and the methods have been applied to complex problems. However, the presentation is deliberately kept at a level comprehensible to a broad audience.

Key words: endogenous uncertainty, catastrophic risk, structured modeling, decision-making support, mathematical modeling, integrated model analysis, optimization.

1 Introduction

In reality everybody has to cope with uncertainty and manage diversified risks in both their private and professional lives. People routinely accommodate uncertainties with diversified rational or accepting responses, in most cases successfully and unconsciously; this proliferates a popular feeling that *on average* uncertainties and risks are easy to handle. Only in those much less frequent situations (when something *unexpected* happens) one realizes that uncertainties and risks play an important role. A closer look at *the unexpected* usually shows that one should have considered such an outcome, and could have implemented measures to at least mitigate the consequences. Analysis of such situations is obviously much easier for personal decision-making than for complex problems where decisions potentially affect large communities and/or have long-term (often irreversible) consequences. Such problems are typically characterized by uncertain time-delays (between decisions and consequences) and/or rather complex relations between decision-making and the resulting consequences (in particular resulting from rare events) that occur at a random time or place.

Rational decision-making should be based on science, i.e. on organized knowledge.[1] Yet science is no different to all other areas of human activities:[2] uncertainties are certainly present in science as well (although often ignored or denied). Unfortunately, rational decision-making processes as well as public policy debates are hindered by inadequate representations, communication and the perception of diverse types of uncertainties and risks. Bertrand Russell summarized this with: *The whole problem with the world is that fools and fanatics are always so certain of themselves, with wiser people so full of doubts.*

The truth is that there are no certain (undoubted, definite) answers/solutions for/to uncertain problems. Nevertheless science can support rational decision-making (also despite uncertainties) by identifying so-called robust solutions (see [15]) that are not only good for most (unknown) futures (i.e. many scenarios describing various futures) but also sensitive to extreme scenarios. This truth is hardly recognized in public debates. People simply prefer (over)-simplified deterministic representations of complex problems, and easy to understand solutions. We discuss just three explanations for this situation. Firstly, in a non-scientific discussion it is virtually impossible to amply present a model adequately representing a complex problem. Thus the discussed problem is typically illustrated by an oversimplified[3] representation of the model, e.g., averages are used instead of (spatial, temporal, social) distributions, only selected scenarios are outlined, and simplified relations between decisions and consequences are considered. Secondly, humans naturally tend to perceive the world in deterministic terms; it is commonly known that perception, especially of "medium" and low probabilities (in the range of 0.3-0.7, and smaller than 0.01, respectively), is rather poor (this also applies to most scientists). Thirdly, public discussions are typically dominated by presentations and defense of arguments justifying one of the extremes:[4] (1) *Do nothing, wait, and react/adapt to the consequences*, or (2) *Plan for the worst case*. During such discussions it is often "forgotten" that in the presence of uncertainties one can easily select two subsets of scenarios that justify (separately) each of these extremes.

This chapter aims at combining the knowledge and experience of researchers working in various fields (including mathematics, operational research, and social science) with practitioners (including researchers) experienced in applying diversified methods of uncertainty treatment in a wide spectrum of fields. One of the many divisions in science is that between the

---

[1] Knowledge is understood here as being composed of facts and rules.

[2] Recall e.g., *In this world, nothing is certain but death and taxes* (Benjamin Franklin).

[3] In the sense of the ability to present all key characteristics of the problem.

[4] Although an extreme solution is rarely the best one, see e.g., [36]. This observation not only can be derived from the properties of the classical two-stage stochastic control problem; in most cases it can be simply justified by a common sense.

deterministic and stochastic views of a problem. This chapter is written from a deterministic perspective, by the author who has developed modeling technology for complex deterministic models, and extensively participated in the development of such models. However, the author has also been collaborating with colleagues working with complex stochastic models, and thanks to their friendly attitude has developed a reasonable understanding of novel methods for the treatment of uncertainties and risks. This has resulted in on-going activities directed at extending the capability of the structured modeling technology developed for deterministic models to handle stochastic models as well. Currently we only have a sort of gangway between the deterministic and stochastic modeling technology; the chances are good however, that this foot-bridge will soon extend into a full size bridge allowing for truly comprehensive problem analysis by integrating pertinent methods developed for both deterministic and stochastic approaches.

2   Context

Decision making is becoming more and more difficult because decision problems are no longer well-structured problems that are easy to solve by intuition or experience supported by relatively simple calculations. Even the same type of problems that used to be easy to define and solve, are now much more complex because of the globalization of the economy, and a much greater awareness of its linkages with various environmental, social, and political issues. Moreover, decision-making is done for the future, which is actually always uncertain. Thus, any decision-maker needs to cope with various uncertainties in order to rationally manage corresponding risks.

Many decision problems have a lot in common, and sets of such problems can be identified by different distinct problem characteristics. The varied contexts of decision problems has important implications on the decision making support (see e.g., [20]) and the development of the corresponding models (see e.g., [30]). The latter in turn implies that mathematical modeling has been, and will continue to be, a combination of science and craft [37].

Rational decision making typically requires, see e.g. [28]:

- a mathematical model adequately representing the relationships between decisions and outcomes (the consequences of applying a decision), including an assessment of the temporal and spatial consequences of implementing a selected decision;
- understanding the uncertainties related to various representations of such relationships;
- a representation of preferential structures (a measure of the trade-offs between various outcomes) of the stake holders (persons and/or institutions affected by the consequences of implementing decisions);

- criteria for the evaluation of various risks related to either implementing a (best at the moment) decision or postponing making a decision (until a possibly better decision can be made); and
- a procedure (conventionally called DMP – Decision Making Process) for selecting *the best* solution (decision), including a process for involving stake holders in the DMP.

It is not practicable to attempt to deal with all these issues for any given decision problem. Each of these elements has a large number of methods and corresponding tools and an attempt to fully exploit the capabilities of many of them for a given problem is doomed to failure. Different decision problems and the associated DMPs have different characteristics, which call for focusing on the implementation of a selection of methods and tools.

There are however some issues common to many such methods and which are therefore interesting for a broad audience. In particular we focus on challenging problems involving endogenous uncertainties and catastrophic risks; the analysis of such problems requires complex models which in turn demand new modeling technology. Therefore, the two sets of corresponding issues, namely those related to (1) uncertainty and risk, and (2) modeling technology are discussed in the remaining part of this chapter.

3   Uncertainty and Risk

3.1   General Comments on Uncertainty

One distinguishes between two types of uncertainty related to a considered phenomena:

- epistemic uncertainty: due to incomplete knowledge (which ranges from deterministic knowledge to total ignorance) of the phenomena,
- aleatory (variability) uncertainty: due to the inherent variability (i.e., natural randomness) of the phenomena, e.g., natural processes; human behavior; social, economic, technological dynamics; and discontinuities (or fast changes) in some of these processes.

The epistemic uncertainty can be reduced provided that there is time and the resources to do so. Also in some cases better characteristics of the aleatory uncertainty can be obtained by additional observations and/or experiments. However, during the time needed for reducing any of these uncertainties, the decision space can be substantially reduced, or even irreversible changes in the system under consideration may occur. This problem is discussed in more detail in [36].

The most common treatment of the uncertainty is through one of the following three paradigms of probability defined as:

- the ratio of favorable events to the total number of *equally likely* events (Laplace),

- the long-run frequency of the event, if the experiment was repeated *many* times (von Mises),
- a measure of a subjective degree of certainty about the event (Ramsey [38], Bayes, Keynes).

The first two paradigms assume that probability is an attribute of the corresponding event (or object), the third one is based on beliefs.

There are at least three pitfalls when using one of the established paradigms for supporting decision making under uncertainty:

- Incorrect calculation of probabilities, e.g. applying the Laplace's paradigm to events that are not equally likely; or violating the assumption of von Mises by: counting frequency from observations of events that occurred under different conditions, or by using a small sample of data, or by interpreting as data results provided by various models based on related data, or by multiple use of the same data each interpreted[5] as independent events. Probability defined as the relative frequency is equal to the limit of an infinite sequence; however, in practice it is rarely proved to what extent the probability actually corresponds to the relative frequency (which in real applications has to be inferred from a finite, often rather small, subset of the infinite sequence). For such cases distribution-free approaches[6] or "uncertain"-distribution methods are actually required.
- Correct probabilities provide a good basis for frequently repeated decision-making provided neither the probability distribution nor payoffs change "substantially"[7] (because this is a condition for a good approximation of an infinite sequence of decisions by a finite subsequence), and one wants to optimize a total expected outcome (defined as a sum of payoffs weighted by their probabilities). However, as demonstrated already in 1739 by Bernoulli's St Petersburg paradox (see for e.g., [3]), maximization of an expected outcome (or utility) is not rational for situations where a decision is made only once, or when, for a sequence of decisions, the consequences of each decision should be evaluated separately.
- The fact that the vast majority of experts is certain about the validity of a statement/theory does not make it true. For example, over 600 years ago it was commonly agreed that the Earth was the center of the Universe.

Of course, the well established paradigms of uncertainty treatment have been successfully used in countless applications in science and industry, and will continue to be a major method for dealing with uncertainty. The above summary aims to highlight that in many other situations the well established methods cannot provide useful results. This is caused by either the nature of

---

[5] Wittgenstein described this as buying several copies of a newspaper to increase a probability that a news is true.

[6] Methods that do not postulate specific a priori distributions.

[7] Note that this condition can hardly be met for observations spanned over long time periods.

the underlying uncertainties or an inadequate set of data. In such situations proper robust approaches have to deal with sufficiently large sets of possible future scenarios rather than with only one (e.g., either negotiated or believed to be most likely) scenario.

Rational analysis of, and decision-making support for, problems involving some types of uncertainties and risks require novel approaches. In particular, proper treatment of endogenous uncertainties and/or catastrophe risks demands new methodology and corresponding software tools. Such problems are characterized by a vast variety of inherent, practically irreducible uncertainties and/or "unknown" risks, and/or by spatial, temporal or social heterogeneity, see e.g., [5,13,17]. The risks often involve events with catastrophic consequences due to either irrecoverable shocks (e.g., insolvency), or a magnitude of impact that may affect at once large territories and communities (e.g., natural or man-made catastrophes).

Such a methodology, called *Robust Decisions*, is presented in [15]. Thus, below we only outline the basic characteristics of endogenous uncertainties and catastrophe risks in order to illustrate the complexity of the corresponding models, which in turn need new modeling technology.

3.2   Endogenous Uncertainty

Traditional models in economics, insurance, risk-management, and extreme value theory require evaluations based on corresponding assumptions and large-enough sets of data. For example, standard insurance theory essentially relies on the assumption of independent, frequent, low-consequence (conventional) risks, such as car accidents, and extensive sets of data about accidents/losses, owners, etc. Thus insurance companies use well established models (exploiting rich sets of historical data) for making decisions on premiums, estimating claims and the likelihood of insolvency. However, such data are not available for new[8] problems or for problems involving rare events. Experiments aiming at collecting data, even if possible, may be very expensive and/or dangerous. Moreover, in many situations, especially in policy-making and management, experiments are simply impossible.

In particular, traditional approaches are not applicable to problems involving catastrophes (understood as rare events with large consequences). Catastrophes typically result in abrupt irreversible changes occurring on extremely large spatial, temporal, and social scales. Large-scale potential catastrophic impacts, and the magnitudes of the uncertainties that surround them in particular, are critically important for the climate-change policy debates and the associated decision-making processes, see e.g., [26,31,34,35,42].

---

[8] *New* also represents old/known types of problems which, however, are not stationary, i.e., whose parameters change over time. This in turn may imply that even large existing sets of data are not adequate for the identification of parameters of models representing such problems.

A more detailed discussion of the relevance of a proper treatment of uncertainty to policy-making is presented in [25].

Traditional risk analysis (based on the concept of expected cost-benefit analysis) practically ignores catastrophic events. Thus, extreme events are treated as improbable during a human lifetime, and consequently are not rationally considered in decision-making processes. However, a 1000-year disaster[9] may actually occur at any time, e.g., next year, or even tomorrow, or in many years, or not within our planning time horizon. Moreover, even ex-post evaluations of extreme events are often difficult. For example, the flood of 1997 in Western Poland has been evaluated (by different scientists) as a 300-, 500-, 1000-, or even 3000-year flood.

The extremal value theory can hardly help in the analysis of catastrophic events because it deals with independent events and assumes that they are quantifiable by an aggregate (a single number, e.g., loss value) [12]. Such a characteristic cannot be provided in a verifiable way for catastrophes. Moreover, it is impossible to properly evaluate complex heterogeneous processes on "average". Such processes have significantly diversified spatial and temporal patterns and induce heterogeneity of losses and gains which make it inappropriate to use average (aggregate) characteristics. For example, on average, residents may even benefit from some climate-change scenarios, while some regions may incur dramatic losses.

3.3   Catastrophe Risks

A common way to deal with risks is to purchase an insurance. However, although this is a very popular approach to risk management, for many risks buying insurance alone is not necessarily a rational solution. By purchasing an insurance one actually buys the rights to a capital for covering the consequences of the insured event; however, it is actually a rather expensive solution (although not commonly recognized as such) because the insurance premium is higher (often much higher) than the expected losses covered by the insurance. Therefore, an insurance (possibly for only part of the risks) should be considered as an element in a diversified risk management portfolio, typically composed of ex-ante (e.g., insurance, mitigation measures, contingent credits, cat-bonds) and ex-post (e.g., loans, diversion of funds) instruments.

Designing a rational portfolio of catastrophe risk management requires concerted interdisciplinary activities needed for developing a system of models that:

- represent a multi-agent decision making structure that typically involves a population exposed to the risks, central and local governments, insurance and finance industry, and possibly also other agents;

---

[9] An extreme event that occurs on average once in 1000 years.

- represent relations between the decisions (composed of elements of the risk management portfolio) and various measures of the consequences;
- typically compensate for the lack of historical data on the occurrence of catastrophes in locations where the effects of catastrophes may never have been experienced in the past.

Such a system of models typically consists of three major modules:

- A catastrophe module that simulates a natural phenomenon using a model based on the knowledge of the corresponding type of event (e.g., earthquakes, floods, hurricanes) represented by a set of variables and the relations between them. The catastrophe models used in IIASA's case studies are based on the Monte Carlo dynamic simulations of geographically explicit catastrophe patterns in selected regions.[10].
- An engineering vulnerability module to estimate the damages that may be caused by the catastrophes.
- An economic multi-agent module that maps spatial economic losses (which depend on implemented loss mitigating and sharing policy options) into gains and losses of the agents (stakeholders), see e.g., [16,17].

Such a system of models supports the integrated analysis of spatial and temporal heterogeneity of diversified characteristics of agents (stake holders) induced by mutually dependent losses from extreme events while also taking into account the diversified (and partially conflicting) objectives of each type of the stakeholders. The model addresses the specifics of catastrophic risks: fragmented data, the need for long term perspectives and geographically explicit models, and a multi-agent decision-making structure. The model, using advanced stochastic optimization techniques, supports the search and analysis of robust risk management portfolios for decreasing regional vulnerability measured in terms of economic, financial, and human losses as well as in terms of selected welfare growth indicators.

A more detailed overview of the integrated approach to catastrophe risk management developed at IIASA and applied to many diversified case studies can be found in [28]. Here we only stress that the corresponding models are complex[11] and their analysis aims to find robust decisions requiring structured modeling approaches.

4 Structured Modeling

The novel methods characterized in Section 3 that are needed to properly treat endogenous uncertainties and catastrophe risks involve complex models. Such models when used for supporting policy-making have to meet addi-

---

[10] A discussion of these models is beyond the scope of this paper but can be found e.g., in [2,4,6,14,18,21,40]

[11] Actually one needs a system of models, each using large sets of data.

tional requirements. The complexity of models combined with such requirements also require new modeling technology. These issues are discussed in this Section.

4.1 Modeling for Science-Based Policy-Making Support

As mentioned in Section 1, models can integrate knowledge pertinent to science-based support of policy-making. Actually, models can also be used for creating knowledge, see Section 4.3. However, the modeling processes which support policy-making have to meet much stronger requirements (than those for models used for research or educational purposes) of: credibility, transparency, replicability of results, integrated model analysis, controllability (modification of model specification and data, and various views on, and interactive analysis of, results), quality assurance, documentation, controllable sharing of modeling resources through the Internet, and efficient use of resources on computational Grids.

Actually such requirements are not new for the modeling communities. Dantzig summarized in [8] the opportunities and limitations of using large-scale models for policy-making. Thanks to the development of algorithms and computing power today's large-scale models are at least 1000-times larger; thus, large-scale models of the 1970s are classified as rather small today. This, however, makes Dantzig's message relevant to practically all models used today, not only for policy-making but also in science and management. Today's models are not only much larger, the modeled problems are also more complex (e.g., by including representation of knowledge coming from various fields of science and technology), and many models are developed by interdisciplinary teams.

The traditional modeling methods and general-purpose modeling tools are developed to deal with one of the standard problem-types through a particular modeling paradigm, and cannot meet the requirements summarized above. This can only be achieved by a qualitative jump in modeling methodology: from supporting individual modeling paradigms to supporting a *Laboratory World*[12] in which various models are developed and used to learn about the modeled problem in a comprehensive way. The truth is that there are no simple solutions for complex problems, thus learning about complex problems by modeling is in fact more important than finding an *"optimal"* solution. Such a Laboratory World requires the integration of various established methods with new (either to be developed to properly address new challenges, or not yet supported by any standard modeling environment) approaches needed for an appropriate (in respect to the decision-making process, and available data) mathematical representation of the problem and the ways of its diversified analyses.

---

[12] Originally proposed by Dantzig, see e.g., [8,24].

The complexity, size, model development process, and requirements for integrated model analysis form the main arguments justifying the need for the new modeling methodology; the standard general-purpose modeling tools are not able to adequately meet such demand. More detailed arguments (including an overview of the standard modeling methods and tools) supporting this statement are available in [29].

4.2 Structured Modeling Technology (SMT)

The development, maintenance and exploitation of models is composed of interlinked activities, often referred to as a *modeling process*. Such a process should be supported by modeling technology that is a craft of the systematic treatment of modeling tasks using a combination of pertinent elements of science, experience, intuition, and modeling resources, the latter being composed of knowledge encoded in models, data, and modeling tools. Thus the key to a successful modeling undertaking is defined by the appropriate choice of *"a combination of pertinent elements"*. This can only be achieved through the long-term and efficient collaboration of researchers advancing disciplinary methodology with those progressing modeling methodology, the latter keeping abreast of recent developments in operations research, see e.g., [29].

The complexity of problems, and the corresponding modeling process are precisely the two main factors that determine the requirements for modeling technology that substantially differ from the technologies successfully applied for modeling well-structured and relatively simple problems. In most publications that deal with modeling, small problems are used as an illustration of the presented modeling methods and tools. Often, they can also be applied to large problems. However, as discussed above, the complexity is characterized not primarily by the size, but rather by: the requirements of integrating heterogeneous knowledge, the structure of the problem, the demand for integrated problem analysis, and the requirements for the corresponding modeling process. Moreover, efficient solving of complex problems requires the use of a variety of models and modeling tools; this in turn will require even more reliable, re-usable, and shareable modeling resources (models, data, modeling tools). The complexity, size, model development process, and the requirements for integrated model analysis form the main arguments in justifying the need for the new modeling methodology.

Unfortunately, modeling resources are fragmented, and using more than one modeling paradigm for the problem at hand is too expensive and time-consuming in practice. The low productivity of model-based work compared with the high productivity of data-based work has already been discussed in [22]. In the case of databases, DBMSs are mature and well-established, and there is a broad agreement on the definitions of the abstract data models, as well as on the operations (e.g., those featured in SQL) to be supported for working with these data. This broad agreement has made it possible to efficiently use data from different sources because DBMS products of high quality are available and widely used. It is therefore strange that professional-quality DBMS techniques are not routinely used in most modeling systems, especially since it is generally agreed that dealing properly with models of a realistic size requires the use of modern DBMS technology; moreover, the DBMS technology has advanced immensely and is now well integrated with the Web. Geoffrion [22] formulated the principles of structured modeling thus providing a methodological framework for the integration of various paradigms. However, the proposed integrating framework has been to a large extent ignored, and most modeling paradigms have been developed somewhat separately.

Continuing progress in the foundations of modeling, and in database management, and new opportunities emerging from the network-based, platform-independent technologies offer a solid background for providing the desired modeling support needed for management, policy makers, research, and education. Arguments supporting this statement are summarized e.g., in [7,9–11,19,23,39]. However, modeling technology is still at the stage where data-processing technology was before the development of DBMS. The data-management revolution occurred in response to severe problems with data reusability associated with file-processing approaches to application development. DBMSs make it possible to efficiently share not only databases but also tools and services for data analysis that are developed and supplied by various providers and made available on computer networks [27]. Data processing was revolutionized by the transition from file processing (when data was stored in various forms and software for data processing had to be developed for each application) to DBMS. The need to share data resources resulted in the development of DBMSs that separate the data from the applications that use the data. The modeling world has not yet learned this lesson: almost every modeling paradigm still uses a specific format of model specification and data handling.

Structured Modeling Technology (SMT) has been developed for meeting such requirements. SMT supports distributed modeling activities for models with a complex structure using large amounts of diversified data, possibly from different sources. A description of SMT is beyond the scope of this paper, therefore we will only summarize its main features:

- SMT is Web-based, thus supporting *any-where, any-time* collaborative modeling.
- It follows the principles of Structured Modeling proposed by Geoffrion, see e.g., [22]; thus it has a modular structure supporting the development of various elements of the modeling process (model specification, handling (subsets of) data, integrated model analysis) by different teams possibly working in distant locations.
- It provides automatic documentation of all modeling activities.

- It uses a DBMS for all persistent elements of the modeling process, which results in efficiency and robustness; moreover, the capabilities of DBMSs allow for the efficient handling of both huge and small amounts of data.
- It assures the consistency of: model specification, meta-data, data, model instances, computational tasks, and the results of model analysis.
- It automatically generates a Data Warehouse with an efficient (also for large amounts of data) structure for:
  - data, and tree-structure of data updates,
  - definitions of model instances (composed of a symbolic specification and a selected data update),
  - definitions of preferences for diversified methods of model analysis,
  - results of model results,
  - logs of all operations during the modeling process.

This conforms to the requirement for persistency for all elements of the modeling process.

- It exploits computational grids for large amounts of calculation.
- It also provides users with easy and context sensitive problem reporting.

Thus SMT supports the entire modeling process composed of:

- Analysis of the problem and a development of the corresponding model (symbolic) specification.
- Collection and verification of the data to be used for calculating the model parameters.
- Definition of various model instances (composed of a model specification, and a selection of data defining its parameters).
- Diversified analyses of model instances.
- Documentation of the whole modeling process.

A more detailed presentation of SMT (including an overview of the standard modeling methods and tools) is available in [29]. Here we only mention that SMT has been developed for the collaborative modeling of complex problems, and has been effectively used for this purpose. In particular, it is being used for the new version of the RAINS model [1], which is a large[13] model composed of submodels of:

- current and future economic activities, energy consumption levels, fuel characteristics, etc.), for emission control options and costs,
- atmospheric dispersion of pollutants,
- environmental and health effects, including sensitivities (i.e., databases on critical loads defined for each at about 5400 grids for several types of environmental indicators).

---

[13] Consisting of over 1,000,000 variables and over 10^7 non-zero coefficients (many coefficients are defined by evaluation of expressions composed of "primary" data).

For handling complex models SMT provides support for the specification of a whole model, and also for the extraction of consistent sub-models that can be analyzed separately by using the data from the common Data Warehouse.

SMT also supports efficient *data processing* which for large models is a very time consuming, and, in fact, most critical element of the modeling process. To achieve the needed functionality the data processing has several characteristics not used in other modeling systems. This includes definitions of not only the trees of data updates, but also composite updates (which are actually parameterized sets of updates).

Such a structured method of interdisciplinary knowledge integration through a mathematical model provides excellent opportunities for effective collaborative work.

4.3   Integrated Model Analysis

Model analysis is probably the least researched element of the modeling process. This results from the focus that each modeling paradigm has on a specific type of analysis. However, the essence of model-based decision-making support is precisely the opposite; namely, to support diversified ways of model analysis, and to provide efficient tools for various comparisons of solutions. Such an approach can be called Integrated Model Analysis.

A typical model for supporting decision-making has an infinite number of solutions, and users are interested in analyzing trade-offs between a manageable number of solutions that correspond to various representations of their preferences, often called the preferential structure of the user. Thus, an appropriate integrated analysis should help users to find and analyze a small subset of all solutions that correspond best to their preferential structures that typically change during the model analysis. SMT provides the computational technology framework for the analysis, but there are three types of problems that call for innovative research:

- the integration of various paradigms of model analysis,
- the extraction of knowledge from large sets of solutions,
- the efficient solution of computational tasks (either resource-demanding, or numerically difficult, or large sets of simple jobs).

We briefly summarize each of them below.

For a truly integrated problem analysis one should actually combine different methods of model analysis, such as: classical (deterministic) optimization (and its generalizations, including parametric optimization, sensitivity analysis, fuzzy techniques), multicriteria model analysis, stochastic optimization and Monte Carlo simulations, classical simulation, soft simulation, and several of its generalizations (e.g. inverse simulation, softly constrained simulation). However, no modeling tool supports such a complete analysis, and the development of separate versions of a model with tools supporting different modeling paradigms is typically too expensive. Thus we aim to find a satisfactory solution to this problem.

The second research challenge is to develop and implement a methodology for a comprehensive analysis of large sets of solutions. One needs to explore the applicability of various data-mining and knowledge engineering techniques, and either adapt some of them, or develop new methods to extract and organize knowledge from large sets of solutions, and supply users with this knowledge in a form that will help further problem analysis.

The third set of research issues is related to the efficient and robust organization of computational tasks typically needed for large-scale models, and includes:

- Efficient support for handling a large number of results, possibly coming from various types of analyses of large models.
- Adaptation of specialized optimization algorithms for badly conditioned problems.
- Support for exploiting the structure of huge optimization problems.
- Support effective use of computational grids.

5   Conclusions

Models can play a key role in the science-based support for policy-making provided they are properly developed, i.e., use appropriate methodology for the treatment of uncertainty and risks, and also appropriate modeling technology. Successful experience with actual applications in diversified areas shows how problems which cannot be dealt with by using traditional approaches, can be successfully coped with.

Any actual policy analysis focuses attention on situations where processes can be affected or controlled by decisions that should be selected in the best possible manner. *The best* however has to be understood differently from the traditional optimization-driven paradigms that aim at computing the optimal solution of a well-defined mathematical programming problem. Optimization still plays a key role in problem analysis but it is now a tool used in integrated problem analysis. In this approach a policy analysis problem is represented by a model consisting of several submodels; such a model contains sets of goals and feasible decisions, the latter implicitly defined by the substantive part of the model, which represents the pertinent knowledge about the problem. In reality these sets are also uncertain and they can be specified through a dialogue of users with a system of models.

The structured modeling technology supports the whole modeling process in a consistent way. By structuring the modeling process one can organize and document knowledge integration and creation much more effectively, especially for problems requiring interdisciplinary work. This includes the distinction of two parts of a model: objective (a representation of pertinent knowledge) and subjective (various representations of user preferences which are typically modified during the model analysis).

Advances in modeling and computational methods [30] allow us to create a "laboratory world" [24], where one can also analyze new policies that have not yet been implemented. Such "learning-by-modeling" dialogue of users with models requires specific methods for finding robust solutions which able properly account for the effects of rare extreme events. These methods also maintain a consistency of outcomes under the changing environment of the "laboratory world" where goals and sets of feasible solutions are subject to changes that reflect the views of users, new knowledge created by model analysis, and experience from diversified implementations.

The diversified experience presented in this chapter comes from two streams of activities: (1) development of novel methods and tools for integrating and creating knowledge pertinent to solving complex problems, and (2) development of structured modeling technology to support the whole modeling process of model-based support for solving complex policy problems. An important challenge for on-going activities is to exploit the synergy of these two streams.

Acknowledgments. Unpublished work by Y. Ermoliev and M. Makowski has been adapted for part of this chapter. Moreover, the author acknowledges the contributions of T. Ermolieva and Y. Ermoliev to the write-up for Section 3. Several ideas exploited in SMT have resulted from many discussions and the joint activities of the author with A. Geoffrion, J. Granat, and A.P. Wierzbicki. Furthermore, the author thanks Y. Ermoliev for sharing over the years his knowledge in a friendly way, and for his constructive comments on a draft of this chapter.

The author gratefully acknowledges all these contributions, but he assumes the sole responsibility for the content of this paper.

References

1. Amann, A., Makowski, M.   Effect-focused air quality management.   In Wierzbicki et al. [41], 367–398 ISBN 0-7923-6327-2.
2. Amendola, A., Ermoliev, Y., Ermolieva, T., Gitits, V., Koff, G., Linnerooth-Bayer, J. (2000) A systems approach to modeling catastrophic risk and insurability. *Natural Hazards Journal 21*, **2/3**, 381–393
3. Anand, P. (1993) *Foundations of Rational Choice under Risk.* Oxford University Press, Oxford
4. Baranov, S., Digas, B., Ermolieva, T., Rozenberg, V. (2002) Earthquake risk management: Scenario generator. Interim Report IR-02-025, International Institute for Applied Systems Analysis, Laxenburg, Austria
5. Chichilnisky, G., and Heal, G. (1993) Global environmental risks. *Journal of Economic Perspectives 7*, **4**, 65–86
6. Christensen, K., Danon, L., Scanlon, T., Bak, P. (2002) Unified scaling law for earthquakes. Proceedings 99/1, National Academy of Sciences, US

7. Cohen, M., Kelly, C., Medaglia, A. (2001) Decision support with Web-enabled software. *Interfaces 31*, **2**, 109–129
8. Dantzig, G. (1983) Concerns about large-scale models. In *Large-Scale Energy Models. Prospects and Potential*, R. Thrall, R. Thompson, and M. Holloway, Eds., vol. 73 of *AAAS Selected Symposium*. West View Press, Boulder, Colorado, 15–20
9. Dolan, E., Fourer, R., More, J., Manson, T. (2002) Optimization on the NEOS server. *SIAM News 35*, **6**, 1–5
10. Dolk, D.(1988) Model management and structured modeling: The role of an information resource dictionary system. *Comm. ACM 31*, **6**, 704–718
11. Dolk, D. (2000) Integrated model management in the data warehouse era. *EJOR 122*, **2**, 199–218
12. Embrechts, P., Klueppelberg, C., Mikosch, T. (2000) *Modeling Extremal Events for Insurance and Finance. Applications of Mathematics, Stochastic Modeling and Applied Probability*. Springer Verlag, Heidelberg
13. Ermoliev, Y., Ermolieva, T., MacDonald, G., Norkin, V. (2000) Stochastic optimization of insurance portfolios for managing exposure to catastrophic risks. *Annals of Operations Research 99*, 207–225
14. Ermoliev, Y., Ermolieva, T., MacDonald, G., Norkin, V. (2000) Stochastic optimization of insurance portfolios for managing exposure to catastrophic risks. *Annals of Operations Research 99*, 207–225
15. Ermoliev, Y., Hordijk, L. Facets of robust decisions. In Marti et al. [32]
16. Ermoliev, Y., Norkin, V. Stochastic optimization of risk functions. In Marti et al. [33]
17. Ermolieva, T. (1997) The design of optimal insurance decisions in the presence of catastrophic risks. Interim Report IR-97-068, International Institute for Applied Systems Analysis, Laxenburg, Austria
18. Ermolieva, T., Ermoliev, Y., Linnerooth-Bayer, J., and Vari, A. (2002) Integrated management of catastrophic flood risks in the Upper Tisza basin: A dynamic multi-agent adaptive Monte Carlo approach. In *Proceedings of the Second Annual IIASA-DPRI Meeting* Integrated Disaster Risk Management: Megacity Vulnerability and Resilience. International Institute for Applied Systems Analysis, Laxenburg, Austria http://www.iiasa.ac.at/Research/RMS
19. ETAN Expert Working Group (1999) Transforming European science through information and communication technologies: Challenges and opportunities of the digital age. ETAN Working Paper September, Directoriate General for Research, European Commission, Brussels
20. French, S., Geldermann, J. (2005) The varied contexts of environmental decision problems and their implications for decision support. *Environmental Science & Policy 8*, 378–391.
21. Froot, K. (1997) *The Limited Financing of Catastrophe Risk: and Overview.* Harvard Business School and National Bureau of Economic Research, Harvard
22. Geoffrion, A. (1987) An introduction to structured modeling. *Management Science 33*, **5**, 547–588
23. Geoffrion, A., Krishnan, R. (2001) Prospects for operations research in the e-business era. *Interfaces 31*, **2**, 6–36
24. Holling, C., Dantzig, G., Clark, W., Jones, D., Baskerville, G., Peterman, R. (1979) Quantitative evaluation of pest management options: The spruce budworm case study. Tech. rep., US Department of Agriculture: Washington Forest Service, Washington, USA

25. Hordijk, L., Ermoliev, Y., Makowski, M. (2005) Coping with uncertainties. In *Proceedings of the 17th IMACS World Congress*, P. Borne, M. Bentejeb, N. Dangoumau, and L. Lorimier, Eds. Ecole Centrale de Lille, Villeneve d'Ascq Cedex, France, 8 ISBN 2-915913-02-1, EAN 9782915913026

26. IPCC (2001) Climate change 2001: The scientific basis. Technical report, Intergovernmental Panel on Climate Change

27. Makowski, M. (2004) Model-based problem solving in the knowledge grid. *International Journal of Knowledge and Systems Sciences 1*, 1, 33–44. ISSN 1349-7030.

28. Makowski, M. Mathematical modeling for coping with uncertainty and risk. In *Systems and Human Science for Safety, Security, and Dependability*, T. Arai, S. Yamamoto, and K. Makino, Eds. Elsevier, Amsterdam, the Netherlands, 35–54. ISBN: 0-444-51813-4

29. Makowski, M. (2005) A structured modeling technology. *European J. Oper. Res. 166*, 3, 615–648. draft version available from http://www.iiasa.ac.at/ marek/pubs/prepub.html

30. Makowski, M., Wierzbicki, A. (2003) Modeling knowledge: Model-based decision support and soft computations. In *Applied Decision Support with Soft Computing*, X. Yu and J. Kacprzyk, Eds., vol. 124 of *Series: Studies in Fuzziness and Soft Computing*. Springer-Verlag, Berlin, New York, 3–60. ISBN 3-540-02491-3, draft version available from http://www.iiasa.ac.at/ marek/pubs/prepub.html

31. Manne, A., Richels, R. (1997) The greenhouse debate: Economic efficiency, burden sharing and hedging strategies. *The Energy Journal 16*, 4, 1–37.

32. Marti, K., Ermoliev, Y., Makowski, M., Pflug, G., Eds. (2006) *Coping with Uncertainty: Modeling and Policy Issues*. Springer, Berlin, Heidelberg, New York

33. Marti, K., Ermoliev, Y., Pflug, G., Eds. (2004) *Dynamic Stochastic Optimization*. Springer Verlag, Berlin

34. Morgan, M., Kandlikar, M., Risbey, J., Dowlatabadi, H. (1999) Why conventional tools for policy analysis are often inadequate for problems of global change: An editorial essay. *Climate Change 41*, 3-4, 271–281.

35. Nordhaus, W., Boyer, J. (2001) *Warming the World: Economic Models of Global Warming*. MIT Press, Cambridge, Mass.

36. O'Neill, B., Ermoliev, Y., Ermolieva, T. Facets of robust decisions. In Marti et al. [32]

37. Paczyński, J., Makowski, M., Wierzbicki, A. Modeling tools. In Wierzbicki et al. [41], 125–165. ISBN 0-7923-6327-2

38. Ramsey, F. (1928) A mathematical theory of savings. *Economic Journal 138*, 543–559

39. Tsai, Y.-C. (2001) Comparative analysis of model management and relational database management. *Omega, 29*, 157–170

40. Walker, G. (1997) Current developments in catastrophe modelling. In *Financial Risks Management for Natural Catastrophes*, N. Britton and J. Oliver, Eds. Griffith University, Brisbane, Australia, 17–35

41. Wierzbicki, A., Makowski, M., Wessels, J., Eds. (2000) *Model-Based Decision Support Methodology with Environmental Applications*. Series: Mathematical Modeling and Applications. Kluwer Academic Publishers, Dordrecht, ISBN 0-7923-6327-2

42. Wright, E., and Erickson, J. (2003) Incorporating catastrophes into integrated assessment: Science, impacts, and adaptation. *Climate Change 57*, 265–286

Part II

Modeling Stochastic Uncertainty

Using Monte Carlo Simulation to Treat Physical Uncertainties in Structural Reliability

D. C. Charmpis and G. I. Schuëller

Institute of Engineering Mechanics, Leopold-Franzens University, Technikerstr. 13, A-6020 Innsbruck, Austria, EU

Abstract. This chapter is concerned with the estimation of the reliability of structures in view of physical uncertainties encountered due to the inherent variability in structural properties and loads. In this respect, methods based on the traditional Monte Carlo simulation method are employed to deal with probabilistically modeled uncertainties. Hence, suitable variance reduction techniques and efficient computational procedures are presented, in order to alleviate the high processing demands associated with Monte Carlo computations and make the overall reliability estimation process more tractable in practice. The focus of this chapter is on statistically high-dimensional problems, which involve large numbers of random variables. The merits of some of the techniques and algorithms described are demonstrated with two application examples.

1 Introduction

The treatment of physical uncertainties has been identified as a research area of great importance and interest within the structural mechanics community [16]. Physical variability is related to the inherent randomness involved in the properties of engineering structures and systems, as well as in the imposed loading and boundary conditions. Since physical uncertainties are a consequence of the randomness in nature, they cannot be controlled by humans. For instance, uncertainties in material properties (modulus of elasticity, mass density, yield strength, etc.) or environmental loads (earthquake, wind, wave, etc.) cannot be reduced using a scientific procedure; one can only accept their existence, then observe and understand them and finally attempt to appropriately treat them in the framework of an engineering process.

Experience shows that several of the aforementioned uncertainty types – depending each time on the particular structural problem at hand – result in severe variations of structural response and therefore directly affect structural safety and reliability. Therefore, physical uncertainties should be taken into account during the modeling of structural systems regardless of the features and the solution accuracy of the numerical method employed to simulate structural behavior. In an effort to cope with such modeled uncertainties, various approaches have been developed incorporating statistical and probabilistic procedures into structural mechanics formulations.

The approaches basically applied for the probabilistic modeling of physical uncertainties in structural mechanics use either simple random variables (e.g. [15]), random processes (e.g. [14]) or random fields (e.g. [22]). In the simple random variable case, a single variable following some probability distribution describes the random characteristics of an uncertain parameter in space and time. In the random field case, the value of an uncertain property (e.g. material property) or loading (e.g. wind pressure on area-like structures) varies across the structural domain according to some correlation pattern. Hence, the adopted probability distribution does not refer to a single random variable, but to several random variables required to adequately represent the spatially correlated random fluctuation of the uncertain property or loading. Accordingly, when the uncertain parameter is time-dependent (e.g. stochastic earthquake loading), a stochastic process with some correlation pattern can be used to model the random variation in time. When random processes and/or fields are employed, the resulting structural investigation is called stochastic analysis. Clearly, the stochastic approach allows for more detailed description of uncertainties than the use of simple random variables. However, stochastic modeling implies that sufficient information on the time or spatial variation of the uncertain parameter is available, in order to be able to define the corresponding random process or field.

A challenging problem encountered when uncertainties are incorporated into structural mechanics applications is the estimation of structural reliability, which is defined as the probability that a structure will respond within acceptable limits. The probability of exceeding these bounds or else the *probability of failure* of the structure can be simply expressed as:

$$P_F = \int_{g(\vec{\theta}) \leq 0} f(\vec{\theta}) \, d\vec{\theta} = \int_{\mathbb{R}^d} I_F(\vec{\theta}) f(\vec{\theta}) \, d\vec{\theta}, \tag{1}$$

where $g(\vec{\theta})$ is the so-called performance or limit state function, $I_F(\vec{\theta})$ is the indicator function of a failure event $F \subseteq \mathbb{R}^d$ whose probability P_F is to be calculated and the vector $\vec{\theta} \in \mathbb{R}^d$ represents the random parameters of the system (i.e. the uncertainties in material properties, loads) with joint probability density function $f(\vec{\theta})$. $g(\vec{\theta})$ is a scalar function associated with F such that $g(\vec{\theta}) \leq 0$ indicates failure of the system, while $g(\vec{\theta}) > 0$ denotes non-failure. Thus, the hyper-surface $g(\vec{\theta}) = 0$, which is called limit state surface, separates the d-dimensional input space of random variables $\vec{\theta} = (\theta_1, \theta_2, \ldots, \theta_d)$ into a failure and a safe domain. Then, the indicator function I_F is defined as:

$$I_F(\vec{\theta}) = \begin{cases} 1 & \text{for } g(\vec{\theta}) \leq 0 \\ 0 & \text{for } g(\vec{\theta}) > 0 \end{cases} \tag{2}$$

and the failure probability may be interpreted as:

$$P_F = \mathrm{E}[I_F(\vec{\theta})], \tag{3}$$

where $\mathrm{E}[\cdot]$ denotes the expectation operator.

The present work is focused on the estimation of structural reliability with techniques based on the traditional Monte Carlo Simulation (MCS) method, which is the most effective and widely applicable approach for handling arbitrary probabilistic or stochastic structural mechanics problems. The severe computational workload associated with MCS implementations for reliability estimation is drastically reduced by introducing suitable variance reduction techniques to minimize the number of simulations required and by alleviating the processing effort of each simulation with the use of efficient computational procedures.

The remainder of this chapter is organized as follows. The simple direct MCS approach is overviewed in Section 2. Section 3 is concerned with variance reduction techniques for low- and high-dimensional problems. Section 4 addresses computational efficiency issues involving solution and parallel processing procedures. Finally, numerical results are presented in section 5 and closing remarks are given in section 6.

2   Direct Monte Carlo Simulation

According to the direct MCS approach, the expectation of the failure probability (3) is estimated by:

$$P_F \approx \hat{P}_F = \frac{1}{n_{sim}} \sum_{i=1}^{n_{sim}} I_F(\overrightarrow{\theta}^{(i)}), \qquad (4)$$

where the samples $\overrightarrow{\theta}^{(i)}$, $i = 1, 2, \ldots, n_{sim}$, are independently and identically simulated following the probability density function $f(\overrightarrow{\theta})$. The convergence rate of the above unbiased estimator is most appropriately measured by the associated coefficient of variation:

$$\delta_{MC} = \frac{\sqrt{\mathsf{Var}[\hat{P}_F]}}{\mathsf{E}[\hat{P}_F]} = \sqrt{\frac{1 - P_F}{n_{sim} P_F}}, \qquad (5)$$

where $\mathsf{Var}[\cdot]$ represents the variance. It is important to note that δ_{MC} is independent of the dimensionality of the random vector $\overrightarrow{\theta}$. It is also independent of the type of the probability density function $f(\overrightarrow{\theta})$ employed, as well as of the type, size and particular configuration of the structural problem considered. The convergence rate of the estimator (4) depends only on the failure probability and the number of samples n_{sim}. Hence, direct MCS is a general method with very wide applicability provided that a deterministic solver is available for the structural problem at hand, in order to determine $g(\overrightarrow{\theta}^{(i)})$ or $I_F(\overrightarrow{\theta}^{(i)})$ by carrying out an analysis for each simulation i.

The main disadvantage of the estimator (4) is its inefficiency in calculating low failure probabilities P_F due to the large number (essentially proportional to $1/P_F$) of samples or equivalently system analyses needed to achieve an acceptable level of accuracy. For instance, in order to calculate a failure probability $P_F = 10^{-4}$ with a coefficient of variation for the estimator $\delta_{MC} = 30\%$, $n_{sim} = (1 - P_F)/(\delta_{MC}^2 P_F) = 111,100$ samples are required following Eq. (5).

3   Variance Reduction Techniques

In an effort to alleviate the disadvantage of inefficiency of direct MCS in calculating low failure probabilities, several more advanced variants have been developed, which reach P_F-convergence with considerably smaller n_{sim}-numbers compared to estimator (4). A popular class of such simulation methods essentially reduce the coefficient of variation (5), therefore they are called *variance reduction techniques*. As opposed to the general applicability of the direct MCS procedure, these techniques focus on efficiency. In this sense, direct MCS is not competitive when compared with methods which succeed to extract and exploit essential properties of the problem considered. However, generality is usually inversely proportional to efficiency. Hence, variance reduction sacrifices generality to some extent, in order to yield results in acceptable computing times.

Contrary to direct MCS, the applicability and effectiveness of variance reduction techniques is typically not independent of the dimensionality of the random vector $\vec{\theta}$, which includes all uncertain parameters of the structural system considered. Thus, one needs to separately refer to problems with relatively small numbers (say up to 20) and large numbers (up to hundreds or even thousands) of random variables, i.e. one has to distinguish between *statistically low- and high-dimensional problems*, respectively. The methods and variants existing today allow the assessment of static/dynamic linear/non-linear reliability estimation problems quite effectively, since failure probabilities can be reliably estimated in low and high dimensions by performing manageable numbers of simulations. Even low probabilities can be satisfactorily dealt with (e.g. $P_F = 10^{-6}$), which would require unacceptably high numbers of simulations (e.g. tens of millions) with direct MCS.

The present section overviews the following variance reduction techniques:

- importance sampling,
- subset simulation and
- line sampling.

These methods, as well as other alternatives, are implemented in software packages ISPUD [10] for low-dimensional problems and COSSAN B [7] for high-dimensional ones. ISPUD can primarily handle reliability problems employing a smaller number of random variables only. COSSAN B can also treat problems with large numbers of random variables encountered e.g. in Stochastic Finite Element (SFE) applications and stochastic dynamics.

3.1 Low-Dimensional Problems

Importance sampling has been one of the most prevalent simulation-based methods for the estimation of structural reliability (see e.g. [18]). The underlying concept of the method is to draw samples of the vector of random parameters $\overrightarrow{\theta}$ from a distribution $f'(\overrightarrow{\theta})$ which is concentrated in the 'important region' of the random parameter space. In other words, $f'(\overrightarrow{\theta})$ needs to be selected in a way that a considerable part of the generated samples falls within the failure domain F. According to this approach, the weight of each realization considered in the important region is appropriately adjusted, in order to avoid the distortion of the original Monte Carlo estimator. For this purpose the probability of failure of Eqs. (1) and (3) is re-expressed as:

$$P_F = \int_{\mathbb{R}^d} I_F(\overrightarrow{\theta}) \frac{f(\overrightarrow{\theta})}{f'(\overrightarrow{\theta})} f'(\overrightarrow{\theta}) d\overrightarrow{\theta}$$

$$= \mathbf{E}_{f'}\left[I_F(\overrightarrow{\theta}) \frac{f(\overrightarrow{\theta})}{f'(\overrightarrow{\theta})} \right]. \tag{6}$$

Then, based on the above expectation with respect to f', the estimator (4) takes the form:

$$P_F \approx \hat{P}_F = \frac{1}{n_{sim}} \sum_{i=1}^{n_{sim}} I_F(\overrightarrow{\theta}^{(i)}) \frac{f(\overrightarrow{\theta}^{(i)})}{f'(\overrightarrow{\theta}^{(i)})}, \tag{7}$$

where the samples $\overrightarrow{\theta}^{(i)}$, $i = 1, 2, \ldots, n_{sim}$, are independently and identically simulated following the probability density function $f'(\overrightarrow{\theta})$. The variance of estimator (7) is given by:

$$\mathbf{Var}[\hat{P}_F] = \frac{1}{n_{sim}} \mathbf{Var}_{f'}\left[I_F(\overrightarrow{\theta}) \frac{f(\overrightarrow{\theta})}{f'(\overrightarrow{\theta})} \right]$$

$$= \frac{1}{n_{sim}} \left(\int_{\mathbb{R}^d} I_F(\overrightarrow{\theta}) \frac{f^2(\overrightarrow{\theta})}{f'^2(\overrightarrow{\theta})} f'(\overrightarrow{\theta}) d\overrightarrow{\theta} - P_F^2 \right). \tag{8}$$

The optimal choice for the importance sampling density $f'(\overrightarrow{\theta})$, which is obtained from Eq. (8) for $\mathbf{Var}[\hat{P}_F] = 0$, is:

$$f'_{opt}(\overrightarrow{\theta}) = \frac{I_F(\overrightarrow{\theta}) f(\overrightarrow{\theta})}{P_F}. \tag{9}$$

Clearly, this is a practically infeasible choice, since P_F is not known a priori. Therefore, a number of techniques have been developed, which approximate the optimal sampling density (9) or construct a different one exhibiting a relatively low variance of the estimator in Eq. (8). Two such approaches are those based on kernel density estimators or design points [19].

Importance sampling, as well as other popular reliability estimation approaches such as first/second order reliability methods (FORM/SORM) and the response surface method, are known to effectively deal with statistically low-dimensional problems only [19]. Therefore, alternative methods have been developed for high-dimensional reliability estimation problems, as described in the next subsection.

3.2   High-Dimensional Problems

In the framework of *subset simulation* [2] P_F is expressed as a product of larger conditional probabilities by defining a decreasing sequence of events (subsets) F_i, $i = 1, 2, \ldots, m$, such that:

$$F_1 \supset F_2 \supset \ldots \supset F_m = F. \tag{10}$$

Then, it holds that:

$$\bigcap_{i=1}^{k} F_i = F_k \ \forall \ k \leq m \tag{11}$$

and the probability of failure P_F can be written as:

$$P_F = P(F_m) = P(F_1) \prod_{i=1}^{m-1} P(F_{i+1}|F_i). \tag{12}$$

With an appropriate selection of F_i, $i = 1, 2, \ldots, m - 1$, the probabilities $P(F_1)$ and $P(F_{i+1}|F_i) \ \forall \ i \geq 1$ can be made sufficiently large so that they can be efficiently estimated through direct MCS. Hence, the original reliability estimation problem is substituted by a series of m problems, each of which can be solved with a small number of simulations. A key ingredient in the overall efficiency of the method is the use of Markov chains, which allows the generation of the samples needed for estimating $P(F_{i+1})$ by using the samples simulated in the previous step for calculating $P(F_i)$. Apart from the original version of subset simulation [2], a number of variants have been developed building upon the basic concept of the method [12,5–7].

The *line sampling* approach [13,19] employs lines instead of points in order to collect information about the probability content of the failure domain. It requires the specification of an important direction, which points towards the nearest region of the failure domain. For this purpose, assuming without loss of generality that the important direction points towards the direction of θ_1, the failure domain F is expressed as:

$$F = \{\vec{\theta} \in \mathbb{R}^d : \theta_1 \in F_1(\vec{\theta}_{-1})\}, \tag{13}$$

where F_1 is a function defined on $\mathbb{R}^{d-1}$ and $\vec{\theta}_{-1} = (\theta_2, \ldots, \theta_d)$ is a $(d$-1)-dimensional vector. Using Eqs. (1) and (13), P_F takes the form:

$$P_F = \underbrace{\int \cdots \int}_{d} I_F(\vec{\theta}) \prod_{i=1}^{d} \phi(\theta_i) d\vec{\theta}$$

$$= \underbrace{\int \cdots \int}_{d-1} \left(\int I_{F_1}(\vec{\theta}_{-1}) \phi(\theta_1) d\theta_1 \right) \prod_{i=2}^{d} \phi(\theta_i) d\vec{\theta}_{-1}$$

$$= \underbrace{\int \cdots \int}_{d-1} \Phi\left(F_1(\vec{\theta}_{-1}) \right) \prod_{i=2}^{d} \phi(\theta_i) d\vec{\theta}_{-1}$$

$$= \mathrm{E}_{\vec{\theta}_{-1}} \left[\Phi\left(F_1(\vec{\theta}_{-1}) \right) \right], \tag{14}$$

where $\phi(x) = (1/\sqrt{2\pi})e^{-x^2/2}$ is the unit-variance Gaussian probability density function and $\Phi(F_1) = \int I_{F_1}(x)\phi(x)dx$ is the probability content of the Gaussian distribution for F_1. Then, based on the above expectation with respect to $\vec{\theta}_{-1}$, the estimator for P_F can be written as:

$$P_F \approx \hat{P}_F = \frac{1}{n_{sim}} \sum_{i=1}^{n_{sim}} \Phi\left(F_1(\vec{\theta}_{-1}^{(i)}) \right), \tag{15}$$

where $\vec{\theta}_{-1}^{(i)}$, $i = 1, 2, \ldots, n_{sim}$, are independent and identically distributed. It is noted that the estimator for P_F by line sampling can be obtained from the estimator (7) of importance sampling with suitable assumptions for θ_1 and $f'(\vec{\theta}_{-1})$ [19]. The coefficient of variation of the estimator (15) is always smaller than δ_{MC} of Eq. (5), which implies that the convergence of the line sampling procedure is always faster than that of direct MCS. Hence, line sampling is considered as a robust reliability estimation method once an important direction has been identified. This direction plays the role of the design point in importance sampling and can be determined analytically for linear systems, but in general gradient computations are needed. It should also be mentioned that line sampling can be combined with the so-called *averaged probability flow* concept for estimating excursion probabilities in stochastic structural dynamics [17].

3.3 Benchmark Studies

As already mentioned in the previous subsections, several reliability estimation methods, procedures and algorithms with various capabilities, accuracy and efficiency have been suggested in the past. Hence, a quantitative comparison of these approaches is considered to be most instrumental and useful for the engineering community. In this respect, a Benchmark study focused on nonlinear stochastic structural dynamics problems was carried out in 1997 [8]. The results of this initiative presented the comparative status of reliability estimation methods suggested until then. Moreover, a more recent partially qualitative comparison of methods was reported in [19].

A new Benchmark study [9], which attempts to assess various recently proposed alternatives for reliability estimation with respect to their accuracy and computational efficiency, was suggested in 2004 and is expected to be completed within 2005. The emphasis of this current study is now on systems which include a large number of random variables. For that purpose three problems have been chosen which cover a wide range of cases of interest in engineering practice and involve linear and non-linear systems with uncertainties in the material properties and/or the loading conditions (static/dynamic). Preliminary results obtained in this new Benchmark study are reported in the Proceedings of a Special Session of the *9th International Conference on Structural Safety and Reliability (ICOSSAR 2005)*, Rome, Italy, 2005 [3].

Reference results obtained with direct MCS allow objective and unbiased comparisons to be made between the various reliability estimation approaches employed in benchmark studies. Thus, the accuracy, the efficiency and the limitations of each approach can be quantitatively determined and the associated advantages and disadvantages can be highlighted. Apart from assessing existing procedures in reliability estimation, the benchmark results can also serve as a reference for the engineering community in order to test new algorithms and computational procedures.

4   Computational Efficiency Issues

The MCS method – either in its simple direct version or in the framework of a variance reduction technique – involves expensive computations due to the successive system analyses required. Consequently, the need for developing efficient computational procedures emerges, in order to accelerate the MCS process and make it more tractable in structural engineering practice.

4.1   Solution Techniques

When the Finite Element (FE) method is applied in the context of an MCS-based technique for the analysis of a structure with uncertain properties, successive linear systems with multiple left-hand sides have to be processed, since the coefficient matrix K representing the stiffness of the structure changes in every simulation. In particular, assuming deterministic loads, each simulation

$i = 1, 2, \ldots, n_{sim}$ involves the solution of a problem of the form:

$$K_i u_i = p, \tag{16}$$

where K_i is the stiffness matrix associated with the ith simulation, u_i is the corresponding vector of unknown nodal displacements and p is the vector of nodal loads. Let K_0 be the stiffness matrix associated with the initial simulation. Then, Eq. (16) can be written as:

$$(K_0 + \Delta K_i)u_i = p, \tag{17}$$

which specifies a set of reanalysis-type or nearby problems. Matrix ΔK_i, which defines the difference between the stiffness matrices K_0 and K_i, is generally small compared to K_0.

Several solution procedures can be applied for solving the series of the systems of the form given in Eqs. (16) or (17). The standard direct method based on Cholesky factorization remains the most popular solution scheme for FE equations. According to this method, the stiffness matrix is factorized usually in the form:

$$K = LDL^t, \tag{18}$$

where D is a diagonal matrix and L is a lower triangular matrix with unit elements on the leading diagonal. FE equations are then solved for the displacements vector with a forward substitution using L, a vector operation involving D and a backward substitution employing L^t. Clearly, the aforementioned factorization and forward/backward substitution procedures need to be performed at each simulation.

A widely known disadvantage of the direct solution approach is its poor performance in solving large-scale problems. This deficiency has led to the development of alternative solution procedures, which are essentially iterative rather than direct and allow the exploitation of the special features of nearby problems encountered in FE reanalyses.

The Preconditioned Conjugate Gradient (PCG) method is such a solution technique, which can be customized to take into account the relatively small differences between stiffness matrices K_i, $i = 1, 2, \ldots, n_{sim}$, avoiding this way the treatment of the n_{sim} systems (16) as stand-alone problems. PCG-customization is localized at the preconditioning matrix $\tilde{K}$ employed to accelerate PCG convergence during the successive FE solutions. Hence, reanalysis problems of the form given in Eq. (17) can be effectively solved using the PCG algorithm equipped with a preconditioner following the rationale of incomplete Cholesky preconditionings. The incomplete factorization of the stiffness matrix $K_0 + \Delta K_i$ can be written as:

$$\tilde{L}_i \tilde{D}_i \tilde{L}_i^t = K_0 + \Delta K_i - E_i, \tag{19}$$

where $\tilde{D}_i$ is a diagonal matrix and $\tilde{L}_i^t$ is a lower triangular matrix with unit elements on the leading diagonal. E_i is an error matrix which is taken as

ΔK_i, therefore the preconditioning matrix becomes the complete factorized initial stiffness matrix: $\tilde{K} = K_0$. The PCG algorithm equipped with this preconditioner throughout the entire MCS process constitutes the powerful PCG-K_0 method for the solution of the n_{sim} nearby problems (17) [4].

4.2 Parallel Computing

Parallel computing is widely appreciated as an effective tool to overcome the computational barriers imposed by CPU speed and memory limitations in sequential processing. The appropriate partitioning of a computational task into subtasks allows the exploitation of powerful parallel computers by executing the produced subtasks on the several available processors concurrently to accelerate the overall computational process. Among the various parallel processing environments available today, a widely used and cost-effective platform for distributed computing is the cluster of networked computers.

Parallel processing is particularly suitable for coping with the excessive computational workloads produced in the context of MCS-based FE analysis (see e.g. [11,4]). There are basically two alternative options for parallelizing the MCS process, in order to take advantage of high performance computing environments like clusters of computers. These two options are realized by partitioning the overall computational process either at the MCS level or at the structural domain level [4]. In the first case, the global set of simulations to be performed is decomposed into subsets, each of which is assigned to a different processor. In other words, several simulations are concurrently conducted by executing the standard sequential MCS-based FE procedure on each processor for a part of the total number of simulations. In the second case, a Domain Decomposition (DD) formulation is applied by partitioning the FE mesh of the structure into non-overlapping submeshes (called subdomains). This way, one simulation is conducted at a time with the structural domain spread over the available processors in the form of subdomains.

Clearly, implementing a DD method is a much more complicated task than the straightforward partitioning at the MCS level. However, the partitioning of simulations leads to efficient MCS-based FE analysis only for moderately large-scale problems, while DD algorithms can deal more effectively with very large-scale problems involving hundreds of thousands or even millions degrees of freedom.

5   Applications

Two application examples are given in this section, in order to demonstrate the merits of some of the techniques and algorithms described in the present chapter. The application areas addressed in the two examples are: (i) stochastic dynamics and (ii) stochastic FE analysis.

5.1   Oscillator with Random Properties Subjected to Random Excitation

The excursion probability of a ten-degree-of-freedom Duffing-type oscillator subjected to random excitation is considered to show the effectiveness and efficiency of a variance reduction technique in estimating low probabilities in high dimensions (see case 2 of problem 2 in the Benchmark study [9]). The governing equation of this problem is expressed with respect to time t as:

$$\vec{M}\ddot{\vec{u}}(t) + \vec{C}\dot{\vec{u}}(t) + \vec{K}\vec{u}(t) = \vec{F}(t) \tag{20}$$

with zero initial conditions, where the mass, damping and stiffness matrices are respectively given by:

$$\vec{M} = \begin{bmatrix} m_1 & 0 & 0 & \dots & 0 \\ 0 & m_2 & 0 & \dots & 0 \\ . & . & . & \dots & . \\ 0 & \dots & 0 & 0 & m_{10} \end{bmatrix}, \tag{21}$$

$$\vec{C} = \begin{bmatrix} c_1 + c_2 & -c_2 & 0 & \dots & 0 \\ -c_2 & c_2 + c_3 & -c_3 & \dots & 0 \\ . & . & . & \dots & . \\ 0 & & \dots & 0 & -c_{10} & c_{10} \end{bmatrix}, \tag{22}$$

$$\vec{K} = \begin{bmatrix} k_1 + k_2 & -k_2 & 0 & \dots & 0 \\ -k_2 & k_2 + k_3 & -k_3 & \dots & 0 \\ . & . & . & \dots & . \\ 0 & & \dots & 0 & -k_{10} & k_{10} \end{bmatrix}. \tag{23}$$

The displacements vector is $\vec{u}(t) = [u_1, u_2, \dots, u_{10}]^t$ and the random excitation $\vec{F}(t)$ has the form $\vec{F}(t) = p(t)[m_1, m_2, \dots, m_{10}]^t$, where $p(t)$ is modeled by a modulated filtered Gaussian white noise. Failure for this system occurs when the maximum absolute relative displacement $\delta u_{\max}$ between two consecutive degrees of freedom over the time interval of interest exceeds some pre-defined critical value.

   The total number of random variables for the Duffing-type oscillator application is 4,030. The random variables consist of 30 structural parameters (i.e. mass, damping and stiffness values for each of the 10 degrees of freedom) and the excitation values at 4,000 time instants considered in the interval $[0, 20s]$ with a step $\Delta t = 0.005$ s. Naturally, the effect of these two groups of random variables on the response of the system and as a consequence on the excursion probabilities is markedly different. Therefore, the original random parameter space is orthogonally decomposed into two subspaces, i.e. each vector $\vec{\xi} \in \mathbb{R}^{4030}$ can be uniquely written as:

$$\vec{\xi} = \vec{\xi}_s + \vec{\xi}_e, \tag{24}$$

where $\overrightarrow{\xi}_s \in \mathbb{R}^{30}$ represents the structural parameters and $\overrightarrow{\xi}_e \in \mathbb{R}^{4000}$ the random excitation. This random parameter space decomposition is exploited to estimate the failure probability P_F for 4 test cases by calculating the averaged probability flow using line sampling computations [20]. Test cases 1 and 2 refer to the excursion probability of the absolute displacement of the 1st degree of freedom, while test cases 3 and 4 examine the excursion probability of the relative absolute displacement of the 10th degree of freedom. Hence, the failure of the oscillator must be monitored for 4 different critical displacement values, each of which corresponds to one of test cases 1-4.

Table 1 reports the results obtained with the procedure mentioned above, while direct MCS results are also provided for comparison purposes. The results of this Table illustrate the accuracy and efficiency of the variance reduction technique. The P_F-values estimated using averaged probability flows with line sampling are close to the corresponding reference values of direct MCS. Moreover, a number of only 330 analyses suffices to obtain P_F-results with the variance reduction technique for any test case; a total number of 360 analyses are required to obtain the results of Table 1, since 30 analyses are initially performed for gradient evaluations. Thus, sufficiently accurate reliability estimations can be calculated by performing manageable numbers of simulations, avoiding this way the millions of analyses required by direct MCS. In addition to this advantage, the small Coefficients Of Variation (COV) obtained for P_F with 10 independent runs demonstrate the robustness of the utilized procedure in effectively and efficiently handling high-dimensional spaces of linear stochastic dynamics problems.

Table 1. Simulation results for the Duffing-type oscillator test problem

Test case	1	2	3	4
Critical displacement (m)	0.057	0.073	0.013	0.017
Averaged probability flows using line sampling				
P_F	9.8E-5	9.7E-7	6.0E-5	4.6E-7
$COV(P_F)$	4.9%	7.6%	11.9%	15.7%
Analyses performed	360	360	360	360
Direct MCS				
P_F	1.0E-4	9.0E-7	6.0E-5	3.0E-7
Analyses performed	29,750,000	29,750,000	29,750,000	29,750,000

5.2   Stochastic FE Analysis of Cylindrical Shell

The cylindrical Scordelis-Lo shell roof of Fig. 1 is used to demonstrate the potential of efficient computational procedures in MCS applications. The two longitudinal y-edges of the shell are L=15.2m long and are assumed to be free, while the two circular edges, each of which is defined as a $80\,°$-arc of a circle with radius R=7.6m, are supported by rigid diaphragms. The shell is subjected to deterministic gravity loading and is assumed to have two uncertain structural properties: the spatial variations of the shell's modulus of elasticity E and thickness t are described by two non-correlated two-dimensional univariate homogeneous Gaussian stochastic fields with mean values E_0=21,000N/mm^2 and t_0=76mm, respectively, and a common coefficient of variation σ=10%. The correlation length values considered for the random fields are b=1.3m and b=6.5m. The objective in this application example is to evaluate the probability P_F that the absolute value of the structure's vertical displacement w_A at node A (mid-point of a free edge of the shell in Fig. 1) exceeds some critical value $w_{A,cr}$ (we assume $w_{A,cr}$=46mm in the case of b=1.3m and $w_{A,cr}$=54mm in the case of b=6.5m). Hence, a direct MCS-based stochastic approach is applied to obtain P_F-results [1] using the spectral representation method to generate stochastic field samples [21] and the local average approach to obtain discretized random field values at the elements' centroids [22].

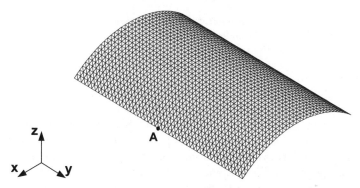

Fig. 1. The structural mesh for the cylindrical shell test problem (35 × 49 nodes)

Meshes of various sizes consisting of triangular shell elements are produced for this stochastic test problem. The finest of these meshes, which contains 35 × 49 nodes resulting in 10,080 active degrees of freedom (Fig. 1), is used as the structural model for carrying out all standard FE calculations. A stochastic mesh, whose size depends on the adopted correlation length b, is also employed in each test run of the shell example for the generation of random field values. The produced stochastic information is transfered from the stochastic to the structural mesh using a computationally inexpensive bivariate interpolation algorithm [4], as visualized in Fig. 2. Since the adopted stochastic mesh is typically coarser than the structural one, the task of generating discretized random field samples during the simulations is considerably accelerated by producing stochastic field values for a smaller number of elements and then appropriately interpolating.

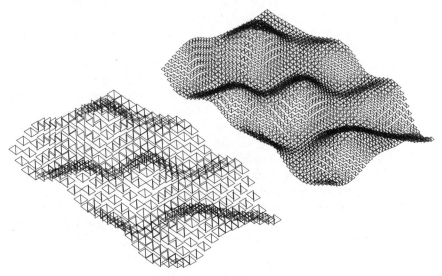

Fig. 2. Cylindrical shell: A Gaussian random field sample (b=1.3m) generated with a stochastic mesh of 17×24 nodes (left) and then interpolated onto the structural mesh of 35×49 nodes (right)

The timing results of Table 2, which are obtained on a cluster of 12 Pentium PCs running the Linux operating system, show the sequential and parallel computational efficiency of the PCG-K_0 solver on n_p PCs in combination with the concept of separate structural and stochastic meshes. The performance of the conventional direct solution approach is also presented for comparison purposes. Clearly, the use of parallel computing by partitioning the overall computational process at the MCS level leads to very efficient stochastic FE analyses for this moderately large-scale problem. For instance, the total processing time needed to conduct 1,900 simulations is reduced from about 1.2 h using standard techniques (sequential direct solution and coinciding structural and stochastic meshes) to just 33 s when computationally more efficient schemes (parallel solution with the PCG-K_0 method, separate structural and stochastic meshes) are employed. It is emphasized that the P_F-results yielded by such efficient schemes in the test runs of Table 2 are practically equal to the reference results obtained with standard techniques. Hence, MCS results of the same accuracy can be achieved in processing times, which are order(s) of magnitude smaller.

Table 2. Simulation results for the cylindrical shell test problem

b (m)	n_{sim}	n_p	Solver	Stochastic mesh	Total time (s)	P_F
1.3	1400	1	Direct	35×49	3119.7	3.0E-2
		1	PCG-K_0	17×24	941.7	3.0E-2
		6	PCG-K_0	17×24	159.7	3.0E-2
		12	PCG-K_0	17×24	80.7	3.0E-2
6.5	1900	1	Direct	35×49	4233.9	7.1E-2
		1	PCG-K_0	5×7	381.6	7.2E-2
		6	PCG-K_0	5×7	64.8	7.2E-2
		12	PCG-K_0	5×7	32.8	7.2E-2

6   Closing Remarks

The present work presents efficient methods and computational procedures for calculating the reliability of structural systems. Although reliability estimation is a rather computationally intensive task, this work demonstrates that it can be effectively and efficiently handled using Monte Carlo-based approaches. However, despite the developments and achievements in this scientific field, a number of issues are still open and call for further research efforts.

Current research needs are primarily concentrated on problems with large numbers of random variables. Typical statistically high-dimensional applications are problems encountered in stochastic finite element analysis (finite element formulations incorporating structural uncertainties, which are modeled using random fields) and in stochastic dynamics (structural analyses under uncertain dynamic loadings, which are described by random processes). Particularly challenging in this respect are problems which are non-linear and time-dependent.

Another area with significant research potential is Reliability-Based Design Optimization (RBDO), in which the design of a structure is optimized for a particular target safety level. More specifically, the aim in RBDO is to minimize the cost of a structure under the condition that certain pre-specified reliability constraints are satisfied. It is clear that reliability estimation algorithms have an important role to play in RBDO implementations.

Finally, as reliability estimation and RBDO are and will be becoming more effective, efficient and mature tools, they can be used to study and handle the effect of uncertainties in several areas of application: earthquake and wind engineering, materials science, fracture and fatigue behavior, etc.

Acknowledgments. This work has been supported by the project *"Modelling Product Variability and Data Uncertainty in Structural Dynamics Engi-*

neering (MADUSE)" of the European Union under contract number MRTN-CT-2003-505164.

References

1. Argyris J., Papadrakakis M., Stefanou G. (2002) Stochastic finite element analysis of shells. Comput. Meth. Appl. Mech. Engrg. **191**, 4781–4804
2. Au S. K., Beck J. L. (2001) Estimation of small failure probabilities in high dimensions by subset simulation. Probabilistic Engrg. Mech. **16**, 263–277
3. Augusti G., Schuëller G. I., Ciampoli M., eds. (2005) Safety and Reliability of Engineering Systems and Structures (Proceedings of ICOSSAR 2005). Millpress, Rotterdam, Netherlands
4. Charmpis D. C., Papadrakakis M. (2005) Improving the computational efficiency in finite element analysis of shells with uncertain properties. Comput. Meth. Appl. Mech. Engrg. **194**, 1447–1478
5. Ching J., Au S. K., Beck J. L. (2005) Reliability estimation for dynamical systems subject to stochastic excitation using subset simulation with splitting. Comput. Meth. Appl. Mech. Engrg. **194**, 1557–1579
6. Ching J., Beck J. L., Au S. K. (2005) Hybrid Subset Simulation method for reliability estimation of dynamical systems subject to stochastic excitation. Probabilistic Engrg. Mech. **20**, 199–214
7. COSSAN (2005) COmputational Stochastic Structural ANalysis, User's Manual. Institute of Engineering Mechanics, Leopold-Franzens University, Innsbruck, Austria
8. IfM (1997) A Benchmark Study on Nonlinear Stochastic Structural Dynamics. http://mechanik.uibk.ac.at/Publications/benchmark.html
9. IfM (2004) Benchmark study on reliability estimation in higher dimensions of structural systems. http://mechanik.uibk.ac.at/Benchmark_Reliability/
10. ISPUD (1997) Importance sampling procedure using design points, User's manual. Institute of Engineering Mechanics, Leopold-Franzens University, Innsbruck, Austria
11. Johnson E. A., Proppe C., Spencer B. F. Jr., Bergman L. A., Szekely G. S., Schuëller G. I. (2003) Parallel processing in computational stochastic dynamics. Probabilistic Engrg. Mech. **18**, 37–60
12. Katafygiotis L. S., Cheung S. H. (2004) Auxiliary Domain Method for solving nonlinear reliability problems. Proceedings (CD) of the 9th ASCE Specialty Conference on Probabilistic Mechanics and Structural Reliability (PMC2004), Albuquerque, New Mexico, USA
13. Koutsourelakis P. S., Pradlwarter H. J., Schuëller G. I. (2004) Reliability of structures in high dimensions, part I: algorithms and applications. Probabilistic Engrg. Mech. **19**, 409–417
14. Lin Y. K. (1976) Probabilistic theory of structural dynamics. McGraw-Hill, New York
15. Nowak A. S., Collins K. R. (2000) Reliability of structures. McGraw-Hill, New York
16. Oden J. T., Belytschko T., Babuska I., Hughes T. J. R. (2003) Research directions in computational mechanics. Comput. Methods Appl. Mech. Engrg. **192**, 913–922

17. Pradlwarter H. J., Schuëller G. I. (2004) Excursion probabilities of non-linear systems. Int. J. Non-Linear Mech. **39**, 1447–1452
18. Schuëller G. I., Stix R. (1987) A critical appraisal of methods to determine failure probabilities. Struct. Safety **4**, 293–309
19. Schuëller G. I., Pradlwarter H. J., Koutsourelakis P. S. (2004) A critical appraisal of reliability estimation procedures for high dimensions. Probabilistic Engrg. Mech. **19**, 463–474
20. Schuëller G. I., Pradlwarter H. J., Koutsourelakis P. S., Charmpis D. C. (2005) Application of line sampling simulation method to reliability benchmark problems. 9th International Conference on Structural Safety and Reliability (ICOSSAR 2005), Rome, Italy
21. Shinozuka M., Deodatis G. (1996) Simulation of multi-dimensional Gaussian stochastic fields by spectral representation. Appl. Mech. Rev. (ASME) **49**, 29-53.
22. Vanmarcke E. (1983) Random fields: Analysis and synthesis. MIT Press, Cambridge, Massachusetts

Explicit Methods for the Computation of Structural Reliabilities in Stochastic Plastic Analysis

I. Kaymaz[1] and K. Marti[2]

[1] Atatürk University, Faculty of Engineering, Dept. of Mechanical Engineering, 25240 Erzurum, Turkey
[2] Federal Armed Forces University Munich, Aero-Space Engineering and Technology, 85577 Neubiberg/Munich, Germany

Abstract. Problems from plastic limit load or shakedown analysis and optimal plastic design are based on the convex yield criterion and the linear equilibrium equation for the generic stress (state) vector σ. The state or performance function $s^*(y, x)$ is defined by the minimum value function of a convex or linear program based on the basic safety conditions of plasticity theory: A safe (stress) state exists then if and only if $s^* < 0$, and a safe stress state cannot be guaranteed if and only if $s^* \geq 0$. Hence, the probability of survival can be represented by $p_s = P(s^*(y(\omega), x) < 0)$.

Using FORM, the probability of survival is approximated then by the well-known formula $p_s \sim \Phi(\|z_x^*\|)$ where $\|z_x^*\|$ denotes the length of a so-called β-point, hence, a projection of the origin 0 to the failure domain (transformed to the space of normal distributed model parameters $z(\omega) = T(y(\omega))$). Moreover, $\Phi = \Phi(t)$ denotes the distribution function of the standard N(0,1) normal distribution. Thus, the basic reliability condition, used e.g. in reliability-based optimal plastic design or in limit load analysis problems, reads $\|z_x^*\| \geq \Phi^{-1}(\alpha_s)$ with a prescribed minimum probability α_s. While in general the computation of the projection z_x^* is very difficult, in the present case of elastoplastic structures, by means of the state function $s^* = s^*(y, x)$ this can be done very efficiently: Using the available necessary and sufficient optimality conditions for the convex or linear optimization problem representing the state function $s^* = s^*(y, x)$, an *explicit* parameter optimization problem can be derived for the computation of a design point z_x^*. Simplifications are obtained in the standard case of piecewise linearization of the yield surfaces.

In addition, several different response surface methods including the standard response surface method are also applied to compute a β-point z_x^* in order to reduce the computational time as well as having more accurate results than the first order approximation methods by using the obtained response surface function with any simulation methods such as Monte Carlo Simulation. However, for the problems having a polygon type limit state function, the standard response surface methods can not approximate well enough. Thus, a response surface method based on the piecewise regression has been developed for such problems. Applications of the methods developed to several types of structures are presented for the examples given in this paper.

Keywords: Reliability analysis, FORM, probability of survival/failure, elasto-plastic mechanical structures, optimizational representations of state functions, direct computation of the β-point, Response Surface Methods (RSM)

1   Stochastic Plasticity Analysis

1.1   Introduction

Problems from plastic limit/shakedown analysis and optimal plastic design are based [5–7] on the linear equilibrium equation

$$C\sigma = P(p, x) \tag{1a}$$

and the convex or piecewise linear convex yield criterion

$$\pi\left(R_i(\sigma_y, x)_d^{-1}\sigma_i | K_i\right) \leq 1, i = 1, \ldots, n_G, \tag{1b}$$

for the stress (state) vector σ composed of the n_0–subvectors $\sigma_i = (\sigma_{ij})_{1 \leq j \leq n_0}$ of stress (state) components $\sigma_{ij}, 1 \leq j \leq n_0$, at the reference or nodal points $X_i, i = 1, \ldots, n_G$, arising from a discretization of the mechanical structure by finite elements (FE). In (1a,b) the following notations are used: C is the $m \times n$ equilibrium matrix with rank $C = m < n$, $P = P(p,x)$ is the external load vector depending on a random vector $p = p(\omega)$ of load parameters and an r–vector x of design variables, including e.g. the load factor $\mu \geq 0$ in limit and shakedown analysis, cf. [8]. Moreover, the feasible stress domain K_i at the point X_i is a closed convex subset of $\mathbb{R}^{n_0}$ containing the origin 0 of $\mathbb{R}^{n_0}$ as an interior point, $R_i = R_i(\sigma_y, x) = (R_{ij}(\sigma_y, x))_{1 \leq j \leq n_0}$ is the vector of material strength parameters $R_i = R_i(\sigma_y, x) = (R_{ij}(\sigma_y, x))_{1 \leq j \leq n_0}$ at X_i, depending on the vector σ_y of yield stresses and the design vector x, and R_{id} denotes the diagonal matrix having the components $R_{ij}, j = 1, \ldots, n_0$, on its main diagonal. Finally,

$$\pi = \pi(z|K_i) := \inf\left\{\lambda > 0 : \frac{z}{\lambda} \in K_i\right\}, z \in \mathbb{R}^{n_0} \tag{1c}$$

is the distance or Minkowski functional of K_i. Hence, $\sigma_i \to \pi(R_{id}^{-1}\sigma_i|K_i)$ is the yield function at X_i; for more details see [9].

1.2   State Function

According to conditions (1a-c), for the safety of structures made of elasto-plastic materials we have [8,10,11] the following criterion:

$$s^* = s^*(R, P) \begin{cases} < 0 : \text{the structure is in a safe (stress) state} \\ \geq 0 : \text{a safe (stress) state is not guaranteed.} \end{cases} \tag{2}$$

The state function (limit state function or performance function) s^* is defined, see [7–9], by

$$\min s \tag{3a}$$

s.t.

$$C\sigma = P \tag{3b}$$

$$\pi(R_{id}^{-1}\sigma_i|K_i) - 1 \le s, i = 1, \ldots, n_G \tag{3c}$$

We may assume that the random ν_σ–vector $\sigma_y = \sigma_y(\omega)$ and the random ν_p–vector p are stochastically independent. Let $\nu := \nu_\sigma + \nu_\theta$, and define the ν–random vector

$$y = y(\omega) := (\sigma_y(\omega), p(\omega)) \tag{4a}$$

In the following, let $x \in D$ denote an arbitrary, but fixed design vector. By means of (4a), the state function s^* can be represented as a function of the parameter vector y and the design vector x:

$$\tilde{s}^*(y, x) = \tilde{s}^*\left((\sigma_y, p), x\right) := s^*\left(R(\sigma_y, x), P(p, x)\right) . \tag{4b}$$

1.3   Computation of the β–Point z_x^*

In the present case of elastoplastic materials, according to (1a-c), the transformed state function $\tilde{s}_T^* = \tilde{s}_T^*(z, x)$ is the minimum value of the convex minimization problem:

$$\min s \tag{5a}$$

s.t.

$$C\sigma = P\left(T_p^{-1}(z_\sigma), x\right) \tag{5b}$$

$$\pi\left(R_i\left(T_\sigma^{-1}(z_\sigma), x\right)_d^{-1}\sigma_i|K_i\right) - 1 \le s, \quad i = 1, \ldots, n_G . \tag{5c}$$

Since (5a-c) is a convex minimization problem in the variables (s, σ), where the problem fulfills the Slater condition, an optimal solution $(\tilde{s}_T^*, \sigma^*) = (\tilde{s}_T^*(z, x), \sigma^*(z, x))$ can be characterized by the **necessary and sufficient** Kuhn–Tucker conditions. Hence, if

$$\tilde{P}(z_p, x) := P\left(T_p^{-1}(z_p), x\right), \tilde{R}_i(z_\sigma, x) := R_i\left(T_\sigma^{-1}(z_\sigma), x\right) , i = 1, 2, \ldots n_G . \tag{6}$$

and $C_{(i)}$ denotes the submatrix of C related to σ_i, then the projection problem can be represented in the following equivalent **explicit** form:

Lemma 1.3.1 The projection problem for the computation of a β–point z_X^* reads (with Lagrange multipliers λ, μ):

$$\min \|z\|^2 \tag{7a}$$

s.t.

$$1 - \sum_{i=1}^{n_G} \mu_i = 0 \tag{7b}$$

$$C_{(i)}^T \lambda + \mu_i \tilde{R}_{id}(z,x)^{-1} \nabla_z \pi \left(\tilde{R}_{id}(z,x)^{-1} \sigma_i | K_i \right) = 0, i = 1, \ldots, n_G \tag{7c}$$

$$C\sigma - \tilde{P}(z,x) = 0 \tag{7d}$$

$$\pi \left(\tilde{R}_{id}(z,x)^{-1} \sigma | K_i \right) - (1+s) \leq 0, i = 1, \ldots, n_G \tag{7e}$$

$$\mu_i \left(\pi \left(\tilde{R}_{id}(z,x)^{-1} \sigma_i | K_i \right) - (1+s) \right) = 0, i = 1, \ldots, n_G \tag{7f}$$

$$\mu_i \geq 0, i = 1, \ldots, n_G, s \geq 0 \tag{7g}$$

A further representation can be obtained if, by piecewise linearization, the convex feasible domain K_i is replaced by a convex polyhedron

$$\tilde{K}_i = \{z \in \mathbb{R}^{n_0} : \tilde{N}_i z \leq 1\}, i = 1, \ldots, n_G \tag{8}$$

with given $\tilde{m}_y \times n_0$, matrices $\tilde{N}_i$ and the vector $1 = 1_{\tilde{m}_y} = (1, \ldots, 1)^T \in \mathbb{R}^{\tilde{m}_y}$.

Then, the state function $\tilde{s}_T^* = \tilde{s}_T^*(z,x)$ is the minimum value function of the LP

$$\min s \tag{9a}$$

s.t.

$$C\sigma = \tilde{P}(z,x) \tag{9b}$$

$$\tilde{N}_i \tilde{R}_{id}(z,x)^{-1} \sigma_i \leq (1+s) 1_{\tilde{m}_y}, i = 1, \ldots, n_G . \tag{9c}$$

Selecting fixed positive strength values $R_{ij}^\circ, j = 1, \ldots, n_0, i = 1, \ldots, n_G$, condition (9c) can be replaced by

$$\tilde{N}_i^0 \tilde{R}_{id}^{0-1} \sigma_i \leq (\tilde{\rho}_{i\,\min}(z, x) + s)\, 1_{\tilde{m}_y^0} , \tag{9c}$$

where $(\tilde{N}_i^0, 1_{\tilde{m}_y^0})$ are given $\tilde{m}_y^0 \times (n_0 + 1)$ matrices and

$$\tilde{\rho}_{i\,\min}(z, X) := \min_{1 \leq j \leq n_0} \frac{R_{ij}\left(T_\sigma^{-1}(z_\sigma), x\right)}{R_{ij}^0} . \tag{10}$$

Consequently, the state function $\tilde{s}_T^* = \tilde{s}_T^*(z, x)$ can be represented then also by the maximum value function of the dual program to (9a-c):

$$\max \tilde{P}(z, x)^T u - \tilde{F}_0(z, x)^T \tilde{u} \tag{11a}$$

s.t.

$$C^T u - \hat{R}(z, x)_d^{-1} N^T \tilde{u} = 0 \tag{11b}$$

$$1_{\nu_G}^T \tilde{u} = 1, \tilde{u} \geq 0 , \tag{11c}$$

where $\nu_G := n_G \cdot \hat{m}_y, \hat{m}_y := \tilde{m}_y, \hat{m}_y := \tilde{m}_y^0$, resp., $R = (R_i)_{1 \leq u \leq n_G}, N$ is composed of the submatrices $\tilde{N}_i, \tilde{N}_i^0, i = 1, \ldots, n_G, U$ is a generalized unit matrix, $\rho_{\min} := (\tilde{\rho}_{i\,\min})$, and

$$\hat{R}(z, x) := \tilde{R}(z, x), \hat{R}(z, x) = R^0, \text{ resp.,} \tag{11d}$$

$$\tilde{F}_0(z, x) = U 1_{\nu_G}, \tilde{F}_0(z, x) = U \tilde{\rho}_{\min}(z, x), \text{ resp..} \tag{11e}$$

Consequently, due to this duality relation, the constraint $\tilde{s}_T^*(z, X) \geq 0$ holds, if and only if there is a vector $(u, \tilde{u})$ of dual variables fulfilling the relations

$$\tilde{P}(z, x)^T u - \tilde{F}_0(z, x)^T \tilde{u} \geq 0 \tag{12a}$$

$$C^T u - \hat{R}(z, x)_d^{-1} N^T \tilde{u} = 0 \tag{12b}$$

$$1_{\nu_G}^T \tilde{u} = 1, \tilde{u} \geq 0 \tag{12c}$$

Here, corresponding to Lemma 1.3.1, we find the next **explicit** characterization of a β–point:

Lemma 1.3.2 Replace the feasible domains K_i by convex polyhedrons $\tilde{K}_i, i = 1, \ldots, n_G$. Then the projection problem for the computation of a β–point z_x^* reads:

$$\min ||z||^2 \tag{13a}$$

s.t.

$$\tilde{P}(z, x)^T u - \tilde{F}_0(z, x)^T \tilde{u} = 0 \tag{13b}$$

$$C^T u - \hat{R}(z, x)_d^{-1} N^T \tilde{u} = 0 \tag{13c}$$

$$1_{\nu_G}^T \tilde{u} = 1, \tilde{u} \geq 0 \tag{13d}$$

Remark 1.3.1 Special representations of the β–point of the above type can be obtained for trusses and frames.

Having a new formulation for the beta-computation for the elastoplastic mechanical structures described above, the classical First Order Reliability Method (FORM) is given in the following section in order to compare the proposed method with the standard FORM approach.

2   First Order Reliability Method (FORM)

In FORM/SORM, the main effort is to solve the constrained optimization problem in defined in $\mathbb{R}^\nu$ by

$$\min ||z||^2 \tag{14a}$$

s.t

$$g(z) = 0 \tag{14b}$$

and

$$g(z) := -\tilde{s}_T^*(z, x) \tag{15}$$

In structural reliability analysis, another method called Response Surface Method (RSM) has emerged to be able to overcome difficulties of using FORM/SORM approaches for problems having an implicit or time-consuming evaluation of limit state function. One of the advantages of using a RSM is to get a better result with the help of any simulation method with the obtained response surface function. Thus, a simple function is replaced in place of the performance function for structural reliability analysis.

3 Response Surface Method

In structural analysis, Response Surface Methods (RSM) are used to approximate (estimate) the complex relationship between the performance of a structure and the variables that affect the performance. Hence, for many applications, the problem is to estimate a response (output) function or mechanism:

$$g = g(\xi) \tag{16}$$

with the input r-vector $\xi = (\xi_1, ..., \xi_r)'$, and an unknown (to be estimated) function g. For the estimation of the unknown (eventually partly known) response function g, observations or estimates $\eta^{(i)} \sim g(\xi_i)$ of the response $g = g(\xi_i)$, $i = 1, .., p$ corresponding to p input r-vectors $\xi^{(1)}$, $\xi^{(2)}$, $...,\xi^{(p)}$ are available. The unknown function g is estimated by approximating g by a polynomial of a certain (low) order s. Hence, if $s = 1$, i.e., if a linear approximation is used, then

$$g(\xi) \approx b_0 + b_1\xi_1 + ... + b_r\xi_r \tag{17}$$

with $r + 1$ unknown coefficients b_k, $k = 0, 1, 2, ..., r$. Consequently, the observations $\eta^{(i)}, i = 1, .., p$ may be represented by

$$\eta^{(i)} = b_0 + b_1\xi_1^{(i)} + ... + b_r\xi_r^{(i)} + \varepsilon_i$$

$i = 1, .., p$. Here, ε_i, $i = 1, .., p$, are error terms including observation/measurement errors as well as approximation errors.

The above p equations are then represented by the matrix equation

$$\eta = Xb + \varepsilon \tag{18}$$

where

$$\eta := (\eta^{(1)}, ..., \eta^{(p)})^T \tag{19}$$

is the p-vector of all observations of g at the p input vectors $\xi^{(i)}, i = 1, .., p$,

$$b := (b_0, b_1, ..., b_r)^T \tag{20}$$

is the $(r + 1)$-vector of the unknown coefficients, X is the $p \times (1 + r)$ matrix with the p rows $X_i = (1, \xi_1^{(i)}, ..., \xi_r^{(i)}), i = 1, .., p$, which is called design matrix, and $\varepsilon = (\varepsilon_1, ..., \varepsilon_p)'$ denotes the vector of all errors, i.e. the observational or measurement errors as well as the (analytical) errors from the approximation of g by a (first order) polynomial. Having no more information, the vector b of unknown coefficients is determined by LSQ-techniques [3,4], i.e. the estimate $\hat{b}$ of b is defined by minimizing the function

$$L(b) := \|\eta - Xb\|^2 \tag{21}$$

Under corresponding rank conditions, the estimator of b is then given by

$$\hat{b} = (X^T X)^{-1} X^T \eta \qquad (22)$$

For reliability problems, Bucher [1] reformed the traditional RSM given above so that the response function can fit around a region that might include the design point in structural reliability analysis, for which the main points of standard RSM technique(s) are described in the following section.

3.1   Standard Response Surface Method

The aim of a standard RSM is to replace the performance function $g(y)$, see (15), by an appropriate approximative response surface function $\bar{g}(y)$ which is mostly given as a second order polynomial as shown in (23) below:

$$\bar{g}(y) = a + \sum_{i=1}^{\nu} b_i y_i + \sum_{i=1}^{\nu} c_i y_i^2 \qquad (23)$$

Here, y_i, $i = 1,2,...,\nu$, denotes the ν physical ("real") variables, as e.g. load factors, material resistance coefficients, cost factors, etc., and the coefficients a, b_i, c_i are to be determined. As mentioned above, cf. (4a), the values of the ν-vector y of physical variables are realizations $y=y(\omega)$ of a certain random vector y(y) having a known probability distribution. As the number of free parameters in (23) is $2\nu+1$, only a few calculations are needed to obtain the coefficients of RSF. In the standard RSM, the suggested way of obtaining these coefficients is interpolation using the points generated along the axes of the physical Y-space, $Y := R^\nu$, which are chosen to be of the form:

$$y_i = \bar{y}_i \pm f_i \sigma_i \qquad (24)$$

in which $\bar{y}_i$ and σ_i are the mean value and standard deviation of random variables $y_i(\omega)$, resp., and f_i is a certain scale factor. Hence, the interpolation points are selected as indicated in Fig. 1. The values of these points are then substituted into the performance function to get the response of $g(y)$. According to the estimation method described at the beginning of Section 3, a first approximation $\bar{g}(y)$ of the true state function g(y) is determined. This first approximate RSF is used then in place of the true performance function to compute a first approximate y_{D1} of the "design point" in Y-space by using one of the structural reliability methods such as the First or Second Order Reliability Method (FORM or SORM) or Importance Sampling. In case of using FORM, the design point is obtained by using (14a,b). For this purpose, the function $\bar{g}(y)$ is transferred to the Z-space of standard normal distributed ν-random vectors:

$$\bar{g}_T(z) := \bar{g}(T^{-1}(z))$$

Then the projection problem is solved with $g(z) := \bar{g}_T(z)$. Let z_{D1} denote the corresponding optimal solution (β-value in Z-space). By back transformation

$$y_{D1} := T^{-1}(z_{D1})$$

one obtains then the corresponding β-point y_{D1} in the physical Y-space. The next iteration will be based on this point which significantly affects the accuracy of the second and final RSF. Having y_{D1}, the new center point y_M for interpolation is obtained on a straight line from $\bar{y}$ to y_{D1} as shown in Fig. 1b, and an explicit formulation of y_M is given as

$$y_M = \bar{y} + (y_{D1} - \bar{y})\frac{g(\bar{y})}{g(\bar{y}) - g(y_{D1})} \tag{25}$$

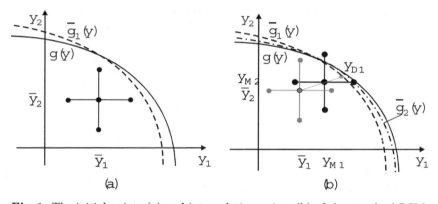

Fig. 1. The initial points (a) and interpolation points (b) of the standard RSM.

In order to use some properties of the structural reliability problems in the construction of the RSF, Kaymaz [2] proposed a new approach, called Adaptive Response Surface Method (ADAPRES), based on weighted regression to get the coefficients of the RSF as explained in the following section.

3.2 ADAPRES: A Response Surface Method for Structural Reliability Analysis

In the standard RSM normal regression is commonly used which gives equal weight to the coefficients of the RSF formed. However, the main effort in the application of the RSM is to form a RSF as close as possible to the limit state function. Therefore, in the following normal regression is replaced by a weighted regression method in which the RSF is formed by giving higher weight to the experimental points closer to the limit state surface.

The main aim in the formation of the RSM is to fit a RSF as closely as possible to the limit state function. In the standard RSM, see above, the coefficients of the RSF using least square method are given by

$$\hat{b} = (X^T X)^{-1} X^T \eta \qquad (26)$$

Here, X denotes the design matrix comprising the experimental points that are generated in the Y-space, and η represents the response vector obtained from the performance function cor responding to the experimental points. In this method the estimation errors are equally weighted. However, a good RSF must be formed such that it describes the performance function well, especially close to the limit state surface given by

$$g(y) = 0 \qquad (27)$$

Therefore, the weighted regression method [13] is utilized to find the coefficients of the RSF, for which the weights are usually determined by allowing the uncorrelated residuals to have different variance, unlike the normal regression, for the error term ε as

$$V\left[\varepsilon\right] = \begin{bmatrix} \sigma_1^2 & \cdots & 0 \\ \vdots & \ddots & \vdots \\ 0 & \cdots & \sigma_r^2 \end{bmatrix} \qquad (28)$$

The above equation indicates that the error terms independently follow probabilistic distributions of different variance. Thus, a weight w_i to each observation is assigned so that $w_1 \sigma_1^2 = \cdots = w_r \sigma_r^2 = 1 \sigma_0^2$ where σ_0^2 is termed the standard deviation of unit weight. The estimate $\hat{b}$ of b is defined by minimizing the function

$$L(b) := (\eta - Xb)^T W (\eta - Xb) \qquad (29)$$

Under corresponding rank conditions, the estimator of b is then given by

$$\hat{b} = (X^T W X)^{-1} X^T W \eta \qquad (30)$$

where W is an $n \times n$ diagonal matrix of weights as:

$$W = \begin{bmatrix} w_1 & 0 & \cdots & 0 \\ 0 & w_2 & \cdots & 0 \\ \vdots & \vdots & \ddots & \vdots \\ 0 & 0 & \cdots & w_r \end{bmatrix} \qquad (31)$$

In general the weights are assigned to observations so that the weight of an observation is proportional to the inverse expected (prior) variance of that observation, $w_i \propto 1/\sigma_{i,prior}^2$. However, in this study, a different approach is proposed to select the weights for the observations since we are seeking to find a RSF close to the limit state function where $g(y) = 0$, which is achieved as follows:

Among the responses of the performance function corresponding to the design matrix the best design is selected based on closeness to a zero value, which indicates that the experimental point is close to the limit state:

$$\hat{y}_{best} = \min_{y \in Y} |g(y)| \tag{32}$$

where min indicates the minimum response value of the performance function evolutions obtained according to the design of the experiments, thus $\hat{y}_{best}$ indicates the value of the absolute minimum performance function response.

The following expression is found to be suitable to obtain the weight for each experiment:

$$w_i = e^{\left(-\frac{\eta^{(i)} - \hat{y}_{best}}{\hat{y}_{best}}\right)} \tag{33}$$

where $\eta^{(i)}$ indicates the ith response from the ith experiment designed according to the design of experiment selected, where i corresponds to the number of the experiment.

The obtained weights are used in the weighted regression to estimate $\hat{b}$ of b as:

$$\hat{b} = (X^T W X)^{-1} X^T W \eta \tag{34}$$

where W is a diagonal matrix as given in (31).

Thus, the RSF to be formed from the weighted regression will have coefficients with greater weights for the points closer to the limit state, thus leads to a better estimate for the reliability index as shown in the examples.

ADAPRES can approximate the performance function around the design point better than the traditional RSM as shown in the examples. However, for the problems having polygon type limit state function, the methods given above can not approximate well enough as will be shown in the examples. Therefore, a response surface method based on the piecewise regression has been developed for such problems, for which the theoretical background is given below.

3.3 A Response Surface Method with a Spline

In piecewise regression, the surface is approximated by subdividing the corresponding ranges $a_i \le y_i \le b_i$ of the physical variables $y_i, i = 1, ..., \nu$, into sufficiently small intervals $[\zeta_j..\zeta_{j+1}]$, with $a = \zeta_1 < ... < \zeta_{r+1} = b$, such that, on each subinterval, a polynomial p_j of relatively low degree can provide a good approximation to the function to be fitted [12]. This can even be done in such a way that the polynomial pieces blend smoothly, hence, guaranteeing that the resulting patched or composite functions $s(x) := p_j(x)$ for $x \in [\zeta_j..\zeta_{j+1}]$ has several continuous derivatives. Such a smooth piecewise polynomial function is called a spline. One of the most widely used spline type is B-spline which is explained in terms of its use in this study in detail below.

In case of only one physical variable $y_1 = y$, i.e. if $\nu = 1$, a B-spline based response surface function $\bar{g}(y)$ is defined by

$$\bar{g}(y) = \sum_{i=0}^{n} \mathbf{P}_i N_{i,k}(y) \tag{35}$$

where $\mathbf{P}_i$ are the control points, k is the order of the polynomial segments of the B-spline curve, $N_{i,k}(y)$ are the normalized B-spline blending functions, which are described by the order k and by a non-decreasing sequence of real numbers $\{t_i : i = 0, ..., n + k\}$ that are called knot sequence. An explicit definition of the blending functions is given as follows

$$N_{i,1}(y) = \begin{cases} 1 \text{ if } t_i \leq y \leq t_{i+1} \text{ and } t_i < t_{i+1} \\ 0 \text{ otherwise} \end{cases} \tag{36}$$

and if $k > 1$,

$$N_{i,k}(t) = \frac{y - t_i}{t_{i+k-1} - t_i} N_{i,k-1}(y) + \frac{t_{i+k} - y}{t_{i+k} - t_{i+1}} N_{i+1,k-1}(y) \tag{37}$$

4 Examples

4.1 Example 1: A Three-Bar Truss

In the following example, a three-bar truss as depicted in Fig. 2 is subjected to a random vertical load, and the material yield strength is also considered as a random variable with normal distribution.

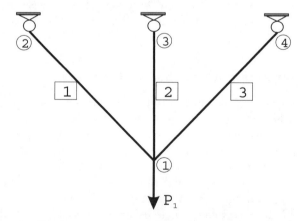

Fig. 2. Three-bar truss with a random load

Table 1. Problem parameters of the three-bar truss structure.

Random parameters		
	Mean	Standard Deviation
Applied force, P (N)	1e5	1e4
Material strength σ_y(N/mm^2)	190	20
Deterministic parameters		
The length of the second bar, l (mm)	1000	
The cross sectional areas of each bar	A(1)=350, A(2)=800, A(3)=350	

The problem parameters for the three-bar truss structure are given in Table 1.

The β-value is determined by the minimization problem as explained in detail in Section 1.3, for which the formulation is represented as follows:

$$\begin{aligned}
&\min \ \|z\|^2 \\
&\text{s.t.} \\
&P(T_p^{-1}(z_p), x)^T v - F_o(T_\sigma^{-1}(z_\sigma), x)^T \tilde{v} \geq 0 \\
&C^T v - H^T \tilde{v} = 0 \\
&1^T \tilde{v} = 1 \\
&\tilde{v} \geq 0
\end{aligned} \tag{38}$$

The new program, called beta_direct, explained in Section 1.3 is used to obtain the reliability results given in Table 2.

Table 2. Reliablilty results for the three-bar truss structure

The method applied	Design Point	P_f	$\beta-$value
Standard FORM	u(1)=1.898 u(2)=-4.907	7.186e-008	5.260
Direct method	u(1)=1.898 u(2)=-4.907	7.186e-008	5.260

As can be seen from Table 2, the proposed direct method gives exactly the same results with that of the standard FORM approach.

Since one of the main aims of this paper is to develop a response surface method for stochastic plasticity analysis, the results given in Table 2 are compared with that of the both standard RSM and the proposed ADAPRES, and the reliability results are given in Table 3. The limit state function and the response function as well as the experimental points are graphically shown in Fig. 3, where the solid line indicates the state function while the dashed line shows the response surface function fitted, and the circles represents the experimental points generated from Central Composite Design.

Table 3. Reliability results from standard_RSM and ADAPRES

	Standard RSM	ADAPRES
selected reliability method	FORM	FORM
probability of failure	3.31e-9	7.190e-8
β−value	5.80	5.260
design points (z-space)	u(1)=1.649	u(1)=1.894
	u(2)=-5.560	u(2)=-4.907

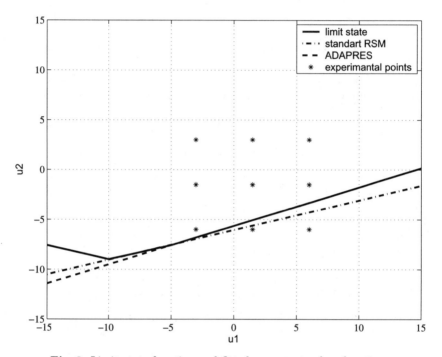

Fig. 3. Limit state function and fitted response surface functions

As seen from Fig. 3, there is a discrepancy between the limit state function and the RSF generated from the standard RSM, which can be caused from the weakness of the classical RSM.

ADAPRES based on the weighted regression is applied to above problem and the following results are obtained, indicating that the results are almost the same as the exact solution obtained from beta_direct given in Table 2.

4.2 Example 2: A 5-Bar Truss Structure

The structure is loaded with two random variables as shown in Fig. 4, and related problem parameters are given in Table 4.

The equilibrium matrix for the given structure is as

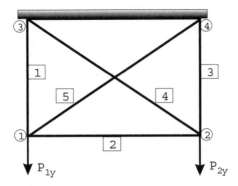

Fig. 4. Five-bar truss structure carrying two random loads

Table 4. Problem parameters for the five-bar truss structure

Random parameters		
	Mean	Standard Deviation
Applied force, P_{y1} (N)	1e5	1e4
Applied force, P_{y2} (N)	1e5	1e4
Deterministic parameters		
Yield Strength, σ_y (N/mm2)	190	
The area of the bars	A(i)=410, i=1..n, n=5	

$$
C = \begin{bmatrix}
0 & -1 & 0 & 1/\sqrt{2} & 0 \\
-1 & 0 & 0 & -1/\sqrt{2} & 0 \\
0 & 1 & 0 & 0 & 1/\sqrt{2} \\
0 & 0 & -1 & 0 & -1/\sqrt{2}
\end{bmatrix}
\tag{39}
$$

The cross sectional areas of each bar are selected as 410. The limit state function and the design point corresponding to the β-value computed are graphically shown in Fig. 5.

The corresponding reliability results obtained from both the standard FORM and the proposed direct method are given in Table 5.

Table 5. Reliability results for the five-bar truss structure

The method applied	Design Point	P_f	β−value
Standard FORM	u(1)=3.298	4.868e-4	3.298
	u(2)=3.277e-12		
Direct method	u(1)=3.298	4.868e-4	3.298
	u(2)=0.000011		

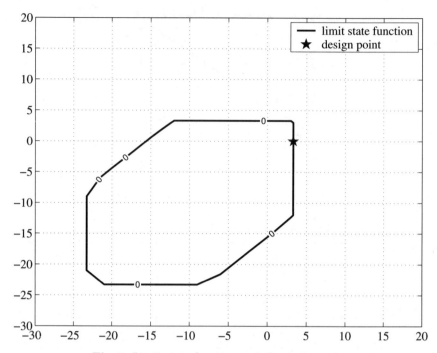

Fig. 5. Limit state function and the design point.

The reliability results given in Table 5 indicate that the proposed direct method can also work for problems having more complex limit state function as depicted in Fig. 5.

Two response surface methods,which are the proposed ADAPRES and the response surface method with spline, are applied to the example to compute the reliability results that are represented in Table 6.

Table 6. Reliability results from ADAPRES and beta ADAPRES_spline

	ADAPRES	**ADAPRES_spline**
selected reliability method	FORM	FORM
probability of failure	2.101e-5	4.868e-4
β−**value**	4.096	3.298
design points (z-space)	u(1)=2.896	u(1)=3.298
	u(2)=2.896	u(2)=0

As the results given in Table 6 indicates, the ADAPRES can not give accurate results when compared to those given in Table 5. However, the reliability results obtained from ADAPRES_spline gives almost the same results with the direct method since the shape of the limit state function given in Fig. 6 is more suitable for the response surface method with spline.

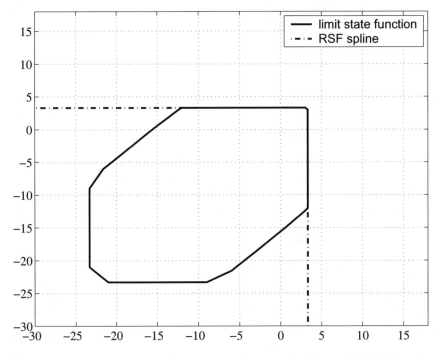

Fig. 6. Limit state function and response surface obtained from ADAPRES_spline.

Since one of the reason of using RSM is to get more accurate results than the approximation methods such as FORM or SORM, this example has been also studied using ADAPRES_spline with MCS, and the results are represented in Table 7.

Table 7. Reliability results and comparison of computational times of MCS and ADAPRES

Applied method	No. of simu-lation	β−value	P_f	Comput. time (sec.)
MCS (exact solution)	1e5	3.167	7.7e-4	1652.7
MCS with ADAPRES_spline	1e5	3.102	9.6e-4	325.47

As can be seen from the results, ADAPRES_spline can also reduce the computational time of the structural reliability analysis of the five-bar truss, which indicates another advantage of the proposed RSM technique. Even though the number of the random variables is two, the reduction in the computational time is very promising when the ADAPRES_spline is used. It also gives better estimate for the reliability results when compared to both the classical and proposed direct method.

5   Conclusions

In the present case of elastoplastic mechanical structures, using the necessary and sufficient optimality conditions for the convex optimization problem representing the (limit) state function $s^* = s^*(y, x)$, an explicit parameter optimization problem has been developed and represented for the computation of a β-point z_x^*. This yields then a considerable reduction of the computational difficulties within FORM that requires computing the projection of the origin to the transformed failure domain. This new technique is applied to several types of structures, such as trusses, and the results are given by comparing with that of FORM. In addition, several different response surface methods including the standard response surface method are also applied to compute a β-point z_x^* in order to reduce the computational time. As shown in the examples, ADAPRES can approximate the performance function around the design point better than the standard response surface method. However, for the problems having polygon type limit state function, the standard response surface methods can not approximate well enough as shown in the examples. Therefore, a response surface method based on the piecewise regression has been developed for such problems and promising results are obtained for the given examples.

Acknowledgments. This research was supported by the Turkish Science Foundation TÜBITAK and the Forschungszentrum Jülich, Germany, under the Project Code of **42.6.I1A.6.A,** which is gratefully acknowledged by the authors.

References

1. Bucher, C.G; Bourgand, U.: Efficient Use of Response Surface Methods. *Report No.9-87*, Institute of Engineering Mechanics, University of Innsbruck, Austria, 1987.
2. Kaymaz, I.; McMahon C. A.: A Response Surface Method Based on Weighted Regression for Structural Reliability Analysis. *Probabilistic Engineering Mechanics 20 (1),* 11-17, 2005.
3. Kleinbaum, D. G.; Kuper, L. L.; Muller K. E.: Applied Regression Analysis and Other Multivariable Methods. PWS-KENT Publishing Company, Boston, 1987.
4. Kreyszig, E.: Advanced Engineering Mathematics. John Wiley & Sons, Singapore, 1993.

5. Marti, K.: Optimal Design of Trusses as a Stochastic Linear Programming Problem. In: Noval, A.S.(ed.): *Reliability and Optimization of Structural Systems*. The University of Michigan Press, Ann Arbor, 231-239, 1999.

6. Marti, K.: Optimal Structural Design under Stochastic Uncertainty by Stochastic Linear Programming Methods. *Journal on Reliability Engineering and Systems Safety (RESS) 72 (3a)*, 165-177, 2001.

7. Marti, K.: Optimal Engineering Design by Means of Stochastic Optimization Methods. In: J. Blachut, H.A. Eschenauer (eds.): *Emerging Methods for Treating Multidisciplinary Optimization Problems. CISM Courses and Lectures 425*, Springer-Verlag, Wien-New York, 107-158, 2001.

8. Marti, K.: Plastic Structural Analysis under Stochastic Uncertainty. *MCMDS 9 (3)*, 303-325, 2003.

9. Marti, K.: Stochastic Optimization Methods in Optimal Engineering Design under Stochastic Uncertainty. *ZAMM 83(12)*, 795-811, 2003.

10. Marti, K.: Stochastic optimization Methods in plastic analysis and optimal plastic design. In: B.H.V. Topping (ed.): *Progress in Civil and Structural Engineering Computing*. Saxe-Coburg Publ.,Stirling, Scotland, UK, 171-189, 2003.

11. Marti, K.: Reliability-based plastic analysis and design. In: Maes, M.A., Huyse, L. (eds.): *Reliabilty and Optimization of Structural Systems*. A.A. Balkema Publishers, Leiden (etc.), 377-383, 2004,

12. McMahon, C. A.; Browne, J.: CADCAM: Principles, Practice and Manufacturing Management.: Addison-Wesley , Harlow, 1998.

13. Myers, R. H.; Montgomery, D. C.: Response Surface Methodology: Process and Product Optimization Using Designed Experiments. John Wiley & Sons, New York, 1995.

Statistical Analysis of Catastrophic Events

J. L. Teugels[1,2] and B. Vandewalle[1]

[1] Katholieke Universiteit Leuven, Belgium
[2] EURANDOM, Technische Universiteit Eindhoven, the Netherlands

Abstract. We make a first attempt to give an extreme value analysis of data, connected to catastrophic events. While the data are readily accessible from SWISS-RE, their analysis doesn't seem to have been taken up. A first set refers to insured claims over the last 35 years; the second deals with victims from natural catastrophes. Together these sets should provide ample proof that extreme value analysis might be able to catch some essential information that traditional statistical analysis might overlook. We finish with a number of cautious remarks.

1 WMO-Release 695

We start with a short summary table that indicates how the number of recorded catastrophes has risen over the second half of the previous century. For background information on catastrophes, see El-Sabh & T.S. Murty, [5]. For information on catastrophes and natural disasters from the point of view of insurance business, see Teugels & Sundt [13]. For the use of statistical procedures within an environmental framework, see W.W. Piegorsch & G. Casella [10].

	50-59	60-69	70-79	80-89	90-99
Number of disasters	20	27	47	63	86
Economic loss (b$)	39.6	71.1	127.8	198.6	607
Insured loss (b$)	0	6.8	11.7	24.7	109.1

Let us make a few comments on this table that can be traced down from WMO Press Release no 695, 2 July 2003 www.wmo.ch/Press695.doc. A natural catastrophe has been described by Munich Re as: *great if the ability of the region to help itself is distinctly overtaxed, making interregional or international assistance necessary.* There is the possibility that in the earlier periods, disasters were not as readily recorded as has been the case during the later decades. Still, the increase in the number of disasters is remarkable. The two other rows are linked. One can safely expect that people, responsible for risk-prone and expensive items, will take appropriate insurance to cope with the economic loss resulting from a catastrophe. Nevertheless, the comparison of the amounts over the last two decades shows a dramatic increase which is not totally predictable from the number of disasters.

2 Extreme Value Statistics

To introduce our statistical methodology we start by an example. We then summarize the most essential aspects of extreme value statistics and apply it to the entire table, as well as to some smaller sub-tables.

2.1 SWISSRE-table of Most Costly Catastrophes

EVENT	TYPE	DATE	LOSS	EVENT	TYPE	DATE	LOSS
WTC-attack	M	11.09.01	21.062	Petro US	M	23.10.89	1.959
Andrew	H	23.08.92	20.900	Fran	H	05.09.96	1.870
Northridge	E	17.01.94	17.312	Fifi	S	18.09.74	1.859
Mireille	T	27.09.91	7.598	X_{13}	ES	04.07.97	1.827
Daria	ES	25.01.90	6.441	Luis	H	03.09.95	1.804
Lothar	ES	25.12.99	6.382	X_4	S	27.04.02	1.707
Hugo	H	15.09.89	6.203	Gilbert	H	10.09.88	1.694
X_{11}	ES	15.10.87	4.839	Isabel	H	18.09.03	1.685
Vivian	ES	25.02.90	4.476	Anatol	ES	03.12.99	1.651
Bart	T	22.09.99	4.445	X_5	S	03.05.99	1.634
Georges	H	20.09.98	3.969	Canada 1	C	17.12.83	1.619
Allison	S	05.06.01	3.261	X_6	S	04.04.03	1.605
X_1	S	02.05.03	3.205	X_7	S	02.04.74	1.600
Piper Alpha	M	06.07.88	3.100	X_8	S	25.04.73	1.527
Kobe, Japan	E	17.01.95	2.973	X_9	S	15.05.98	1.512
Martin	ES	27.12.99	2.641	Loma Pieta	E	17.10.89	1.479
Floyd	H	10.09.99	2.597	Celine	H	04.08.70	1.463
X_{12}	ES	06.08.02	2.548	Vicki	T	19.09.98	1.435
Opal	H	01.10.95	2.526	Petro France	M	21.09.01	1.405
US	F	20.10.91	2.288	Canada 2	C	05.01.98	1.384
X_2	S	06.04.01	2.277	X_{10}	S	05.05.95	1.366
X_3	S	10.03.93	2.220	Grace	H	29.10.91	1.346
Iniki	H	11.09.92	2.090				

The table contains data collected by SWISS-RE, one of the leading reinsurance companies. The figures can for instance be found on the web-site www.swissre.com/INTERNET/pwswpspr.nsf. The data refer to insured losses over the period 1970-2003. *EVENT* refers to the catastrophe itself. Each one of them is recognizable by the *DATE* that can be found in the third columns. When the event received a specific name, then this name has been used to specify the event; if not then a symbol of the type X_i has been used instead. The figures under *LOSS* in the fourth columns refer to insured losses in millions of US dollars. These figures exclude liability losses and they have been indexed to 2003.

The second columns are meant for further use and give a first possible classification of the *TYPE* of event. We used the following abbreviations: M:

man-made disaster, H: hurricane (US-Caribbean region), E: earthquake, T: typhoon (Far East region), ES: European storm, S: US-based storm, C: cold spell, F: forest fire, X_i: alternative for storm without name.

Before we make an analysis of the above table it is instructive to remark the following: The insured losses of the first three events clearly stand out. Their joint total is more than that of the next dozen in the list. This kind of observation is typical when one deals with extreme values. For this reason we include a bit of information on *extreme value statistics*.

2.2 The Maximum

We give a quick survey on how to approximate the distribution of

$$X_{n,n} := \max\{X_1, X_2, \ldots, X_n\} \, ,$$

the *maximum* of a sample by an *extreme value distribution*. For a thorough treatment see the recent book by Beirlant e.a. [2]. This is obtained as follows: Find sequences of normalizing constants $\{a_n\} > 0$ and of centering constants $\{b_n\}$ such that

$$P\left\{ \frac{X_{n,n} - b_n}{a_n} \le x \right\} \to G(x)$$

where $G(x)$ is a non-degenerate distribution. If such an expression can be found, then the distribution of the maximum in the sample can be approximated by

$$P\{X_{n,n} \le y\} \approx G(b_n + a_n \, y) \, .$$

The understanding of the above expression is that once the explicit form of G is known to the insurer, he can get an idea about the distribution of the largest claim he can expect in a sample from an otherwise unknown distribution. Under very weak conditions (satisfied by all classical distributions from statistics) the above expression can be validated.

It turns out that the possible distributions on the right hand side come from a one-parameter family $\{G_\gamma; \gamma \in \Re\}$ of *extreme value distributions*. Each one of them is fully characterized by one single parameter γ which is called the *extreme value index*. The latter parameter needs to be estimated from the data and with a bit more work, even confidence intervals can be obtained. The explicit link between a distribution F (with potentially many parameters) and its corresponding extreme value distribution G_γ (with one single parameter) is technically complicated but possible. We refer the reader to the literature, for example Beirlant e.a. [2].

The tail behavior of the extreme value distribution heavily depends on the sign of γ. We point at some of the highlights.

(i) **Pareto-Fréchet case**: $\gamma > 0$

The explicit form $G_\gamma(x) = \exp{-(1 + \gamma x)^{1/\gamma}}$ is usually referred to as of *type II*. The distribution has infinite right tail. It is useful when modelling data that come from a distribution where there might be fear of the non-existence of moments. For example, if $\gamma > 1$ then there is no finite mean; if $\frac{1}{2} < \gamma < 1$ then there is a mean, but no finite variance.

The most classical example is the Pareto-distribution $1 - F(x) = x^{-\frac{1}{\gamma}} (x > 1)$ often used in insurance and in economics. Distributions of the type $1 - F(x) \sim x^{-\frac{1}{\gamma}} \ell(x)$ with the function ℓ slowly varying are called of *Pareto-type*. They even fully characterize all distributions for which the above approximation by an extremal distribution with positive γ is actually valid.

(ii) **(Extremal) Weibull case**: $\gamma < 0$

This extreme value distribution $G_\gamma(x) = \exp{-(1 + \gamma x)^{1/\gamma}}$ (for $1 + \gamma x > 0$) is often called of *type III*. It is bounded to the right by the value $-\gamma^{-1}$. As a consequence, all of its moments are finite since the right tail of the distribution is exponentially bounded.

Here the most famous examples are formed by all beta-distributions, in particular the uniform distribution.

(iii) **Gumbel case**: $\gamma = 0$

This distribution has a particularly simple form $G_0(x) = \exp(-e^{-x})$ and is known as the *Gumbel distribution* or of *type I*. While is looks a central case from the parametric point of view, many traditional distributions lead to the Gumbel distribution as an approximation for the maximum. The Gumbel distribution itself is somewhat heavy tailed, heavier than exponential in any case. Still, all of its moments are finite.

Among the many distributions that lead to the Gumbel distribution, we mention the normal, the log-normal, the exponential and gamma distributions and many others.

Among the most important statistical assignments in extreme value analysis is the estimation of the extreme value index γ and the construction of confidence intervals. In general, the practitioner should only use the largest (say k) values of the sample as these are the only measurements that refer to the tail behavior of the distribution. The estimation of the value of k is part of the statistical assignment. But it often happens - and the reader can verify this in almost all the forthcoming examples - that there is a sizable number of different k-values leading to comparable estimates of γ.

Over the last decades one has seen a plethora of potential estimators for γ and k. Some apply only to the case where $\gamma > 0$, while others do not make this prior assumption. For more information we again refer to Beirlant e.a. [2].

2.3   Application to the Swiss-Re Loss Table

Here are a few illustrations on how extreme value analysis is applied to the Swiss-Re loss table.

The Entire Table

By way of illustration it is worthwhile to apply the statistical procedure to the entire table above. If one assumes that the 45 different losses from catastrophes can be considered as coming from the same distribution with a positive γ, then one can use the existing statistical know-how to estimate γ. The graph below (Figure 1) illustrates the performance of four different estimators when one gradually includes more and more of the data.

As a first estimator, the Hill [8] estimator was considered (black full line), which can basically be seen as a slope estimator for the linear part of the Pareto quantile plot as discussed below. A second regression estimator was considered in the use of the Zipf [14] estimator (grey dash-dotted line), as introduced in this context by Schultze and Steinebach [12] and Kratz and Resnick [9]. Furthermore, also two bias reduced estimators are considered, a first one (grey dash-dot line) corresponding to an exponential regression model as introduced by Beirlant *et al.* [1] and a second one (grey dashed line) corresponding to a recently developed extension to the generalized Pareto distribution (Beirlant *et al.*, [3]). In the abscissa one allows k, the number of included values, to run from 1 to an ultimate 45.

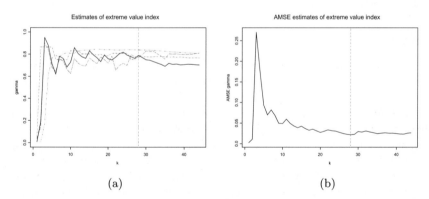

(a)   (b)

Fig. 1. (a) Estimates of the tail index γ for four different well-known estimators, among which the Hill-estimator (full line) and (b) estimates of corresponding asymptotic mean squared errors for the Hill-estimator, along with the minimizing k-value (vertical line).

As one could predict, the estimates show a very strong volatility in the estimation when k is low. This is due to the fact that the number of used data is too small to give any precise estimation. However, when k increases the graphs settle down and show a remarkable stability. This is not only true for the classical Hill-estimates but also for the other three. This is a strong indication that the data are coming from a very heavy tailed distribution.

For each of the four estimators one can of course select a specific k–value that one prefers on the basis of additional statistical precautions. For example, one may choose that value of k for which the asymptotic mean squared error is minimal. The fundamental reason for this choice is that this k–value strikes a balance between the bias (usually large for k large) and the variance (usually large for k small). With this criterium, the Hill-estimator for instance, leads to a k–value of 28 and corresponding estimate .7904 of γ. The least one can say is that near that k-value, the values for the four different estimators are close to each other. Moreover, they all end up around the value .8 indicating that the data might be coming from a distribution with a finite mean but without a finite variance.

Of course, one can argue that the independence among the Swiss-Re data can hardly be discussed. However it is less obvious to consider the data as coming from the same distribution. For this reason we have classified the data according to their type. Note nevertheless that for a reinsurance company such a further subdivision is far less obvious than for an environmentalist who is interested in losses from hurricanes. Whatever, the above analysis has some interest and has been included mainly for illustrative purposes. Moreover, the reader should realize that our analysis, based on just 45 data, still shows remarkable stability. Looking at sub-tables will automatically make all our statements statistically less accurate.

US-Caribbean Hurricanes

We now apply extreme value analysis to a subset of the total table. While the number of data in such subsets will be much smaller, we avoid the risk to be criticized for carelessly amalgamating data.

In Figure 2, it is seen that if we pick out the 12 specific cases from type H, then an estimate around the value 1 seems appropriate for all four estimators as mentioned above. A Pareto-type distribution can be expected to give reasonable results. This indicates that extremely costly hurricanes can still be expected over the given region. Of course, a time dependent examination of a hopefully much larger set would eventually reveal changes in the heaviness of the tail. To illustrate even further that a Pareto-distribution might provide a proper model we have included a $Q - Q$–table. The rationale behind such tables is that if the data are indeed coming from the predicted distribution, then their sample values should be situated close to a straight line.

At least one more comment is in order. The Pareto-fit endorses our opinion that - if the data come from a Pareto distribution as constructed above - then we can roughly estimate the probability of an even larger loss, conditional on already exceeding the threshold present in the data. This is illustrated in

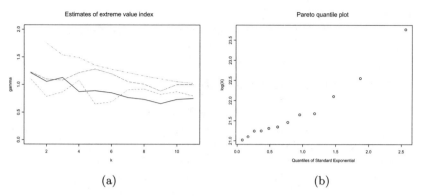

(a)   (b)

Fig. 2. (a) Estimates of the tail index γ for the same four different well-known estimators and (b) Pareto quantile plot for the US-Caribbean hurricane data.

Figure 3. As a first estimator, the Weissman (1978) estimator was considered (black full line), which is based on the Hill estimator as a slope estimator for the ultimate linear part in a Pareto-quantile plot. Furthermore, also two bias reduced estimators are considered, a first one (grey dash-dot line) correspond-ing to the previously mentioned exponential regression model as introduced by Beirlant *et al.* (1999) and a second one (grey dashed line) corresponding to the recently developed extention to the generalized Pareto distribution (Beirlant *et al.*, 2004).

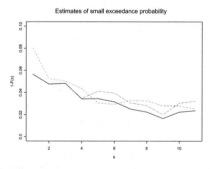

Fig. 3. Estimates for three different estimators, among which the Weissman es-timator (full line), of the probability to attain an even larger loss then already observed.

If we compare the US-Caribbean hurricanes with European storms, then we deduce that European storms seem to have a much smaller γ, perhaps even a negative number. This would indicate that European storms are well modelled by a distribution for which the corresponding distribution has a right upper bound. However, the number of relevant data is too small to perform such an analysis with sufficient confidence.

3 SWISS-RE Casualties Table

On the same Swiss-Re web-site we can find data that refer to numbers of victims in natural disasters. Here are the data concerning the 38 worst catastrophes. (From the original table we dropped two man-made disasters.) For this table we only used as abbreviations C (cyclone), E (earthquake), F (flood), H (hurricane), S (snow) and V (volcanic eruption). Before we analyze the above table and some of its subsets, we make a number of observations.

3.1 The Table Itself

PLACE	TYPE	DATE	VICTIMS	PLACE	TYPE	DATE	VICTIMS
Bangladesh	F	14.11.70	300000	India	E	30.09.93	9500
China	E	28.07.76	250000	Honduras	H	22.10.98	9000
Bangladesh	C	29.04.91	138000	Philippines	E	16.08.76	8000
Peru	E	31.07.70	60000	Kobe (Japan)	E	17.01.95	6425
Gilan (Iran)	E	21.06.90	50000	Philippines	E	05.11.91	6304
Armenia	E	07.12.88	25000	Pakistan	E	28.12.74	5300
Tabas (Iran)	E	16.09.78	25000	Ecuador	E	05.03.87	5000
Colombia	V	13.11.85	23000	Nicaragua	E	23.12.73	5000
Guatemala	E	04.02.76	22000	Indonesia	E	30.06.76	5000
Izmit (Turkey)	E	17.08.99	19118	Fars (Iran)	E	10.04.72	5000
Gujarat	E	26.01.01	15000	Algeria	E	10.10.80	4500
India	C	29.10.99	15000	Afghanistan	E	30.05.98	4000
India	F	01.09.78	15000	Iran	S	15.02.72	4000
Mexico	E	19.09.85	15000	Van (Turkey)	E	24.11.76	4000
India	F	11.08.79	15000	Vietnam	T	01.11.97	3840
India	F	31.10.71	10800	India	F	08.09.92	3800
Venezuela	F	15.12.99	10000	China	F	01.07.98	3656
Bangladesh	C	25.05.85	10000	Taiwan	E	21.09.99	3400
India	C	20.11.77	10000	Reunion	C	16.04.78	3200

- All but four of the events refer to either earthquakes (22) or cyclones-floods (12) catastrophes. These separate categories require further attention. Some countries seem particularly vulnerable for one of these types with the India/Bangladesh region and Iran as particularly vulnerable.
- Only one event appears on both Swiss-Re tables, namely the Great Hanshin earthquake in Kobe in 1995. This single fact clearly illustrates that the concept of risk can get totally different interpretations when looked at it from different perspectives.
- The most surprising observation however is the rounding of the numbers of victims in the fourth columns. It goes without saying that these numbers are rough, even very rough estimates of the actual numbers. The effect on the extreme value analysis will be obvious as the latter will have to be based on rough and imprecise data.

- As in the case of losses, we will again assume that the data in the subsets come from the same distribution and are independent.

3.2   Extreme Value Analysis

We apply our extreme value techniques to the entire table first. We then look at two different subsets.

The Entire Table

Figure 4 shows the estimation of the extremal index for the entire table. The optimal k-value lies at 38 and the corresponding estimate for γ equals 1.197. Using appropriate statistical procedures from extreme value analysis one can obtain confidence bounds for this estimate. By way of example we have drawn the 95% confidence bounds for the entire range of k-values. For our specific choice of $k = 38$ the confidence interval is $(0.913, 1.745)$.

As was the case with the insured losses, also here one might criticize the assumption that all data are independent and come from the same distribution. Independence is probably not a big issue but mixing earthquakes with floods is far less obvious. For this reason we make a separate analysis of these two sub-tables.

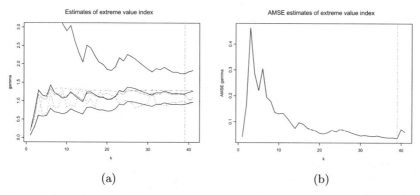

(a)   (b)

Fig. 4. (a) Estimates of the tail index γ for the same four estimators as before, among which the Hill-estimator (full line) and (b) estimates of corresponding asymptotic mean squared errors for the Hill-estimator, along with the minimizing k-value (vertical line).

Casualties From Earthquakes Worldwide

Figures 5 contain some first graphical illustrations on the earthquake data. The figure on the right indicates why k is chosen to be equal to 20, involving almost all of the 22 individual data points. Based on this value we get dramatic estimates $\hat{\gamma} \sim 1.207$ indicating that under this model not even a first moment exist. The recent tsunami of 26.12.04 resulting from an earthquake and with some estimated 275000 victims illustrates this statement.

But also Figure 6 gives more illustrative insight into the earthquake data. On the left the estimated value of γ is used to draw a Pareto-quantile plot. Knowing that only 22 points have been used the fit between the data and the one-parameter Pareto-model is reasonable.

The figure on the right illustrates how one might estimate the small exceedance probability. Given the data and k we estimate the probability that a similar or even higher number of casualties would show up. As was the case with the estimation of γ also the estimation of this exceedance probability can be done using a variety of procedures. The most classical one has been developed by Weissman. For information on such procedures we refer to Beirlant e.a. [2].

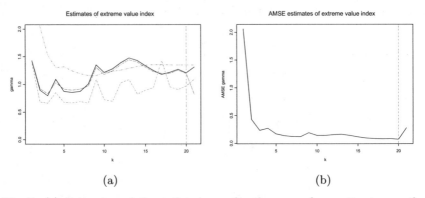

(a)   (b)

Fig. 5. (a) Estimates of the tail index γ for the same four estimators as before, among which the Hill-estimator (full line) and (b) estimates of corresponding asymptotic mean squared errors for the Hill-estimator, along with the minimizing k-value (vertical line).

Casualties From Cyclones

As a second sub-table we combine the casualty data for floods and cyclones as the connection between the two events is obvious. From the 14 data we have to deduce that the situation is even more dramatic than for the earthquake data. The estimated value of γ is $\hat{\gamma} \sim 1.5$ indicating that even worse catastrophes can be expected under the current conditions. On the right, we again illustrate

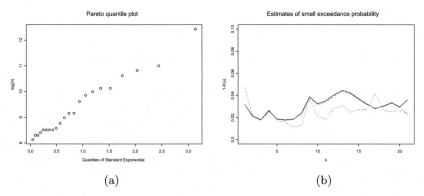

(a)   (b)

Fig. 6. (a) Pareto quantile plot for the number of casualties from earthquakes worldwide and (b) estimates for three different estimators, among which the Weissman estimator (full line), of the probability to attain an even larger number then already observed.

the estimates of exceedance probabilities.

The reader can easily notice that the estimates on the left are far less stable than for some of the previous tables. After all, they are based on merely 14 values. However, the main message from the graph should be that, whatever estimator one uses, all estimates end up far above the value 1. Hence, the underlying distribution is far from having a finite mean.

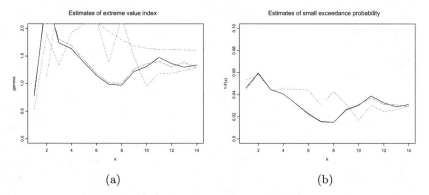

(a)   (b)

Fig. 7. (a) Estimates of the tail index γ for the same four estimators as before, among which the Hill-estimator (full line) and (b) estimates for three different estimators, among which the Weissman estimator (full line), of the probability to attain an even larger number then already observed.

4   Concluding Remarks

We like to finish this contribution with a number of comments.

- It is quite clear that the above applications of extreme value analysis are a very first attempt to get a global insight into the statistical modelling of catastrophic risks. For a number of natural events one has more elaborate historical tables that should be used to make a deeper and more profound analysis. We mention for example the table with earthquake data as provided by Pisarenko and Sornette [11]. Also for hurricane losses larger and time-dependent data should be consulted.

- One serious and rather obvious drawback in some of the examples is the omission of a time coordinate. One normally takes it for granted that insured losses are increasing over time. This is not only due to inflation. For example data provided by Munich Re Group suggest an exponential growth in economic and insured losses over time.

- It is tempting to draw conclusions from the last few tables in connection with the number of victims. In particular, the estimates of γ for the cyclone data could be interpreted as indications of changes in climate or global warming. However, in order to draw such conclusions a much more elaborate and detailed investigation needs to be undertaken. We might hope that our first analysis provides sufficient background to go deeper into the strengths and weaknesses of extreme value analysis.

- One conclusion should stand out: it is necessary that statisticians should be involved in the worldwide analysis of environmental data, especially when attempts are made to use or abuse data in one or the other direction. Topics like global warming and climate change should therefore appear on the subject lists of statistical conferences, not only on those of nature-preserving groups or political forums. For more on this item, see El-Shaarawi & Teugels, [7].

- Certain catastrophes happen at a single site while their aftereffects spread over time and over space. The propagation of the effects depends among others on the intensity of the catastrophe, the characteristics of the surrounding medium, the drop-off effect of the catastrophe, etc. Sometimes one can measure certain quantities directly like in the case of hurricanes. More often, one is forced to look only at secondary effects like in the case of insured losses. An integrated approach of this kind of catastrophes might be revealing. For an example from earthquake analysis, see Brillinger [4].

Acknowledgments. The first author likes to thank Abdel El-Shaarawi for many passed and forthcoming discussions on the raised issues. The authors thank the referees for a careful reading of the manuscript.

References

1. J. Beirlant, G. Dierckx, Y. Goegebeur & G. Matthys Tail index estimation and an exponential regression model. *Extremes*, 2: 177–200, 1999.
2. J. Beirlant, Y. Goegebeur, J. Segers & J.L. Teugels *Statistics of Extremes: Theory and Applications* J. Wiley & Sons, 2004.
3. J. Beirlant, E. Joossens & J. Segers Discussion of "Generalized Pareto Fit to the Society of Actuaries' Large Claims Database" by A. Cebrian, M. Denuit & P. Lambert. *North American Actuarial Journal*, 8: 108–111, 2004.
4. D.R. Brillinger Three environmental probabilistic risk problems. *Statistical Science*, 18: 412–421, 2003.
5. M.I. El-Sabh & T.S. Murty *Natural and Man-Made Hazards Proc. Intern. Symp. Rimouski, Quebec*, D. Reidel Publishing Company, 1987.
6. A.H. El-Shaarawi & W.W. Piegorsch (ed.) *Encyclopedia of Environmetrics* 4 volumes, J. Wiley & Sons, Chichester, 2002.
7. A.H. El-Shaarawi & J.L. Teugels Environmental statistics: Current and future *Proc. 2004 ISI Special Conf., Daejon*, forthcoming, 2004.
8. B.M. Hill A simple general approach to inference about the tail of a distribution. *Annals of Statistics* 3: 1163–1174, 1975.
9. M. Kratz & S.I. Resnick The qq-estimator of the index of regular variation. *Communications in Statistics: Stochastic Models*, 12: 699–724, 1996.
10. W.W. Piegorsch & G. Casella (ed) Statistics and the Environment. *Statistical Science*, Special Issue, 2003.
11. V.F. Pisarenko & D. Sornette Characterization of the frequency of extreme events by the generalized Pareto distribution *Pure and Applied Geophysics*, 160: 2343–2364, 2003.
12. J. Schultze & J. Steinebach On least squares estimation of an exponential tail coefficient. *Statistics and Decisions*, 14: 353–372, 1996.
13. J.L. Teugels & B. Sundt (ed). *Encyclopedia of Actuarial Science* 3 volumes, J. Wiley & Sons, Chichester, 2004.
14. G.K. Zipf *Human Behavior and the Principle of Least Effort: An Introduction to Human Ecology.* Addison-Wesley, 1949.

Scene Interpretation Using Bayesian Network Fragments

P. Lueders[1]

University of Hamburg, Germany

Abstract. We present an approach to probabilistic modelling of static and dynamic scenes for the purpose of scene interpretation and -prediction. Our system, utilizing Bayesian Network Fragments as relational extension to Bayesian networks, provides modelling in an object-oriented way, handling modular repetitivities and hierarchies within domains. We specify a knowledge-based framework, which maintains both partonomy- and taxonomy-hierarchies of entities, and describe an interpretation method exploiting these. The approach offers arbitrary reasoning facilities, where low level perceptive information as well as abstract context knowledge within scenes can be either given as evidence or queried.

1 Introduction

Objective of our work is to develop an integrated vision system which provides reasoning facilities on plans and intentions in partially observed scenes. The system shall handle possibly uncertain evidence on a hierarchy of information entities ranging from perceptive input of a tracker over representations of objects and basic processes to aggregations of entities and context concepts. The conceptual framework of this system was described in [13].

Basically reasoning methods are needed, which can be applied at domains exhibiting unknown number of entities and uncertainties on entity features and relations. Besides Scene interpretation other application cases are examined, e.g. as user surveillance systems (see related works in the end of this paper).

Plain Bayesian networks [18] provide appropriate reasoning methods handling uncertainty and arbitrary evidence/query combinations, but are limited to modelling of propositional domains [7]. Relational extensions to Bayesian networks were introduced [7,12,16] to model complex domains exhibiting relations between entities, concept hierarchies and modular repetitivity. Our work is based on this previous work, representation elements mostly resemble those in [12].

We present a method to exploit partonomy and taxonomy hierarchies in scene modelling and describe an interpretation algorithm, which is based on constrained hypotheses of scene entities and provides answers to queries on features and existence of entities within scenes.

The remainder of the paper is structured as follows. Chapter 2 presents the representation our framework is based on, chapter 3 describes the interpretation algorithm, chapter 4 outlines related work and chapter 5 finally summarizes our presented approach and provides perspectives for future work.

2   Representation

This section starts with presenting the domain-independent underlying structure of our framework. Subsequently we describe the integration of the framework into our scene-modelling domain.

Basic elements within our framework are Bayesian network fragments [7,10,12]. A *Bayesian network fragment* (BNF) $F = (R, O, I, G, C)$ consists of:

- a finite set R of *resident attributes* or *resident random variables*, where for each random variable X, $X \in R$, there is a local distribution D_X, which may be an unconditional probability distribution or a conditional probability distribution (CPD) depending on the values of other random variables, which are parents of X in the fragment graph (described below);
- a finite set O of *output attributes*, with $O \subseteq R$;
- a finite set I of *input attributes*, disjoint to R, or random variables, whose values correspond to output random variables of other fragments;
- a *fragment graph* G, where G is an directed acyclic graph (DAG), whose nodes are indexed by the random variables in $R \cup I$, and random variables in I correspond to root nodes in G;
- a finite set C of *input fragments*, whose values represent other fragments, which have attributes, whom input attributes of the fragment F correspond with.

An input fragment represents a binary relation on fragments; if the value of an input fragment type A of fragment X is Y (denoted as $X.A = Y$), the relation $A(X, Y)$ holds. A fragment X, having input fragment Y, can be input fragment of Y too. The dot-notation can be extended to *attribute chains* $A_1.A_2.\cdots.A_k$, denoting the composition of the relations $A_1, \ldots, A_k$. Input attributes identify their corresponding attribute of an other fragment via attribute chains.

The probability model of a fragment is specified by the local probability models of its attributes. In the proposed framework fragments are present as *classes* and *instances*. An input fragment of a fragment has an associated *type*, i.e. fragment class specification. A type is a ternary relation; if the type-value of type F on input fragment A in fragment X is Y, then the relation $F(X, A, Y)$ holds. The probability model of a fragment is associated with its class. An instance of a class corresponds to a domain entity of the appropriate type and derives its probability model from its class.

Fragment classes are organized in a *class hierarchy* or *taxonomy*. A subclass inherits the probability model of its superclass and can override or extend it. With this inheritance modelling, equal partial probability models of different fragment classes can be represented by the probability model of a common superclass. The set of output-variables of a class fragment $F' = (R', O', I', G', C')$ must at least contain the output variables of its superclass $F = (R, O, I, G, C)$, $O \subseteq O'$.

Fragment classes reside within a *knowledge base* and represent concepts of domain entities. During the interpretation process fragment instances are plugged together to Bayesian networks, which will be called *compound networks* or, while modelling of scenes, *scene graphs*. Resulting compound networks represent relations between domain entities, in our setting basically *partonomies* of scene entities. Arrangements of domain entity BNFs are specific for different scenes or situations, thats why these networks were introduced in [11] as *situation specific networks*.

To ensure, that the local distributions in Bayesian network fragments define well-defined probability distributions in compound Bayesian networks, an assumption is required, which generalizes the acyclicity condition for Bayesian networks: the node orderings in all fragment graphs must be consistent with a global total ordering on random variables [12].

2.1   Scene Modelling

As domain for scene interpretation we choose place-cover settings. In a simple introductory scenario one could ask for the probability of the existence of a cup-cover given the positions of observed cup and saucer objects on a table. One could look for a joint probability distribution (JPD) on 3 random variables A, B, C, representing the position of the cup, the position of the saucer and the existence of the cover.

The variables can be seen as attributes, which carry belief on features and existence of entities. In a partonomy one could maintain belief on features and existence of an aggregate C given evidence of features and existence of their parts A,B, thus leading to distributions $P(C|AB)$ as basic knowledge representation. According to the chain rule of probabilities the JPD is given by e.g. $P(ABC) = P(C|AB)P(B|A)P(A)$.

We consider features of entities to be mutually independent, if nothing is known about a combining relation (here about the existence of the cover). In other words, in scene specific partonomy modelling, dependencies between parts A, B are given in relation to aggregations C. Therefore we apply pruned factorizations as $P(ABC) = P(C|AB)P(A)P(B)$. Figure 1 (left) depicts a Bayesian network corresponding to this factorization, where cover-existence is modelled by a binary random variable, the positions of cup and saucer are represented as discrete variables, which may take up 4 position values.

Considering the context of describing features and existence of domain entities, the variables of the probability distribution are represented within Bayesian network fragments, which correspond to the entities, they are attributes for. Figure 1 (left) shows 3 BNFs, marked by gray rectangles, each of which having only one resident attribute.

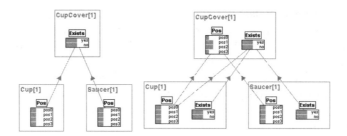

Fig. 1. Example showing the modelling conception

The discussion was limited until this point on a scenario consisting of one aggregate-entity 'CupCover' with its parts 'Cup' and 'Saucer', and will be now generalized. The observation-probability of other aggregate-entities can be defined in a similar way by features of their parts. A concept 'Cover' could e.g. depend on the position of a 'CupCover'; the existence of 'Cup' and 'Saucer' may depend on features of observed blobs. In the example setting we extend the model by inserting existence-variables into the BNFs of 'Cup' and 'Saucer' and a position-variable into the fragment of 'CupCover' (figure 1 right).

We further add edges between the existence-node and between the position-nodes, assuming variable-dependence. The existence of such edges is concept-specific (the position of 'CupCover' e.g. depends on the position of 'Cup': they are equal, if defined by center of gravity) and is defined within the fragment graph of the corresponding BNF. A fragment thus contains *feature-nodes* (here only positions) and an *existence-node*, which represents the observation-probability of the corresponding domain entity. Existence-nodes in network fragments were described by [12] in a similar context, where existence uncertainty of entities was modelled.

In the above example only one entity feature, the position, was described. In advanced settings we expect further features to be involved, e.g. observation times, directions, colors and others. In the result we would get complex conditional probability distributions for existence variables. We therefore try to factorize the existence-distribution, if this is possible, by introducing new variables. A conditional distribution which conditions on features of input entities and contributes to the belief of existence of a concept can be seen as constraint-check element. We may e.g. check constraints for positions and observation times of entities separately to reason on existence of aggregate entities.This factorization can only be done, if constraint checks independently contribute to the existence probability of an entity. Input variables for CPDs of existence-variables of fragments in our domain thus can be so called *constraint-variables* and existence-variables of input entities. Figure 2 illustrates the insertion of a constraint-variable 'SuitPos' at the 'CupCover'-BNF.

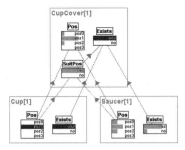

Fig. 2. Scene graph with constraint-variable and inference result

We model feature-variables of entities not only by discrete random variables as in the above example, but also by continuous variables with e.g. Gaussian distributions. Thus we avoid discretization of natural continuous domains of feature-variables representing e.g. positions and times. Resulting scene graphs will be Bayesian networks, containing both discrete and continuous random variables, which are described in [6] as *hybrid networks*. The handling of different distribution types is simplified by separate constraint-variables for different types of features, because their binary discrete distributions are only conditioned on distributions having the certain type of the particular features.

Figure 2 illustrates a result of probabilistic inference application in a scene graph, which is constructed of the three BNFs, corresponding to the entities of the above example. The observation of existence and position of a cup and the existence of the cup-cover was given as evidence. The query results e.g. the expected position and a high existence-probability of a saucer. In figure 2 observed and queried variables are illustrated by black and gray value tables respectively.

Probabilistic inference within scene graphs was processed using an implementation of *importance sampling* (*likelihood weighting*), a method for approximate inference in Bayesian networks, which can be directly applied to networks containing both discrete and continuous random variables [6].

The introductory setting exemplified the modelling of a static scene. Our approach is also applied to dynamic scenes, where scene entities, modelled by BNFs, are not only static objects or aggregations, but also actions or processes. Figure 3 depicts a scene graph in a simple dynamic setting, where the scene of placing a cup-cover is modelled.

The process of placing the cover is represented in relation to two subprocesses of placing a saucer and a cup. Feature-variables of BNFs represent place-positions, start- and end-times of corresponding process-entities. Figure 3 illustrates a query result, where the start-time and position of the saucer-placing was given and features of a temporal succeeding cup-placing were inferred.

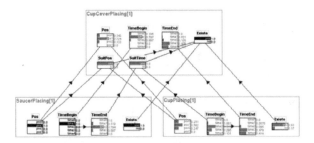

Fig. 3. Scene graph of a dynamic setting example

The compound Bayesian networks examplified above had a hyper tree structure, i.e. the graphs of fragments was tree-structured. In general we allow cycles in the graph of fragments, e.g. to model a 'transport'-concept depending on a 'move'- and a 'touch'-process, which both are connected with the same 'object'-entity. Such cycles are represented by additional equality constraints within BNFs, which are defined by attribute chains in dot-notation. The 'transport'-fragment would then e.g. retain the constraint 'move.object=touch.object'.

2.2 Incorporating Taxonomy Information

A taxonomy structure of domain concepts can be modelled as Bayesian network of binary random variables, where nodes represent the probability of an entity being of a subclass, given, that the entity is of a certain superclass (a similar approach was described in [17]). Figure 4 depicts a taxonomy network of a simple example domain.

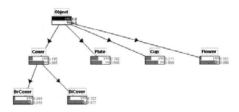

Fig. 4. Taxonomy tree of an example domain

Compound networks of BNFs represent partonomies and other relations between different entities, while a taxonomy graph models class-type relations of a single entity. We integrate taxonomy information into the framework by copying the taxonomy graph into the fragment graphs of entities and unifying the existence-node of each fragment with the node of the taxonomy structure, the fragment class corresponds with. The taxonomy structure thus must contain a node for every fragment class residing in the knowledge base.

3   Scene Interpretation

The above discussion regarded to scene graphs as Bayesian networks, that were composed of Bayesian network fragments. Given a set of BNFs, which reside in the knowledge base and represent concept classes, there exist different possibilities of combining instance BNFs to a scene graph. In our domain a compound of BNFs can be seen as interpretation or explanation of a scene. The challenge is to find a composition of BNFs, which explains a (partially) observed scenario best.

We now describe an incremental interpretation algorithm. Given scene observations are placed as instance BNFs of corresponding scene entities into the scene graph, where relevant random variables are set with the observation evidence. Given the set of observation BNFs in the scene graph, the algorithm instantiates new BNFs as hypotheses. The hypothesized BNFs may be concepts, whose input fragments are (partially) existing in the scene graph (bottom up direction), as well as missing input fragments of an existing concepts (top down direction).

Other steps of the interpretation algorithm are given by *specialization* and *generalization* of entities, which exist in the scene graph. Since output attributes of subclasses at least contain the output attributes of the corresponding superclass, we possibly have to instantiate new input fragments during specialization and cut of input fragments in the generalization step. By utilizing taxonomy operations the interpretation algorithm is able to reason on different abstraction levels of scene entities.

Each operation of the hypothesizing algorithm generates a new scene graph. Different scene graphs are evaluated and compared by a metric. The value of the metric increases with the existence-probability of instance BNFs within the scene graph, since higher probabilities of existence of hypothesized concepts reflect a better interpretation or explanation of a scene. We thus measure the accuracy of hypothesized scene graphs by multiplying the existence-probabilities of all BNFs, that the graph contains. The lower the metric value, the more speculative the scene graph is.

When hypothesizing top-down, i.e. hypothesizing missing input fragments of existing BNFs, we may instantiate new BNFs, which again have input fragments missing. To compute the metric of a scene graph we need a valid probability model, i.e. all input attributes and input fragments must be instantiated. While hypothesizing we had to 'unroll' the scene graph completely. To maintain an incremental algorithm we adapt the concept of *default distributions* [16], where CPDs of variables with not yet existing parents are temporarily replaced by non-conditional distributions. We introduce *default BNFs* $F' = (R', O', I', G', C')$ for every BNF $F = (R, O, I, G, C)$ within the knowledge base, where $O' = O$, $I' = \{\}$ and $C' = \{\}$.

In top-down hypothesizing also BNFs, which already exist in the scene graph can be bound for missing input fragments. These BNFs must fit the class-type, the input fragment is associated with. In addition these BNFs must be unbound, i.e. not yet set as input fragment to another BNF, or match additional equality constraints of BNFs.

In the following the interpretation algorithm is presented as pseudo code.

- Instantiate fragments according to observational evidence and/or prior context knowledge within a scene graph S. Set $H = \{S\}$.
- Repeat until a termination criterion is fulfilled:
  Set $H' = \{\}$.
  For every $S \in H$ do
    - Add S to H'.
    - *bottom up step*. For every $F \in$ KB (knowledge base) with at least one input fragment $F' \in S$: create S' by copying S and instantiating default BNF F. Add S' to H'.
    - *top down step*. For every default BNF $F \in S$ create S' by copying S and substituting F with the corresponding non-default BNF F'. For all possible instantiations of input fragments of F', with allowed $F'' \in S$ or new $F'' \in$ KB: create S'' by copying S' and instantiating the input fragment configuration and add S'' to H'.
    - *generalization step*. For every BNF $F \in S$ having superclass X, where holds: if F is input fragment to another BNF, X must still match class-type condition, else X must be more specific then the root-class. Create S' by substituting F with instantiation of BNF F' of class X. Add S' to H'.
    - *specialization step*. For every BNF $F \in S$ having subclass X: Create S' by substituting F with instantiation of BNF F' of class X. If F' has no unbound fragments, add S' to H'. If F' has unbound fragments, for all possible instantiations of unbound input fragments with allowed $F'' \in S$ or new $F'' \in$ KB: Create S'' by copying S' and instantiating the input fragment configuration. Add S'' to H'.
  Set $H := H'$.
  Compute metric for all $S \in H$.
  Order and prune H according to the metric.

Within the interpretation algorithm loop a number of scene graphs is maintained within an agenda H. Since the agenda is pruned of graphs with low metrics, the algorithm can be characterized as *beam search*. After the termination criterion is fulfilled, e.g. after the maximal metric value stopped rising, the scene graph with the best metric is returned as optimal interpretation of a scene given the evidence.

Feature distributions of hypothesized entities in explanations may be uniform or of high variance. This originates from having graph structure like $A \rightarrow C \leftarrow B$, where parts A, B influence aggregate C. If there is no knowledge about C the variables A and B are mutually independent, but entering evidence into C renders them dependent (explaining away-structure). In our framework while evidence of observed parts influences probability of hypothesized aggregats (if constraints match), hypothesized parts may remain independent, having uniform or prior distributions.

Thus to infer features of hypothesis concepts we proceed as follows. Within the graph of the optimal explanation for given observations we put evidence into all yet unobserved existence-nodes. Subsequently random variables of hypothesis concepts depend on observed random variables and after processing probabilistic inference, feature values can be read from the graph.

Figure 5 illustrates a first step of an example interpretation task, where positions and shape of two blobs were given and the features of other possible scene-entities were queried.

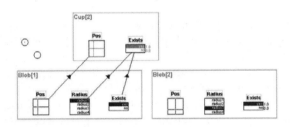

Fig. 5. Bottom-up hypothesis

The result of a run of the interpretation algorithm is shown in figure 6. Within this best scene-explanation e.g. the existence and features of a third

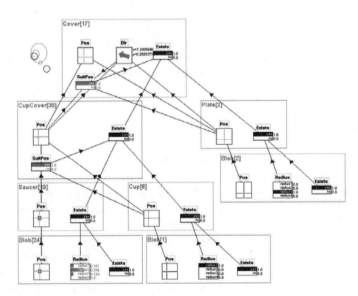

Fig. 6. Features of a hypothesized scene-graph

blob are inferred, which possibly might guide a tracker in a process of selective perception to gather more evidence on the corresponding area of the scene.

Figure 6 also illustrates the use of continuous random variables. The direction of the cover is represented by an angular Gaussian distribution and positions by bivariate Gaussians within the table-plane. Mean and standard deviation of position distributions are depicted at the nodes and in the upper-left part of the picture to point out the spatial configuration of entities within the hypothesized scene.

4   Related Work

BNF-representations were applied to model various relational open world domains e.g. units within a military battlefield [11] or entities within road traffic scenes [7].

An approach for dynamic situation modelling is described in [4]. The probabilistic framework features apllyiance of symbolic probabilistic inference and a taxonomic representation, that differs from ours. As use case for this framework an intrusion detection system is put forth, where multiple inputs from sensors are given and concept entity-types and states need to be inferred.

The authors of [3] present a method using Multi-Entity Bayesian network fragments (MEBNFs), which are in short BNFs enriched by first-order-logic elements [2], to detect human threatening behavior in computer networks.

Relating to measuring scene- or situation assessments, the authors of [1] present an approach within a military domain and suggest inter alia a multiplication of enitity-likelihoods to score a situation. Some of the same authors point out in [12], that decision theory might play a key role to measure reliability of a BNF-model.

There are numerous approaches to scene interpretation which are based on rule based systems and logics. In [14] the usage of description logics is examined, the conceptual representations of enitities are similar to those used here.

5   Conclusion

We described a framework and methods for probabilistic modelling of scenes. Our knowledge-based approach exploits taxonomy hierarchies of scene concepts as well as partonomies and other relations of scene entities.

The scene interpretation algorithm provides results on arbitrary scene-queries, ranging from questions on basic entities given abstract scene knowledge to explanations of scenes given perceptive evidence.

In future work we will evaluate our approach using more complex and other domains. We will examine methods, where probability models are (partially) specified in an unsupervised data-driven process, i.e. where structure and parameters of Bayesian network fragments are learned by statistics [15].

In relational probabilistic modelling with BNFs a frame-based knowledge representation is maintained, where frames correspond to fragment classes and frame-slots to both resident variables and input fragments [8]. We will evaluate other methods of reasoning with frame-based knowledge-bases, e.g. using description logic (DL) systems, and try to combine different abductive and deductive interpretation steps. Paralles between RPM and probabilistic logics, RPM learning and inductive logics programming (ILP) which are outlined in [5] will be examined for integration into the presented framework.

Acknowledgments. We would like to thank Bernd Neumann and other members of the Cognitive Systems Laboratory at the University of Hamburg for inspiring and helpful discussions relating to this work.

References

1. Suzanne M. Mahoney and Kathryn Blockmond Laskey and Ed Wright and Keung Chi Ng (2000) Measuring performance for situation assessment. MSS National Symposium on Sensor Data Fusion,
2. Kathryn Blackmond Laskey (2003) MEBN: A Logic for Open-World Probabilistic Reasoning. Department of Systems Engineering and Operations Research, Georg Mason University. MS4A6.
3. Kathryn Laskey and Ghazi Alghamdi and Xun Wang and Daniel Barbara and Tom Shackelford and Ed Wright and Julie Fitzgerald (2004) Detecting Threatening Behavior Using Bayesian Networks. Conference on Behavioral Representation in Modeling and Simulation.
4. Bruce D'Ambrosio and Masami Takikawa and Daniel Upper (2003) Representation for Dynamic Situation Modeling. Information Extraction and Transport, Inc.
5. Vitor Santo Costa and David Page and Maleeh Qazi and James Cussens (2003) CLP(BN): Constraint Logic Programming for Probabilistic Knowledge. International Conference on Uncertainty in Artificial Intelligence.
6. Uri N. Lerner (2002) Hybrid bayesian network for resoning about complex systems. Department of Computer Science, Stanford University.
7. Daphne Koller and Avi Pfeffer (1997) Object-Oriented Bayesian Networks. Thirteenth Annual Conference on Uncertainty in Artificial Intelligence (UAI-97), 302–313.
8. Daphne Koller and Avi Pfeffer (1998) Probabilistic frame-based systems. Fifteenth National Conference on Artificial Intelligence (AAAI-98), 580–587.
9. Avi Pfeffer and Daphne Koller and Brian Milch and Ken T. Takusagawa (1999) SPOOK: A system for probabilistic object-oriented knowledge representation. Fifteenth Conference on Uncertainty in Artificial Intelligence (UAI-99), 541–550.
10. Suzanne M. Mahoney and Kathryn Blackmond Laskey (1997) Network Fragments: Representing Knowledge for Construction Probabilistic Models. Thirteenth Annual Conference on Uncertainty in Artificial Intelligence (UAI-97).
11. Suzanne M. Mahoney and Kathryn Blackmond Laskey (1998) Constructiong Situation Specific Belief Networks. Fourteenth Annual Conference on Uncertainty in Artificial Intelligence (UAI-98), 370–378.

12. Suzanne M. Mahoney and Kathryn Blackmond Laskey (2003) Knowledge and Data Fusion in Probabilistic Networks. Journal of Machine Learning Research Vol. 4.
13. Bernd Neumann (2002) Conceptual Framework for High Level Vision. Department of Informatics, University of Hamburg, FBI-HH-B-241/02.
14. Bernd Neumann and Ralf Moeller (2004) On Scene Interpretation in DL. Department of Informatics, University of Hamburg, FBI-B-257/04.
15. H. Langseth and T. Nielsen (2003) Fusion of domain knowledge with data for structural learning in object oriented domains. Journal of Machine Learning Research, Vol. 4, 339–368.
16. Olav Bangso and Pierre-Henri Wuillemin (2000) Object Oriented Bayesian Networks: A Framework for Topdown Specification of Large Bayesian Networks and Repetitive Structures. Aalborg University, CIT-87.2-00-obphw1.
17. Raymond D. Rimey and Christopher M. Brown (1994) Control of Selective Perception Using Bayes Nets and Decision Theory. International Journal of Computer Vision, Vol 17, 173–209.
18. J. Pearl (1988) Probabilistic Reasoning in Intelligent Systems: Networks of Plausible Inference.

Part III

Non-Probabilistic Uncertainty

General Equilibrium Models
with Discrete Choices in a Spatial Continuum

M. Keyzer[1], Y. Ermoliev[2], and V. Norkin[3]

[1] Centre for World Food Studies (SOW-VU), Free University,
De Boelelaan 1105, 1081 HV Amsterdam, The Netherlands,
M.A.Keyzer@sow.vu.nl
[2] International Institute for Applied Systems Analysis (IIASA),
Laxenburg, Austria, *ermoliev@iiasa.ac.at*
[3] Institute of Cybernetics of the Ukrainian Academy of Sciences, 03680 Kiev,
Ukraine, *norkin@i.com.ua*

Abstract. The treatment of spatial characteristics through probability distributions makes it possible to use stochastic optimization methods and to obtain efficiency results and competitive equilibrium prices for general equilibrium models with discrete choices in spatial continuum. Along these lines, and combining results from stochastic optimization with principles established by Aumann and Hildenbrand for economies with continuum of traders the paper develops a practical modeling framework that can combine the spatially distributed aspects of land-use with processes such as market clearing or telecommunication investments concentrated at specific points. It also presents associated stochastic algorithms for numerical implementation. We discuss both a general equilibrium version in which all consumers meet their own budget, and a welfare maximizing version with transfers adjusting among consumer groups for which we formulate a dual approach that solely depends on a finite number of prices.

Keywords: Spatial modelling, Continuum of agents, Discrete choice, General equilibrium, Welfare maximization, Dual welfare function, Stochastic tâtonnement.

1 Introduction

At the interface of geography and economics, the practical relevance of applied policy models has often been limited by their lack of empirical detail in representing the distribution of spatial and social characteristics of the economy under study. Elaborate household surveys have been conducted and detailed geographic information systems were set up, but the databases are rarely used in regional or national models, due to the relatively high level of social and spatial aggregation that is required to keep the analysis tractable. The situation is even less satisfactory when it comes to dealing with spatial distributions of uncertainties, which are either neglected altogether or dealt with through a small number of alternative states of nature.

This paper specifies a general equilibrium model which can be used for land use planning. The model allows to combine the spatially distributed aspects of land-cover with processes such as market clearing or telecommunication investments concentrated at specific points. Assuming that individuals in society are located on a joint distribution over physical space, social characteristics and random events, the model incorporates spatial and stochastic distributions of landuse cover jointly with discrete choices, say, about the (discrete) market the produce should be shipped to, and can be extended to allow for investments at these market points from which the whole region can benefit. Thus, we interpret "spatial" in a broad sense, and the geographical coordinates may only be two among the many coordinates of the space. This representation would seem to offer the natural setting for representation of spatial characteristics as it often significantly simplifies, e.g., convexifies, the problem, allowing for discrete choices and other non-convexities. The paper on the one hand shows how stochastic optimization techniques apply to spatial analysis and on the other hand how the treatment of non-convexities may become easier in a spatial context.

Our discussion proceeds in three parts, the first of which deals in section 2-5 with a general equilibrium model in Arrow-Debreu format, evaluating the excess demand as the integral over spatial and other characteristics of the net demand by individual consumers and producers. The specification of this general equilibrium model has several distinguishing features.

First, as we find it unrealistic to postulate that there exists at every point in the continuum a market where supply and demand are matched, our major assumption is that goods are shipped from producers to consumers in the continuum via a finite number of market points where prices are formed. Consequently, the market excess demand function is finite dimensional, as in the standard model, albeit that it is now to be evaluated as the integral over individual net demand rather than as the sum over a finite number of agents. Second, we deal with discrete choice as sole source of non-convexity but since virtually the most important non-convexity can be approximated in the way, this hardly imposes a limitation. Third, we allow for full satiation of consumers. When consumers are atomless, their demand could remain infinite at equilibrium. This seems unrealistic and creates unnecessary complications. It is avoided by allowing for satiation. Fourth, the critical step in ensuring that non-convexities may be bypassed is to guarantee that all agents making the same discrete choice are sufficiently different. For this, it suffices to require smoothness of the density with respect to a single characteristic.

Representation in a continuum is often used in mathematical modeling, to bypass non-essential effects associated with non-convexities of real world processes. For example, in control theory (see e.g. Alekseev, Tihomirov, Fomin, 1979), the assumption of continuous time convexifies the attainable sets, and leads to Pontryagin's maximum principle. Kantorovich (1942) studied classical transportation problems in a spatial continuum. In economics Aumann

(1964, 1966) and Hildenbrand (1970, 1973) were first to study a continuum of agents within general equilibrium theory. Aumann and Hildenbrand assume continuity of consumer preferences and hence continuity of the corresponding utility functions but relax the usual concavity requirements on utilities. Treating consumers as atomless and distributed according to a smooth density enables to prove existence of competitive equilibria, i.e. the existence of endogenously generated prices at which aggregate demand does not exceed supply, while consumers and producers take prices as given and maximize utility and profits, respectively, according to optimization problems that may exhibit nonconvexities. Their proofs essentially rely on the assumption that because all individuals are sufficiently different to ensure that agents whose demand or supply exhibits a discontinuity at the prevailing prices have measure zero and can be disregarded. However, this assumption is nonconstructive, in the sense that it is introduced after derivation of individual behavior from preferences and technology. This makes it difficult actually to build a model that meets the requirement, and may be one reason for the class of models not to have found numerical application so far, another being that the traditional computational approaches require discretization of infinite dimensional models that could destroy continuity of the aggregate excess demand. Our approach, in a way similar to Anderson et al. 1992, constructively introduces requirements through the utility and production functions themselves. Finally, our convexity requirements are strict, ensuring that individual net demands are almost everywhere single- rather than multivalued.

The second part of our paper (Section 6) addresses the fact that it may be difficult to develop an applied model in which the budget of every individual in the continuum has to be elaborated separately. Hence, we turn to the cases with a finite number of social classes whose member consumers share a common budget, and to the case in which there is only one such class and which can be dealt with as welfare program. It appears that the dual of this program is more tractable than the primal, since it depends on the (finite-dimensional) price vectors only and yields excess demand as its gradient and hence satisfies the Weak Axiom of Revealed Preference.

In Section 7, we turn to computation. Here the main point is that treating the distributions of characteristics as probability distributions enables us to apply stochastic tatonnement procedures on excess demand along the lines described in Ermoliev et al. (2000). These are in effect stochastic equivalents of the classical Walrasian tatonnement but whereas the classical tatonnement at each step adjusts the market price for every commodity on the basis of its aggregate excess demand, the stochastic variant only activates a random sample of agents from an infinite set. The tatonnement indicates a direction of price change, purely on the basis of the net demand within this sample.

In sum the paper proceeds as follows. Section 2 introduces the distribution of spatial and social characteristics. Producer and consumer behavior under discrete choice are described in sections 3 and 4, respectively. Existence of a competitive equilibrium and of a solution to the spatial welfare model is established in section 5 and 6, respectively. Section 7 discusses the stochastic tatonnement procedure and its application. Illustrative examples are given to ease the understanding of the technicalities required for a rigorous presentation of our computational approach.

2   The Continuum of Agents: Distribution of Spatial and Social Characteristics

The description of our economy starts with the specification of the spatial and social characteristics of the households living in it. For example, one may consider the empirical distribution of characteristics as compiled from a household survey. Each answer in the survey questionnaire defines one characteristic, while the frequency of answers specifies the distribution of these characteristics in the sample. If a characteristic relates to an exogenous variable of the analysis (e.g. previous occupation of the respondent, or geographical location), it can be treated as part of a vector $x \in X \subset R^m$. Assuming that the survey was well designed, and representative, it is possible to infer from this sample an estimate of the distribution $G(x)$ at the level of the population. In our computational procedure we view $G(x)$ as a probability measure on an appropriate probability space for which we use the following formal general concepts. Let $L_q(X)$ denote the Banach space of integrable in power q functions on X for some $q, 1 \leq q < \infty$. The multifunction $A : x \to A(x) \subset R^n$ is called (Borel)-measurable if it has a Borel graph in $X \otimes R^n$, (see Aumann, 1965, Hildenbrand, 1974, and Castaing and Valadier, 1977, for the concept of measurable multifunctions).

   Assumption 2.1. *Let x be an m-dimensional real vector of characteristics and $x \in X$, where X is a compact in R^m . The distribution $G(x)$ defines a measure on X, and this measure $G = G_1 \otimes G_y$ on $X = X_1 \otimes Y$ is a product of the absolutely continuous (with respect to Lebesgue measure) measure G_1 on $X_1 \subseteq R$ and the σ-additive and complete measure G_y on Y.*

   We note that continuity is only required for a single characteristic, say, element x_1 . This is important because survey data often comprise a large number of discrete characteristics, such as farm/non-farm or male/female. It is always possible to introduce an artificial, continuous variable, say, x_{m+1} , that creates a "pseudo"-continuum and only serves to eliminate discontinuities that might arise from non-convexities. In this paper, all characteristics are taken to describe the spatial or social diversity of agents. This makes it safe to treat X as a compact set, and also ensures that for a continuous in c, Borel measurable in x function $u(c, x)$ and distribution $G(x)$, the integrability over X is assured and the function $U(c) = \int_X u(c, x) dG(x)$ is continuous. By Lebesgue's dominance convergence theorem $U(c)$ is continuous if $u(c, x)$ is continuous in c and majorated by some integrable function, for example, if $u(c, x) \leq \bar{u}$ for some given $\bar{u}$. However, some components of x could also be taken to represent uncertain events such that the compactness of X is no longer guaranteed. Before formulating the producer model with setup costs and discrete decisions in general terms, we present a simple example that introduces the approach to eliminate discontinuities. This is essentially based on the following lemmas that ensure the non-stationarity w.r.t. x_1 of the value function $F(x)$ of the decision problems, i.e. when $F(x^1) \neq F(x^2)$ for $x^1 \neq x^2$.

Lemma 1. *Assume that (i) the problem $F(x) = \sup_z \{f(z,x)|h(z,x) \leq 0\}$ has a solution for any $x \in R^1$; (ii) the function $f(z,x)$ is strictly increasing in x; and (iii) $h(z,x)$ is nonincreasing in x. Then, the function $F(x)$ is strictly increasing in x.*

Proof. Choose $x^1 < x^2$, $F(x^1) = f(z(x^1), x^1)$, and $h(z(x^1), x^1) \leq 0$. Since $f(z(x^1), x^1) < f(z(x^1), x^2)$, while $h(z(x^1), x^2) \leq 0$, it follows that $F(x^1) < F(x^2)$.

Lemma 2. *Assume that (i) the problem $F(x) = \sup_z \{f(z,x)|h(z,x) \leq 0\}$ has a solution for any $x \in R^1$; (ii) the function $f(z,x)$ is non-decreasing in x; (iii) $h(z,x)$ is strictly decreasing in x; and (iv) for any (z, x) there exists an arbitrary small Δz such that $f(z + \Delta z, x) > f(z, x)$ (for example, $\Delta z = \varepsilon \nabla_z f(z,x) \neq 0$, $\epsilon > 0$). Then, the function $F(x)$ is strictly increasing in x.*

Proof. Choose $x^1 < x^2$ and $F(x^1) = f(z(x^1), x^1)$, and $h(z(x^1), x^1) \leq 0$. Now $z(x^1)$ is an internal point of the set $\{z|h(z, x^2) \leq 0\}$, since by (iii) $h(z(x^1), x^2) < h(z(x^1), x^1) \leq 0$. Furthermore, by (iii) and assumption (iv), there exists a value Δz such that $h(z(x^1) + \Delta z, x^2) \leq 0$ and $F(x^2) \geq f(z(x^1) + \Delta z, x^2) \geq f(z(x^1) + \Delta z, x^1) > f(z(x^1), x^1) = F(x^1)$.

Finally, the next lemma gives a sufficient condition for the level set of a partially nonstationary function to have zero measure. This is the main regularity property that makes it possible to neglect the discontinuities in response functions after integration.

Lemma 3. *Assume that (i) function $f(x,y) : X \otimes Y \to R^1$ is measurable in (x,y) on a product of measurable sets $X \subseteq R^1, Y \subseteq R^m$; (ii)$f(x,y)$ is nonstationary in variable x , i.e. $f(x^1, y) \neq f(x^2, y)$ for any $x^1 \neq x^2 \in X$ and $y \in Y$; (iii) measure $G = G_x \otimes G_y$ on $X \otimes Y$ is a product of a σ-additive and complete measure G_y on Y and (iv) absolutely continuous (with respect to Lebesgue measure) measure G_x on X. Then $G\{(x,y)|f(x,y) = 0\} = 0$.*

Proof. For any $y \in Y$ by (ii) the set $\{x|f(x,y) = 0\}$ consists of no more than one point. By (iv), $G_x\{x|f(x,y) = 0\} = 0$. And by the Fubini theorem (e.g. Kolmogorov and Fomin, 1981):

$$G_x\{x|f(x,y) = 0\} = \int_Y G\{(x,y)|f(x,y) = 0\}dG(y) = 0.$$

In subsequent sections we often use the following important fact. Let

$$V(p) = \int_X \max_{d \in D} v(p, d, x) dG(x),$$

where $D \subset R^n$ is a compact set, $v(p, d, x)$ is convex and continuous in p, continuous in d and integrable in x, its subdifferential $\partial_p v(p, d, x)$ is bounded for all $d \in D$ by an integrable in x function. Then, by well known results on subdifferentiation of integral functions and the differentiation of a maximum function (see e.g. Clarke, 1983, Levin, 1985), and for $co\{\cdot\}$ denoting the convex hull, the following result holds.

Lemma 4. *The subdifferential $\partial V(p)$ of $V(p)$ is expressed as follows:*

$$\partial V(p) = \int_X \partial_p \max_{d \in D} v(p, d, x) dG(x) = \int_X co\{\partial_p v(p, d, x)| d \in d(p, x)\} dG(x),$$

where $d(p, x) = \arg\max_{d \in D} v(p, d, x)$. In addition, if $v(p, d, x)$ is continuously differentiable in p with gradient $v_p(p, d, x)$, then $d(p, x)$ is single valued for any p and $V(p)$ is continuously differentiable with gradient

$$V_p(p) = \int_X v_p(p, d, x)|_{d=d(p,x)} dG(x).$$

Proof. See Clarke (1983), Levin (1985).

Example: Single output and input commodity.

Suppose that households are, in fact, "producers" and are distinguished by a characteristic x, say, geographic location and distributed over an area according to the smooth distribution function $G(x)$. The firm at spot x produces a single output commodity, using a single input commodity, according to a strictly concave production function $f(v, x)$, with setup costs $g_0(x)$, where v denotes input use. The firm maximizes the discontinuous profit function:

$$\pi(p, x) = \max_{v \geq 0} [p_1 f(v, x) - g(v, x) - p_2 v], \tag{1}$$

where $g(v, x) = 0$ if $v = 0$ and $g(v, x) = g_0(x)$ if $v > 0$, while p_1 and p_2 are the given prices of the output and input, respectively, and $p = (p_1, p_2)$. This defines a discontinuous input demand $v(p, x)$ such that $v(p, x) = 0$ if $\pi(p, x) = 0$, and $v(p, x) > 0$ if $\pi(p, x) > 0$. Since $f(0, x) = 0$, this discontinuous problem can also be rewritten as the mixed-integer program:

$$\pi(p, x) = \max_{v \geq 0, \delta = 0, 1} \delta [p_1 f(v, x) - g(v, x)] - p_2 v.$$

Furthermore,

$$\Pi(p) = \int \pi(p, x) dG(x) = \int [\max(\bar{\pi}(p, x), 0)] dG(x), \tag{2}$$

for $\bar{\pi}(p, x) = \max_{v \geq 0} (p_1 f(v, x) - g_0(x) - p_2 v)$, assuming that this function is integrable. Now if $\bar{\pi}(p, x)$ is non-stationary with respect to x, then $G\{x :$

$\bar{\pi}(p,x) = 0\} = 0$, implying that the points at which a switch takes place can be neglected in the integration. Consequently, the following properties hold. First, aggregate profit $\Pi(p)$ is continuously differentiable and convex in p. Second, the aggregate output and input coincide with the one obtained after integration of input demand in the original problem (1), and, by Hotelling's lemma (Varian, 1992), are equal to the negative of the derivative of the profit function (see also Lemma 4).

To our knowledge, this approach has not found practical application so far, presumably because of the difficulties in dealing with maximization problems generally involving, multi-dimensional integrals, as in (2). Stochastic quasi-gradient procedures – to be discussed in section 7 – enable us to deal with the maximization of multidimensional integrals without having to evaluate them explicitly or to approximate them, which might undermine the convexity properties.

3   Producer Behavior

Our next example considers a set of marketplaces indexed λ, located at x^λ, with $\lambda = 1, ..., L$. We suppose that N commodities are traded at these marketplaces and fetch a price p^λ. Hence, these are $n = N \times L$ prices in the economy. Let $p \in R^n_+$ denote the vector of stacked prices of all marketplaces partitioned into $(p^1, ..., p^\lambda, ..., p^L)$. Next, we introduce the production model with discrete characteristics, representing H technology types indexed h, J firm types, indexed j, K commodities, indexed k. At x, every firm of type j maximizes profits, at given prices p solving:

$$\pi_j(p,x) = \max_{y_j^h, \delta_j^h} \sum_h \delta_j^h \left(p y_j^h \right)$$

$$\text{s.t.} \sum_h \delta_j^h \mathrm{H}_j^h(\mathrm{y}_j^h, \mathrm{x}) \le 0, \tag{3}$$

$$\sum_h \delta_j^h = 1, \delta_j^h \in \{0, 1\},$$

where $y_j^h = (y_j^{h,1}, ..., y_j^{h,\lambda}, ..., y_j^{h,L})$ denotes net supply of firm j at x using technology h; $\pi_j(p,x)$ is the optimal profit; and x has a distribution $G(x)$ satisfying assumptions 2.1. By definition, this profit is equal to the sum of the value of net supplies at the different locations: $\pi_j(p,x) = \sum_\lambda p^\lambda y_j^{h,\lambda}$. Hence, the firm can in principle buy and sell at every market place λ, taking charge of the transportation costs of outputs to, and of inputs from this market. Clearly, the technology index h might also be associated to a particular configuration of marketplaces at which the producer trades. For notational convenience we do not in the sequel refer explicitly to the marketplace.

Every producer chooses one technology, represented by a transformation function $H_j^h(\cdot)$. The transformation function may have a positive value at $y_j^h = 0$, so as to reflect that setup costs must be incurred before any production can take place, but we also assume that it is feasible to close down the factory, i.e. that there is a technology h for which the transformation function is non-positive at $y_j^h = 0$. In the Example, the associated transformation function can be defined as:

$$H_j^1(y_{1j}^1, y_{2j}^1, x) = y_{1j}^1 - f(-y_{2j}^1, x) \quad \text{and} \quad H_j^2(y_{1j}^2, y_{2j}^2, x) = 0.$$

We note that model (3) has discrete decision variables. This reflects an indivisibility and hence a non-convexity in production. Alternatively, this indivisibility can be expressed in the space of products but on nonconvex and, in general, disconnected sets, as follows:

$$Y_j^h(x) = \{y \in R^n | H_j^h(y, x) \le 0\}, \quad Y_j(x) = \cup_h Y_j^h(x).$$

Hence, model (3) can also be written as the maximization of the profit function py on the generally nonconvex and possibly disconnected set $Y_j(x)$ but we maintain a representation with discrete choice because this permits to eliminate the discontinuity at aggregate level in a constructive way.

Assumption 3.1 (Transformation). *Every firm j with technology h has transformation functions $H_j^h : R^n \times X \to R$, $H_j^h(y_j^h, x)$, and every such function satisfies the following properties: (i) it is continuous and strictly quasiconvex in y_j^h, measurable in x; (ii) for each j it has possibility of inaction $H_j^h(0, x) \le 0$ for some h ; (iii) $\sup_{y_j^h}\{\|y_j^h\| : H_j^h(y_j^h, x) \le 0\} \le \bar{\gamma}_j^h(x) \in L_2(X)$, all j, h.*

Measurability in (i) is a far weaker requirement than continuity and enables us to accommodate abrupt changes in technological conditions over the space of characteristics. Condition (iii) generates a scalar $\bar{\gamma}_j^h(x)$, which is the upper bound on feasible output. Now we can re-define the profit functions in the following way:

$$\pi_j(p, x) = \max_h \pi_j^h(p, x) \tag{4}$$

for

$$\pi_j^h(p, x) = \max_{y_j^h}\{py_j^h | H_j^h(y_j^h, x) \le 0\}. \tag{5}$$

We remark that if we replace the technology constraint $H_j^h(y_j^h, x) \le 0$ by the full set production constraints and balances, it becomes possible to calculate the price $p_j^h(x)$ at location x, as a shadow price to the program. The following assumption is the key step to ensure that the aggregate net supply is G-a.s. a continuously differentiable function.

Assumption 3.2 (Regularity). *For any positive p and fixed $h \ne h'$:*

$$G(x | \pi_j^h(p, x) = \pi_j^{h'}(p, x) = \pi_j(p, x)\} = 0.$$

For a two-dimensional vector x, this assumption means that the boundaries between regions choosing different technologies are lines of zero surface.

This illustrates how the optimization model can be used to generate a zoning map $h_j(p, x) = \arg\max_h \pi_j^h(p, x)$, defined so as to maximize $\pi_j^h(p, x)$, the value of land. Clearly, it is possible to impose legal restrictions on this zoning, expressed as the index set, say, to keep land under natural vegetation. The model to be presented can be used to analyze both the direct effect of such restrictions, and the indirect effect via the adjustment of prices.

Assumption 3.2 is satisfied if for all pairs h, h', the difference between the profit functions is nonstationary with respect to one characteristic, say, x_1 whenever h and h' are both maximal in (5) (see Lemma 1). It is mild since it only requires that two competing best technologies should not lead to profits that coincide everywhere within any sub-region, while the underlying best supplies do not. The requirement can be considered constructive as it can always be enforced by including in the transformation function of Assumption 3.1, an additional perturbation that differentiates between h and h'. For this, we can define a nonnegative perturbation function $\varepsilon_j^h(x_1)$ that is measurable and nonstationary in x_1 and enters as: $H_j^h(y_j^h - \varepsilon_j^h(x_1), x) \le 0$.

Proposition 1 establishes continuous differentiability of the aggregate profit function, and hence single-valuedness and continuity of aggregate net supply.

Proposition 1 (Aggregate net supply). *Let the distribution of characteristics and the transformation function satisfy assumptions 2.1, 2.2 and 3.1, 3.2. Then, the aggregate profit $\Pi_j(p) = \int_X \pi_j(p, x)dG(x)$, where $\pi_j(p, x) = \max_h \pi_j^h(p, x)$, of firms in group j is continuously differentiable, convex, non-negative and homogeneous of degree one in p and the aggregate net supply mapping $Y_j(p) = \partial \Pi_j(p)/\partial p$ is continuous and homogeneous of degree zero in p; $Y_j(p) = \int_X \sum_h \delta_j^h(p, x)y_j^h(p, x)dG(x)$, where $\delta_j^h(p, x), y_j^h(p, x)$ solve problem (3).*

Proof. The proof proceeds in three steps. (1) The profit function $\pi_j^h(p, x)$ is continuously differentiable and convex in p. This follows from the Maximum Theorem (see for instance, Varian, 1992) and assumption 3.1(i)); $\partial \pi_j^h(p, x)/\partial p = y_j^h(p, x)$, where $y_j^h(p, x)$ is a solution of (5), that is measurable due to assumption 3.1(i). Function $\pi_j^h(p, x)$ is measurable in x as the optimal value of the optimization problem whose feasible set is measurable in x (assumption 3.1(i), see Castaing and Valadier(1977)). (2) The profit $\pi_j(p, x) = \max_h \pi_j^h(p, x)$ of firms in group j is, almost everywhere w.r.t. $G(x)$, continuously differentiable, and convex in p, it is nonnegative and homogeneous of degree one; by assumption 3.1(i) and 3.2, the function is continuously differentiable in p almost everywhere in x; convexity in p follows from (1); homogeneity follows from the definition of problem (3); non-negativity from the possibility of inaction 3.1(ii); the almost everywhere property follows from the regularity assumption (see lemmas 1-3). By the rule of subdifferentiation of maximum function (see for instance, Rockafellar, 1973):

$$\partial \pi_j(p, x) = co\{\partial \pi_j^h(p, x)/\partial p | h \in \arg\max_h \pi_j^h(p, x)\}$$

$$= co\{y_j^h(p, x) | h \in \arg\max_h \pi_j^h(p, x),$$

where $co\{\cdot\}$ denotes the convex hull. Since all $\pi_j^h(p,x)$ are measurable, $\pi_j(p,x)$ inherits this property, and multifunction $\partial\pi_j(p,x)$ is also measurable (see Castaing and Valadier, 1977). (3) By assumption 3.1(iii), profit function $\pi_j(p,x)$ and subdifferential $\partial\pi_j(p,x)$ are bounded by integrable functions, so $\Pi_j(p) = \int_X \pi_j(p,x)dG(x)$ is well defined and by Lemma 4

$$\partial\Pi_j(p) = \int_X \partial\pi_j(p,x)dG(x) = \int_X co\{y_j^h(p,x)|h \in \arg\max_h \pi_j^h(p,x)\}dG(x).$$

The subdifferential $\partial\pi_j(p,x)$ is G-a.s. single valued, hence subdifferential $\partial\Pi_j(p)$ is single valued and continuous, and thus $\Pi_j(p)$ is a convex, continuously differentiable function. Choose a measurable function $\bar{h}(p,x) \in \arg\max_h \pi_j^h(p,x)$ and define

$$\delta_j^h(p,x) = \begin{cases} 1, & h = h(p,x), \\ 0, & h \neq h(p,x), \end{cases}$$

all h,

$$Y_j(p) = \int_X \sum_h \delta_j^h(p,x)y_j^h(p,x)dG(x).$$

Remark that since $\sum_h \delta_j^h(p,x)y_j^h(p,x) \in co\{y_j^h(p,x)|h \in \arg\max_h \pi_j^h(p,x)\}$, it follows that $Y_j(p) \subseteq \partial\Pi_j(p)$, and because of single valuedness of $\partial\Pi_j(p)$, we finally obtain $Y_j(p) = \partial\Pi_j(p) = \partial\Pi_j(p)/\partial p$.

4   Consumer Behavior

As indicated in the introduction, the representation of consumers by means of a continuum offers two major advantages. It allows to include detailed empirical distributions of consumer characteristics, as obtained through georeferenced household surveys and censuses, and it permits to deal with discrete choices, which seems important since consumers in general buy goods in discrete quantities, such as one car, two pairs of shoes, and they face discrete personal choices as to which town, province or country they want to live in, the job they will apply to, and so on. To describe consumer behavior, we distinguish r consumer groups, indexed i, coinciding with one of the discrete characteristics, say, $i = x_m$ of the households in the survey. To represent discrete choices, we introduce the option for the consumer to migrate to alternative destinations, indexed s, each with a specific utility function and income. The destination might be a physical location, a specific career or a lifestyle. Migration might be highly temporary and only reflect a shopping visit to the city, or permanent. Like for producers, we assume that all consumer purchases take place at the marketplace and that transportation appears as a separate commodity demand.

Assumption 4.1 (Endowments). *Each consumer x of group i owns fixed commodity endowments $e_i^s(x) \in R^n$ after choosing destination s, such that* *(i)* $e_i^s(x) \in L_2(X)$; *(ii)* $e_1^s(x) \geq e_1(x)$ *and* $\int_X e_1(x)dG(x) > 0$; *(iii) for every* x *there exists a destination* $s(x)$ *such that* $e_i^{s(x)}(x) \geq 0$, *with at least one strict inequality.*

Hence, the stocks available, say of skilled labor, depend on the destination chosen. Furthermore, this specification can be used to describe purchases of indivisible commodities and that setup costs of migration could be treated in this way. The consumer preferences are characterized by a utility function that can also differ across to reflect variations in lifestyle.

Assumption 4.2 (Utility). *Each consumer* x *of group* i *has, for every* s, *a utility function* $u_i^s : R^n \times X \to R$, *such that it is (i) Borel in* (c, x), *for some* $\bar{c}_i^s(x) \in L_2(X)$ $u_i^s(\bar{c}_i^s(x), x) \in L_1(X)$, *(ii) continuously differentiable with respect to consumption vector* $c \in R_{++}^n$ G-*a.s. in* x, *(iii) strictly concave in* $c \in R_{++}^n$ G-*a.s. in* x, *where* R_{++}^n *is the strictly positive orthant; (iv)* $u_i^s(0, x) = 0$; *and (v)* G-*a.s. in* x, $\partial u_i^s(c, x)/\partial c_k \geq 0$ *for* $c \leq \bar{c}_i^s(x)$ *with at least one strict inequality and* $\partial u_i^s(c, x)/\partial c_k < 0$ *whenever* $c_k > \bar{c}_{ik}^s(x)$; *(vi) for* $i = 1$, $u_i^s(c, x) = \tilde{u}_i(c)$ *is increasing in* c *with* $\partial \tilde{u}_1(c)/\partial c_k \to +\infty$ *for* $c_k \downarrow 0$; *(vii) for* $i = 1$, $\bar{c}_{1,k} \geq \int_X \left(\sum_j \max_h \bar{\gamma}_{j,k}^h(x) + \sum_i \max_s e_{i,k}^s(x) \right) dG(x)$.

The Borel measurability requirement in (i) is weaker than a continuity requirement in (c, x). We do not impose continuity with respect to x, in order to maintain all flexibility with respect to possibly abrupt changes in consumer properties in the space of characteristics. Condition (v) defines an individual satiation level $\bar{c}_i^s(x) \in L_2(X)$. Utility is non-satiated as long as all consumption falls below this level, but it is nonincreasing in any commodity for which consumption exceeds it. This guarantees boundedness (even out of equilibrium) and hence integrability of the demand by any member in state s. Assumption (vi) expresses that there is one (possibly very small) consumer group whose utility function is increasing everywhere but does not vary with either the state s or the location x. The requirement on the derivative guarantees positive consumption of all commodities. Condition (vii) indicates that for this consumer group the satiation level is so high that it exceeds maximal potential supply (see also assumption 3.1 (iii)). The integrability of its demand is not an issue because all members are identical. Imposing these relatively tight requirements on group 1 enables us to maintain weak assumptions for all other groups.

Consumer x of group i owns endowments $e_i^s(x)$ and receives a fixed share $\theta_{ij}(x)$ of the profits of firms in group j; hence the identity $\sum_i \int_X \theta_{ij}(x)dG(x) = 1$ must hold. Thus, consumer income $r_i^s(p, x)$ consists before transfers of the value of commodity endowments $pe_i^s(x)$ plus profits:

$$r_i^s(p, x) = pe_i^s(x) + \sum_j \theta_{ij}(x)\Pi_j(p). \tag{6}$$

Assumption 4.3 (Transfers). *Each consumer* x *of group* i *receives transfers* $t_i(p, x) \in R$, *such that (i)* $t_i(p, x) \in L_1(X)$; *(ii)* $t_i(p, x)$ *is continuous* *and homogeneous of degree one in* p *(iii)* $\max_s (r_i^s(p,x)) + t_i(p,x)$ *is positive for all* p; *and (iv)* $\sum_i \int_X t_i(p,x)dG(x) = 0$.

We note that conditions (iv) on zero balance of transfers can be implemented in several ways. For example, if $t_i(p,x) = 0$ for each x then all consumers make their decisions independently relying only on commonly known prices. This leads to Walrasian equilibrium for continuum of traders as in Aumann (1964, 1966) and Hildenbrand (1970, 1973). Alternatively, the specification $\int_X t_i(p,x)dG(x) = 0$ supposes that incomes are redistributed by means of transfers within a corresponding consumer group ("family") i that serves as a risk pool. In this case, the group as a whole may be also represented by a consolidated consumer with a utility function aggregated by means of social welfare weights $\alpha_i^s(x)$ that convert individual utilities into money metric (see, e.g., Ginsburgh and Keyzer, Chapter 2, 2002) and are implemented by allowing for the transfers among consumers within the groups. In Section 6 below, we allow for transfers among groups:

$$U_i(\{c^s(\cdot), \kappa^s(\cdot)\}) = \sum_s \kappa^s \int_X \alpha_i^s(x)u_i^s(c^s(x),x)dG(x)$$

to be maximized over $\{c^s(\cdot), \kappa^s(\cdot)\}$ subject to a consolidated budget

$$\sum_s \kappa^s p \int_X c^s(x)dG(x) \leq \sum_s \kappa^s \int_X r_i^s(p,x)dG(x).$$

This problem can be decomposed by means of a single Lagrange multiplier into individual (x-specific) utility maximization problems (7) with given (endogenous) transfers. Clearly, this also defines the special case in which all consumer budgets are consolidated into a single one.

Now the model of consumer x in group i reads:

$$u_i^*(p,x) = \max_{c^s \geq 0, \kappa^s \in \{0,1\}} \sum_s \kappa^s u_i^s(c^s,x)$$
$$\text{s.t. } \sum_s \kappa^s pc^s \leq \sum_s \kappa^s r_i^s(p,x) + t_i(p,x) \qquad (7)$$
$$\sum_s \kappa^s = 1,$$

for given $t_i(p,x)$. Observe that in view of the satiation assumption 4.2(v), program (7) has a bounded solution, even for $p_k = 0$, that determines an optimal destination $\kappa_i^{*s}(p,x)$ as well as a (nonnegative) optimal consumption $c_i^{*s}(p,x) \leq \bar{c}_i^s(x)$. Because of the strict concavity of utility, $c_i^{*s}(p,x)$ is single valued and continuous for all $p \geq 0$. As in the case of the producer problem we can reformulate the consumer problem (7) in terms of continuous variables only, but now with, in general, piecewise continuous utility functions and nonconvex consumption sets, by defining consumption sets

$$C_i^s(p,x) = \{c \in R_+^n | pc \leq r_i^s(p,x) + t_i(p,x)\}, \quad C_i(p,x) = \cup_s C_i^s(p,x),$$

as well as the index sets $S_i(c, p, x) = \{s | c \in C_i^s(p, x)\}$, and the piecewise continuous utility functions

$$U_i(c, p, x) = \max_{s \in S_i(c, p, x)} u_i^s(c, x).$$

Now problem (7) is equivalent to the maximization of $U_i(c, p, x)$ over $c \in C_i(p, x)$.

Next, we reformulate problem (7) in a more convenient form. For given destination s, consumption can be determined from:

$$u_i^{*s}(\mathrm{p}, \mathrm{x}) = \max_{c^s \geq 0} u_i^s(c^s, x)$$

$$\text{s. t. } pc^s \leq r_i^s(p, x) + t_i(p, x). \tag{8}$$

Note that by nonsatiation assumption 4.2(v), for $s \neq 1$, utility and consumption will at all non-negative prices, be bounded for every s and all x and for satiation level $\bar{c}$ high enough, the budget constraint will hold with equality. Now the functions $u_i^*(p, x)$ can also be determined as

$$u_i^*(p, x) = u_i^{*s_i}(p, x) = \max_s u_i^{*s}(p, x),$$

while $c_i^{*s}(p, x) = c_i^{s_i}(p, x)$, $\kappa_i^{*s}(p, x) = 1$ for $s = s_i$ and $c_i^{*s}(p, x) = 0$, $\kappa_i^{*s}(p, x) = 0$ for $s \neq s_i$. As for production, the maximization can be used to generate a zoning map $s(p, x)$ describing the assignments for every location x. Likewise, it is possible to impose restrictions $S(x)$ on land use, and to analyze their direct effect as well as their price induced effect. The following assumption, similar to (4), ensures that the aggregate consumption is G-a.s. a continuously differentiable function.

Assumption 4.4 (Regularity). *For $s \neq 1$, $s \neq t$, and any positive p:*

$$G(x | u_i^{*s}(p, x) = u_i^{*t}(p, x) = u_i^*(p, x)) = 0.$$

As for assumption (4) on profits, this assumption is satisfied if for all pairs that correspond to maximal utility, the difference between both value functions is nonstationary with respect to x_1. The property is easily constructed via a perturbation function $\varepsilon_i^s(x_1)$ that is measurable and nonstationary in x_1 and enters utility as: $u_i^s(c_i^s, x) = \tilde{u}_i^s(c_i^s, x) + \varepsilon_i^s(x_1)$.

Proposition 2 (Aggregate consumption). *Let the distribution of characteristics and the utility, endowment and transfer functions satisfy assumptions 2.1, 4.2-4.4. Then, the aggregate consumer demand and the aggregate endowment supply are continuous and homogeneous of degree zero in p.*

Proof. Define $S_i(p, x) = \{s : u_i^{*s}(p, x) = u_i^*(p, x)\}$ and

$$C_i^*(p, x) = \{c_i^{*s}(p, x) | s \in S_i(p, x)\}, \quad E_i^*(p, x) = \{e_i^s(x) | s \in S_i(p, x)\}.$$

Multivalued mappings $S_i(p, x)$, $C_i^*(p, x)$, $E_i^*(p, x)$ are:

(i) measurable in x for all $p \geq 0$;
(ii) closed valued and even single valued for almost all x;
(iii) homogenous of degree zero in p for almost all x;
(iv) upper semicontinuous in $p \geq 0$ for almost all x;
(v) bounded by a function that is integrable in x.
In conditions (i), (iv), (v) upper semicontinuity is preserved after integration over x. Hence, mappings $C_i^*(p) = \int_X C_i^*(p, x) dG(x)$, $E_i^*(p) = \int_X E_i^*(p, x) dG(x)$ are upper semicontinuous in $p \geq 0$, single-valued by (ii) and thus continuous. Since $S_i(p, x) = S_i(\lambda p, x)$ for any $\lambda > 0$, $C_i^*(p, x) = C_i^*(\lambda p, x)$, $E_i^*(p, x) = E_i^*(\lambda p, x)$ hence $C_i^*(p) = C_i^*(\lambda p)$, $E_i^*(p) = E_i^*(\lambda p)$, mappings $C_i^*(p)$, $E_i^*(p)$ are homogenuous in $p \geq 0$ of degree zero.

5   Existence of a Competitive Equilibrium

Having specified supply and demand, we can readily obtain the aggregate net supply and demand of consumers and producers in the general equilibrium model (3), (7) with given transfers, as the integral values:

$$C_i^*(p) = \int_X \left(\sum_s \kappa_i^{*s}(p, x) c_i^{*s}(p, x) \right) dG(x), \tag{9}$$

$$E_i^*(p) = \int_X \left(\sum_s \kappa_i^{*s}(p, x) e_i^s(x) \right) dG(x), \tag{10}$$

$$Y_j^*(p) = \int_X \left(\sum_h \delta_j^{*h}(p, x) y_j^{*h}(p, x) \right) dG(x), \tag{11}$$

where $\kappa_i^{*s}(p, x)$, $c_i^{*s}(p, x)$ solve consumer problem (7) and $\delta_j^{*h}(p, x), y_j^{*h}(p, x)$ solve producer problem (3). A competitive equilibrium is characterized by a price vector $p \in P$ such that

$$Z^*(p) = 0, \tag{12}$$

for

$$Z^*(p) = \sum_i C_i^*(p) - \sum_i E_i^*(p) - \sum_j Y_j^*(p),$$

where P denotes the price simplex $P = \{p \geq 0, \sum_k p_k = 1\}$. This price normalization is needed because $Z^*(p)$ is homogeneous of degree zero in p.

Proposition 3 (Competitive equilibrium). *Let the distribution of characteristics and the utility, endowment and transfer functions satisfy assumptions 2.1, 2.2 and 4.1-4.4, and the transformation functions satisfy assumptions 3.1, 3.2, then model (9)-(12) has an equilibrium, with positive prices.*

Proof. By propositions 1 and 2 for all $p \in P$, excess demand $Z^*(p)$ is continuous and homogeneous of degree zero in prices. And by assumption 4.3, equation (6) and nonnegativity of profit in proposition 1, it satisfies Walras Law $(pZ^*(p) = 0$ for all $p \in P$). Furthermore, by assumptions 4.1(ii) and

4.2(v), consumers $i = 1$ have positive income at all prices and demand more than can be supplied if any price drops to zero. Then, by standard arguments (see, e.g. Arrow and Hahn, 1971, chapter 1) there exists an equilibrium, and price must be positive since, by assumption 4.2 (vii) excess demand could not be nonnegative otherwise.

6 Spatial Welfare Optimum: a Dual Approach

While the spatial competitive equilibrium determines prices for a specified transfer function, through the spatial welfare optimum to be considered in this section the transfers are determined on the basis of a welfare program with fixed positive weights $\alpha_i(x)$ on the various consumers. This welfare program maximizes the weighted sum over groups i of the integral over x of individual utilities multiplied by the destination factor κ_i^s and summed over s. The resulting welfare program is hard to handle numerically in a straightforward manner, because it is defined in functional space. Therefore, we propose to formulate the equivalent dual welfare program, that essentially replaces the budget constraint from the model of the previous section by a fixed welfare weight, from which the transfers and the solution of the original program follow.

Thus, for given positive marginal utility of expenditure $\mu_i(x) = 1/\alpha_i(x)$, i.e. equal to the inverse welfare weight we can maximize the surplus of consumer (i, x):

$$w_i^o(p, x) = \max \sum_s \kappa_i^s [u_i^s(c^s, x) - \mu_i(x)(pc^s - r_i^s(p, x))]$$

$$\text{s. t.} \quad c^s \geq 0, \quad \kappa^s \in \{0, 1\} \quad \sum_s \kappa^s = 1, \tag{13}$$

with optimal surplus $w_i^o(p, x)$, consumption $c_i^{os}(p, x)$ and switches $\kappa_i^{os}(p, x)$. By construction, $\kappa_i^{os}(p, x) = 0$ for all s except some s_i, $\kappa_i^{os_i}(p, x) = 1$. By assumptions 4.1-4.3, this problem has a bounded solution. Program (13) defines the (i, x)-specific subproblem:

$$w_i^{os}(p, x) = \max_{c \geq 0} \{u_i^s(c, x) - \mu_i(x)(pc - r_i^s(p, x))\}. \tag{14}$$

By assumptions 4.2(v),(vi), $w_i^{os}(p, x)$, are well defined for all $p \geq 0$, since satiation ensures that consumption will not exceed $\bar{c}_i^s(x)$, this program attains its optimum. The same applies to function

$$w_i^o(p, x) = \max_s w_i^{os}(p, x) = w_i^{os_i}(p, x),$$

that, by (14), is equal to the sum of the consumer surplus $(u_i^{s_i}(c_i^{os_i}(p, x), x) - \mu_i(x)pc_i^{os_i}(p, x))$ and the producer surplus $r_i^{s_i}(p, x)$, multiplied by $\mu_i(x)$, where $s_i = s_i(p, x)$ is the optimal state for member x of group i and $c_i^{os_i}(p, x)$ denotes the optimal consumption in this state. This value can be interpreted as the self-earned utility since for $\mu_i(x)$ such that the revenue balance with expenditure it coincides with individual utility. The associated regularity assumption is:

Assumption 6.1 (Regularity). *For $i \neq 1$ and any positive p:*

$$G\{x|w_i^{os}(p,x) = w_i^{ot}(p,x) = w_i^o(p,x); s \neq t\} = 0.$$

This regularity assumption on consumer surplus can be enforced constructively in the same way as for assumption 4.4 for utility itself. For every i, we can now define associated income transfers:

$$t_i^o(p,x) = \sum_s \kappa_i^{os}(p,x)(pc_i^{os}(p,x) - r_i^s(p,x)), \tag{15}$$

i.e., the transfer that closes the gap between target expenditure and available revenue at prevailing prices (Lagrange multipliers of the welfare program). The following proposition establishes that the Second Theorem of welfare economics also applies in our case. The dual social welfare function can be defined as

$$W(p) = \int_X W^o(p,x)dG(x) \tag{16}$$

for

$$W^o(p,x) = \sum_i \alpha_i(x)w_i^o(p,x). \tag{17}$$

Assumption 6.2 (Welfare weight normalization and nonnegligibiliy of consumer 1). *(i) $\alpha_i(x)$ are nonnegative integrable functions and $\int_X \sum_i \alpha_i(x) dG(x) = 1$; (ii) $\alpha_1(x) \equiv \alpha_1 > 0$.*

By assumptions 4.1(i), 4.2(v),(vi), and 6.2, functions $\alpha_i(x)w_i^o(p,x)$ are integrable and hence function $W(p)$ is well defined for $p \geq 0$.

Proposition 4 (Equilibrium with transfers). *Let the distribution of characteristics and the utility, endowment and transfer functions satisfy assumptions 2.1, 2.2 and 4.1-4.4, 6.1, and the transformation functions satisfy assumptions 3.1, 3.2, while welfare weights satisfy assumption 6.2, then the solution of*

$$\min_{p \geq 0} W(p) \tag{18}$$

defined as in (16)-(17) supports a competitive equilibrium (12) with transfers (15), with unique and positive optimal prices.

Proof. *Part 1. Existence and uniqueness of optimal prices.* Function $W(p)$ is convex because $w_i^{os}(p,x)$ is convex in p (see e.g. Avriel, 1976, Theorem 5.1). Since by assumption 4.2(iv), we have $u_1^s(0,x) = 0$, it follows that $w_1^{os}(p,x) \geq pe_1^s(x)$ and $\alpha_1 w_1^o(p,x) \geq p\alpha_1 \sum_s \kappa_1^{os}(p,x)e_1^s(x)$. By assumption 4.1(ii)

$$\int_X \alpha_1 w_1^o(p,x)dG(x) \geq p\alpha_1 \int_X e_1(x)dG(x) \to +\infty \quad \text{if} \quad p \to +\infty. \tag{19}$$

By assumptions 3.1(ii), 4.2(iv) all $w_i^\circ(p, x) \geq 0$, hence since by assumption 6.2 consumer 1 is nonnegligible, it follows from (16) and (19) that the convex function satisfies $W(p) \geq 0$ and

$$W(p) \to +\infty \quad \text{if} \quad p \to \infty.$$

Hence, $W(p)$ achieves its minimum. Thus, we have the following representation:

$$
\begin{aligned}
W(p) &= \sum_i \int_X \max_s [\alpha_i(x) u_i^s(c_i^{os}(p, x), x) - (pc_i^{os}(p, x) - r_i^{os}(p, x))] dG(x) \\
&= \sum_i \int_X \max_s [\alpha_i(x) u_i^s(c_i^{os}(p, x), x) \\
&\quad - (pc_i^{os}(p, x) - pe_i^s(x) - \sum_j \theta_{ij}(x) \Pi_j(p)] dG(x) \\
&= \sum_i \int_X \max_s [\alpha_i(x) u_i^s(c_i^{os}(p, x), x) \\
&\quad - p(c_i^{os}(p, x) - e_i^s(x))] dG(x) + \sum_j \Pi_j(p) \\
&= \sum_i \int_X \max_s [\max_{c \geq 0} (\alpha_i(x) u_i^s(c, x) - p(c - e_i^s(x)))] dG(x) + \sum_j \Pi_j(p).
\end{aligned}
$$

Now by regularity assumption 6.1, function $W(p)$ is differentiable for $p \geq 0$, and by Lemma 4 and Proposition 1:

$$\partial W / \partial p = -Z^\circ(p) = -\left(\sum_i C_i^\circ(p) - \sum_i E_i^\circ(p) - \sum_j Y_j^\circ(p)\right),$$

where C_i°, E_i°, Y_j° are defined analogously to (9)-(11) with $*$ replaced by $\circ$. A stationary point p^* of the convex function $W(p)$, clears the market, i.e. corresponds to nonnegative excess demand:

$$\partial W / \partial p = -Z^\circ(p^*) \geq 0, \quad p^* Z^\circ(p^*) = 0. \tag{20}$$

Now, by assumption 4.2(vii), consumer $i = 1$ has demand strictly below satiation level and by 4.2(vi), positive consumption. By (14) $\partial u_1 / \partial c_1 = \mu_1 p^*$, and hence $p^* > 0$ and $Z^\circ(p^*) = 0$. Moreover, because of strict concavity of utility, differentiability and the boundary property 4.2(vi) of the derivative, $w_1^{os}(p, x)$ is strictly convex (it is a Legendre transformation, see Avriel, 1976, p. 109), and since by assumption 6.2 consumer $i = 1$ has positive measure, and $w_i^{os}(p, x)$ is convex for $i \neq 1$, this property carries over to $W(p)$. Hence, p^* is unique.

Part 2. Equilibrium with transfers. We show that for $\mu_i(x) = 1/\alpha_i(x) > 0$ a stationary point of $W(p)$ is equivalent to a competitive equilibrium (12) with transfers (15) , where $\mu_i(x)$ is the value of the Lagrange multiplier associated to the budget constraint of the individual consumer problem. Let $\{\kappa_i^{os}(p, x), c_i^{os}(p, x)\}$ be a solution of (13), i.e. for all $c^s \geq 0$, $\kappa^s \in \{0, 1\}$, $\sum_s \kappa^s = 1$, we have

$$\sum_s \kappa^s (u_i^s(c^s, x) - \mu_i(x)(pc^s - r_i^s(p, x))$$

$$\leq \sum_s \kappa_i^{os}(p, x)(u_i^s(c_i^{os}(p, x), x) - \mu_i(x)(pc_i^{os}(p, x) - r_i^s(p, x)). \tag{21}$$

Then,

$$\sum_s \kappa^s u_i^s(c^s, x) \leq \sum_s \kappa_i^{os}(p, x) u_i^s(c_i^{os}(p, x), x)$$

for all $c^s \geq 0$, $\kappa^s \in \{0,1\}$, $\sum_s \kappa^s = 1$, and such that

$$\sum_s \kappa^s p c^s \leq \sum_s \kappa^s r_i^s(p, x) + t_i(p, x), \qquad (22)$$

where $t_i(p, x)$ are defined by (15). Since $\{\kappa_i^{os}(p, x), c_i^{os}(p, x)\}$ satisfies (22), it also provides a solution to (7) with transfers (15) (implicitly dependent on $\mu_i(x)$). Conversely, for given $\mu_i(x)$, solution $\{\kappa_i^{os}(p, x), c_i^{os}(p, x)\}$ and transfers (15) inequality (21) can be rewritten in the form

$$\sum_s \kappa^s(u_i^s(c^s, x) - \mu_i(x) \sum_s \kappa^s(pc^s - r_i^s(p, x) - t_i(p, x)) \leq$$
$$\leq \sum_s \kappa_i^{os}(p, x) u_i^s(c_i^{os}(p, x), x),$$

i.e. $\mu_i(x)$ is a Lagrange multiplier for a budget constraint in (7). The same applies to producer decisions. Obviously, transfers sum to zero.

This proposition shows that the minimum of $W(p)$ uniquely defines a competitive equilibrium with transfers. Such a competitive equilibrium is known to be Pareto efficient in terms of the aggregate utility of every group i, and more generally of any group of consumers with positive measure: no group could achieve higher utility without any group being worse off. As is well known the Pareto-concept only applies to sets of positive measure, since it would be possible to provide unlimited quantities to the individual atomless consumer without affecting the aggregate commodity balance. We observe that in conditions of Proposition 4, since commodity balances hold and solutions of (13) are specialized in the optimum, despite the nonconvexities (14)-(17) imply that dual welfare is equal to primal (social) welfare:

$$W(p^*) = \sum_i \int_X \alpha_i(x) u_i^{s_i(x)}(c_i^{s_i(x)}(p^*, x), x) dG(x)$$

$$= \max_{c_i^s(\cdot), \kappa_i^s(\cdot), y_j^h(\cdot), \delta_j^h(\cdot)} \sum_{i,s} \int_X \alpha_i(x) \kappa_i^s(x) u_i^s(c_i^s(x), x) dG(x), \quad (23)$$

where maximum is taken over measurable functions $c_i^s(x) \geq 0$, $\kappa_i^s(x) \in \{0,1\}$, all i, s, and $y_j^h(x), \delta_j^h(x) \in \{0,1\}$, all j, h, subject to constraints

$$\sum_{i,s} \kappa_i^s(x) c_i^s(x) \leq \sum_{i,s} \kappa_i^s(x) e_i^s(x) + \sum_{j,h} \delta_j^h(x) y_j^h(x);$$

$$\sum_h \delta_j^h(x) H_j^h(y_j^h(x), x) \leq 0, \text{ all } j;$$

$$\sum_s \kappa_i^s(x) = 1, \text{ all } i;$$

$$\sum_h \delta_j^h(x) = 1, \text{ all } j.$$

We also note that in this model the welfare weights define price normalization, so that there is no scope for further normalization on the simplex p, and $Z^o(p)$ is not homogeneous of degree zero in prices, unlike $Z^*(p)$.

Clearly, use of the welfare program with fixed welfare weights has the disadvantage that it does not impose restrictions on transfers. Yet, we note that trade balances at fixed prices can be incorporated within the technology set, as a technique to transform imports into exports.

We also note that in this model the price generated at discrete market points can be interpreted as an ex ante variable that translates via the marginal utility in the spatial continuum into an ex post price at location x. In a similar spirit, the approach is naturally extended to allow for (non-rival, say, telecommunication) investments at the market points so as to improve the productivity of (rival) inputs in the spatial continuum.

Finally, the minimization of $W(p)$ as it is specified in (14), (16), (17) belongs to the class of so-called stochastic minimax problems (see Ermoliev, 1988). In the following section we shall use this fact to develop a stochastic tâtonnement procedure for searching equilibrium.

7   Deterministic Versus Stochastic Welfare Tâtonnement

If it was easy to evaluate excess demand, a deterministic price adjustment procedure could be used to compute equilibrium prices. Specifically, Arrow and Hurwicz (1958) have proved that if the excess demand $Z(p)$ satisfies the Weak Axiom of Revealed Preference (WARP)

$$p^* Z(p) > 0 \text{ for all } p^* \in P, \ p \in P \text{ such that } Z(p^*) = 0, \ Z(p) \neq 0,$$

then Walrasian tâtonnement can be used. For excess demand as defined by the general equilibrium model of proposition 3, and prices on the simplex $P = \{p \geq 0 | \sum_k p_k = 1\}$, the property can only be proved to hold in very special cases. Yet, any excess demand $Z^\circ(p) = -\partial W / \partial p$ associated to a welfare optimum satisfies this condition, with price on a compact set $P = \{0 \leq p \leq \bar{p}\}$. There is in this case no scope for normalization on the simplex, since price normalization already follows from the welfare weights. Starting from given $p(1) = p_1$, one could specify the algorithm:

$$p(t + 1) = \Pi_P[p(t) + \rho_t Z^\circ(p(t))], \quad t = 1, 2, ..., \tag{24}$$

where Π_P is the projection operator on P and step-size multipliers ρ_t are sufficiently small. However, the difficulty in applying this tâtonnement rule to our model is that, due to the integrals, at each step of tâtonement procedure (24) one has to mine and process the information related to all x, making computation of excess demand very hard and necessarily inaccurate. In fact, the procedure presupposes that there is a central planner who is able to compute aggregate excess demand without error, and hence has to possess all information about all points x. Suppose on the contrary that we possess at every iteration t, a statistical estimate of $Z^\circ(p(t))$. Then one might expect that, if this estimate is asymptotically unbiased, the iteration process will eventually converge to an equilibrium. The proposed stochastic Walrasian tâtonnement process builds on this idea. The key observation is that in (13) aggregate excess demand is the expected value of the total net demand $z°(p,x)$ of all consumers with characteristic x, if we treat $G(x)$ as distribution of random events. The stochastic tâtonnement process uses a sequence of independent random drawings $x(t)$ from the distribution $G(x)$, and starting from a given $p(1) = p_1 \in P$ adjusts p(t) according to:

$$p(t+1) = \Pi_P[p(t) + \rho_t z°(p(t), x(t))], \quad t = 1, 2, , \tag{25}$$

for

$$z°(p(t), x) = \sum_{i,s} \kappa_i^{os}(p, x) c_i^{os}(p, x) - \sum_{i,s} \kappa_i^{os}(p, x) e_i^s(x)$$
$$- \sum_{j,h} \delta_j^h(p, x) y_j^h(p, x).$$

We note that the evaluation of excess demand for price adjustment process (25) is only required for agents located at a sequence of points x . Nonetheless, this process converges in the limit.

Proposition 5 (Convergence of stochastic tâtonnement to a welfare equilibrium). *Let the assumptions of proposition 3 hold. Then for step-sizes* ρ_t *such that:*

$$\rho_t \geq 0, \quad \sum_t \rho_t = \infty, \quad \sum_t \rho_t^2 < \infty, \tag{26}$$

process (15) converges, with probability 1, to an equilibrium price.

Proof. See Ermoliev et al. (2000), taking into account that $Z°(p)$ satisfies WARP as a subgradient of the convex function $W(p)$.

In fact, the rule $\rho_t = const/t$ satisfies requirement (26). As argued in Ermoliev et al. (2000), in case WARP does not hold, process (25) requires additional shocks for convergence. The stochastic Walrasian tâtonnement process adds to the intuitive appeal of the classical tâtonnement the property of full decentralization. In the classical process (24) there is an auctioneer who adjusts prices in proportion to the prevailing excess demand whose calculation requires all agents to communicate their net trades. In the stochastic version, at any given point during the iteration process, only a random collection of consumers have to communicate their intentions. However, this purely dual approach has the limitation that, to avoid solving problems (23) in functional space it is required that explicit demand $c_i^{os}(p, x)$ and net supply $y_j^h(p, x)$ be available in closed form. In practice only the primal functions will be available for calculations. In other words, $c_i^{os}(p, x)$, $y_j^h(p, x)$ are solutions of internal problems that require internal iterations and cannot be obtained without errors. Hence, in (25) estimates of the gradient of $W(p)$ in (18) are subject to errors, say $\varepsilon(t)$ at iteration t. Consequently, at every iteration the Weak Axiom only holds with a certain accuracy. Nonetheless, convergence of (25) is ensured if $\varepsilon(t) \to 0$, which amounts to a relatively mild requirement since the change in p tends to zero by construction, making it easier to achieve accuracy. The condition is based on the fact that the approximation of $Z°(p(t))$ calculated in this case is the so-called $\varepsilon(t)$ -subgradient of $W(p)$.

Finally, we may mention that several modifications of the stochastic tâtonnement process could be envisaged. For example, if the continuum is two-dimensional only (purely spatial), it is possible to approximate the derivative of the integral of the dual welfare by Monte Carlo sampling for given prices, i.e. to take a mean over a sufficient number of stochastic gradients prior to any price adjustment, and apply regular gradient algorithms. In between both approaches, one may consider further smoothing of the stochastic gradient, by Cesaro-averaging (Nemirovski et al., 1977). The point made here is merely that once the spatial equilibrium problem has been cast in the framework of stochastic optimization a rich arsenal of stochastic optimization techniques becomes available to analyse its properties and solve it numerically.

Acknowledgments. The authors thank the participants of the IFIP/IIASA/ GAMM workshop on Coping with uncertainty (December 13-16, 2004, IIASA) and the anonymous referees for their useful comments.

References

1. Alekseev, V.M., Tihomirov V.M., Fomin S.V. (1979) Optimal Control. Nauka, Moscow.
2. Anderson, S.P., A. De Palma and J.-F. Thisse (1992) Discrete choice theory of product differentiation. MIT-Press, Cambridge
3. Arrow, K.J., Hahn F.H. (1971) General Competitive Analysis. Holden Day, San Francisco
4. Arrow, K.J., Hurwicz L., Uzawa H. (1958) Studies in Linear and Nonlinear Programming. Stanford University Press, Stanford
5. Aumann, R. (1964) Markets with a continuum of traders. Econometrica **32**, 39–50
6. Aumann , R. (1965) Integrals of set valued functions. J. of Math. Analysis and Applications **12**, 1–12
7. Aumann, R. (1966) Existence of competitive equilibrium with a continuum of traders. Econometrica **34**, 1–17
8. Avriel, M. (1976) Nonlinear Programming: Analysis and Methods. Prentice Hall, New Jersey.
9. Castaing, Ch., Valadier M. (1977) Convex anaysis and measurable multifunctions. Lecture Notes in Mathematics **580**. Springer, Berlin
10. Clarke, F.H. (1983) Optimization and Nonsmooth Analysis. Wiley, New York
11. Drezner, Z. (ed.) (1995) Facility location analysis: survey of applications and methods. Springer, New York
12. Ermoliev, Yu.M. (1988) Stochastic quasigradient methods. In: Yu. Ermoliev R.J.B. Wets, eds. Numerical Techniques for stochastic optimization. Springer, Berlin
13. Ermoliev, Yu., Keyzer M.A., Norkin V. (2000) Global convergence of the stochastic tâtonnement process. Journal of Mathematical Economics **34**, 173–190

14. Fujita M., and Thisse J.F. (2002) Economics of agglomeration: cities, industrial location and regional growth. Cambridge University Press, Cambridge
15. Ginsburgh, V., Keyzer M.A. (2002) The structure of applied general equilibrium models. MIT-Press, Cambridge, MA
16. Hansen, P., Labbé, M., Peeters, D., Thisse J.-F. (1987) Facility Location Analysis. Fundamentals of Pure and Applied Economics **22**, 1–70
17. Hildenbrand, W. (1970). Existence of equilibria for economies with production and a measure space of consumers. Econometrica **38**, 608–623
18. Hildenbrand, W. (1973) Core and equilibria of a large economy. Princeton University Press, Princeton
19. Kantorovich, L.V. (1942) On the transportation of mass, Doklady Academii Nauk of USSR **37**, 227–229
20. Keyzer, M.A., Ermoliev Yu. (1999) Modeling producer decisions in a spatial continuum. In: J.-J. Herings, G. van der Laan and A.J.J. Talman (eds.): Theory of Markets and their Functioning. Elsevier Science, Amsterdam
21. Kolmogorov, A.N., Fomin S.V. (1981) Elements of theory of functions and functional analysis. Nauka, Moscow
22. Levin V.L. (1985) Convex analysis in the spaces of measurable functions and its application in mathematical economics. Nauka, Moscow
23. Rust, J. (1997) Randomization to break the curse of dimensionality, Econometrica **65**, 487–516
24. Varian, H. (1992) Microeconomic Analysis. Norton, New York

Sequential Downscaling Methods for Estimation from Aggregate Data

G. Fischer, T. Ermolieva, Y. Ermoliev, and H. Van Velthuizen

Institute for Applied Systems Analysis, Laxenburg, Austria

Abstract. Global change processes raise new estimation problems challenging the conventional statistical methods. These methods are based on the ability to obtain observations from unknown true probability distributions, whereas the new problems require recovering information from only partially observable or even unobservable variables. For instance, aggregate data exist at global and national level regarding agricultural production, occurrence of natural disasters, on incomes, etc. without providing any clue as to possibly alarming diversity of conditions at local level. "Downscaling" methods in this case should achieve plausible estimation of local implications emerging from global tendencies by using all available evidences.

The aim of this paper is to develop a sequential downscaling method, which can be used in a variety of practical situations. Our main motivation for this was the estimation of spatially distributed crop production, i.e., on a regular grid, consistent with known national-level statistics and in accordance with geographical datasets and agronomic knowledge. We prove convergence of the method to a generalized cross-entropy maximizing solution. We also show that for specific cases this method is reduced to known procedures for estimating transportation flows and doubly stochastic matrices.

Keywords: Cross-entropy, minimax likelihood, downscaling, spatial estimation.

1 Introduction

The analysis of global change processes requires the development of methods, which allow for dealing in a consistent manner with data on a multitude of spatial and temporal scales. Although GIS provides detailed geo-physical information, the socio-economic data often exist only at aggregate level. Integrated analysis of economic and environmental impacts of global changes raises a number of new estimation problems for downscaling and upscaling of available data to ensure consistency of biophysical and economic models. For example, aggregate data on national income does not reveal possibly alarming heterogeneity of its concentration among a small fraction of population or within, say, risk-prone regions of a country. We often need to derive information about the occurrence of disasters and induced potential losses in particular locations from information of their occurrence at global or regional levels. Aggregate regional annual concentrations of pollutants may be well within norms, whereas concentrations in some locations may exceed vital levels for a short time and cause irreversible damages.

The estimation of global processes consistent with local data and, conversely, long-term local implications emerging from global tendencies challenge the traditional statistical estimation methods. These methods are based on the ability to obtain observations from unknown true probability distributions. For the new estimation problems, which can be termed downscaling and upscaling estimation problems (see also [2] discussing other downscaling and upscaling problems), we often have only very restricted samples of real observations. Additional experiments to obtain more observations may be expensive, time consuming, dangerous or simply impossible. For example, although we can estimate total "departures" or "arrivals" of passengers in transportation systems, the estimation of passenger flows between different locations requires expensive origin-destination surveys and in many cases the data does not exist [6]. Similar situations occur with projections of migration flows, estimation of flows in communication systems, and trade flows. The paucity or lack of historical data is especially limiting for regions, which are subject to rapid changes (new developments, shocks, instabilities).

The aim of this article is to develop a recursive sequential downscaling method, which can be used for a large variety of practical situations. Our main motivation for this is the spatially explicit estimation of agricultural production, which is outlined in Section 2.1 and Section 5. In this problem we deal with "downscaling", i.e. attribution of known aggregate national or sub-national crop production and land use to particular locations (grid cell; pixel). Sections 2.2, 2.3 outline also the main idea of the sequential downscaling method of Section 3 by using simple known procedures for estimating transportation flows (e.g., migration flows, combining purely probabilistic prediction with available data on total demands and capacities of locations) and transition probabilities.

Section 3 develops a sequential downscaling method for iterative rebalancing estimates to satisfy general balance equations with unobservable and observable variables. We prove the convergence of this method to the solution maximizing a cross-entropy function. For specific transportation constraints this method reduces to the procedure proposed in the 1930s by the Leningrad architect G.V. Sheleikhovskii for estimating passenger flows. The convergence of Sheleikhovskiis method to the solution maximizing a cross-entropy function was established by Bregman [1] using complex and lengthy analysis of specific mappings and projections arising in the case of the transportation constraints. For recent developments and further references on these pioneering ideas see [5]. Our analysis for general constraints is based only on duality theory, which significantly simplifies proofs and clarifies the convergence properties. This opens up a way for various modifications and extensions, e.g., to situations with uncertainties when the available higher-level information is imprecise or involves stochastic elements.

Section 4 outlines connections between the maximum entropy principle, widely used (see e.g., [3], [12]) for the new estimation (downscaling) problems and the fundamental maximum likelihood principle of statistical estimation theory. We show that the maximum entropy principle can be viewed as an extension of the maximum likelihood principle, the so-called minimax likelihood principle. Therefore, the convergence of downscaling methods to solutions maximizing a cross-entropy function can be considered as an analog of the asymptotic consistency [16] analysis in traditional statistical estimation theory.

Section 5 describes a practical application and results of numerical calculations, with a fast convergence of the proposed basic procedure and its possible modifications. Section 6 concludes. As an important topic for future research, it emphasizes the need for incorporating the downscaling methods within the overall decision making problems, i.e., similar to the existing stochastic optimization theory.

2   Downscaling Problems: Motivating Examples

Let us consider situations, very common in regional studies, when direct observations of uncertain parameters on local levels are practically impossible and the estimation of their spatially explicit representation requires a downscaling procedure making use of information at a higher, more aggregate level. The problem of Section 2.1, in fact, motivated the development of discussed in Section 3 sequential downscaling procedure. Sections 2.2, 2.3 outline the main idea of this procedure by using simpler special cases of the problem.

2.1   Spatial Estimation of Agricultural Production

In general, the available information can be summarized as follows (see also Section 5). Extent of arable land a_i, in a pixel i, $i = \overline{1, m}$, is estimated from land cover satellite images. The degree and extents of suitable area for different crops in a pixel comes from FAO/IIASA crop suitability studies [10], [11]. There is also (computer-simulated) spatial information on the attainable yield d_{ij} of crop j, $j = \overline{1, n}$ in pixel i. From statistics, the price p_j of crop j, the value V_j of crop production j in a country, i.e. the total production of crop j multiplied by its price, the crop-wise sown area and production are available. Let x_{ij} be desirable estimates of crop j production in pixel i. This leads to the following estimate $v_{ij} = p_j d_{ij} x_{ij}$ of crop production value j in pixel i. Since production value V_j of crop j in the country is known from statistics, $\sum_{i=1}^{m} v_{ij} = V_j$, we have equations

$$\sum_{j=1}^{n} x_{ij} = a_i, i = \overline{1, m}. \tag{1}$$

$$\sum_{i=1}^{m} d_{ij} x_{ij} = b_j, j = \overline{1, n}, \tag{2}$$

where $b_j = V_j/p_j$.

By introducing new variable y_{ij} characterizing area shares by crop j in pixel i, i.e., $x_{ij} = a_i y_{ij}$, constraints (1), (2) can be written as the following

$$\sum_{j=1}^{n} y_{ij} = 1, i = \overline{1, m}, \tag{3}$$

$$\sum_{i=1}^{m} a_{ij} y_{ij} = e_j, j = \overline{1, n}, \tag{4}$$

where $a_{ij} = d_{ij} a_i$. This modification of constraint (1), (2) allows the use of entropy-like arguments.

There will usually be an infinite number of feasible solutions x_{ij}, $i = \overline{1, m}$, $j = \overline{1, n}$ satisfying equations (3) and (4). Therefore, to find a unique solution requires application of some additional principles. A key idea is to use some additional prior information on crop-specific area distribution, i.e., a prior distribution q_{ij} of crop j in pixel i. This prior can be based upon available crop distribution maps and other ancillary information, such as agro-climatic, biophysical, terrain and soil, demographic and farming systems characteristics (see discussion in Section 5). In any case, regardless of availability and detail of ancillary information, the prior can even be a (least informative) uniform distribution [18]. If a prior distribution $q_{ij} > 0$, $i = \overline{1, m}$, $j = \overline{1, n}$ is available, then a rather natural way to derive the estimates is from the minimization of the function

$$\sum_{j=1}^{n} \sum_{i=1}^{m} y_{ij} ln \frac{y_{ij}}{q_{ij}}, \tag{5}$$

subject to (3), (4), where (5) defines the so-called Kullback-Leibler distance [13] between distributions y_{ij} and q_{ij}.

Remark 1. Function $-\sum_{i,j} y_{ij} ln \frac{y_{ij}}{q_{ij}}$ is termed the cross-entropy, i.e., the minimization of (5) defines the cross-entropy maximizing estimates. Since $\sum_{i,j} x_{ij} ln \frac{x_{ij}}{q_{ij}} = \sum_{i,j} a_i y_{ij} ln \frac{y_{ij}}{q_{ij}} + \sum_i a_i ln a_i$, therefore the minimization of function $\sum_{i,j} x_{ij} ln \frac{x_{ij}}{q_{ij}}$ subject to equations (1), (2) is equivalent to minimization of a generalized (a weighted) cross-entropy $\sum_{i,j} a_i y_{ij} ln \frac{y_{ij}}{q_{ij}}$.

An alternative approach, which we take in this paper, is to derive a sequence of estimates y_{ij}^0, y_{ij}^1, y_{ij}^2, ... from an appropriate behavioral principle (as Sheleikhovskiis procedure) and to prove their convergence to a cross-entropy maximizing solution. For instance, a general tendency in farming is to allocate a crop j to pixels with maximum production values $p_j d_{ij}$ (or similar, such as maximum net revenue or maximum net present value in case of perennial crops or forestry activities). However, the straightforward application of such a rule to equations (1), (2) will, in general, lead to an overestimation or underestimation of aggregate known production values V_j, $j = \overline{1, n}$, i.e., situations when condition (2) is not fulfilled. Thus, these rule-based initial estimates require a sequential balancing procedure, which is developed in Section 3. Let us illustrate the main idea of the procedure by using two important special cases.

2.2 Estimation of Interzonal Flows

There can be different types of flows requiring estimation or/and projection procedures. It may be immigration or trade flows between different regions, flows of passengers or goods in transportation systems, or flows of messages in communication systems. Purely statistical projections often require expensive and time consuming origin-destination surveys; the necessary historical information may not exist [6]. Yet, the inconsistency of purely statistical estimate of interdependent flows is an inherent problem in the economics [14]. In particular, this is a key issue in situations when land use patterns are changing, e.g., due to new development or "shocks" in some locations. Since standard statistical procedures often do not take into account such available information as "demands" for departures from locations i, $i = \overline{1, m}$, and "capacities" of locations j, $j = \overline{1, n}$, to accommodate inflows. As a result, they may overestimate or underestimate the actual movements between locations.

The downscaling methods attempt to estimate flows among given locations in a way consistent with available data on the expected total number of "departures" a_i from locations i and arrivals b_j in location j.[1] For unknown (to be estimated) flows x_{ij} clearly $\sum_{j=1}^{n} x_{ij} = a_i$, $i = \overline{1, m}$, $\sum_{i=1}^{m} x_{ij} = b_j$, $j = \overline{1, n}$, i.e., we have a particular case of constraints (1), (2) with $d_{ij} = 1$, $i = \overline{1, m}$, $j = \overline{1, n}$. Assume also that there is a prior probability q_{ij} for a passenger from i to choose the destination j. For example, some behavioral models (see, e.g., [8], p. 414) define q_{ij} proportionally to a "distance" r_{ij} from i to j, $q_{ij} = r_{ij} / \sum_j r_{ij}$.

Consider the following iterative estimation procedure:

(i). If a passenger from location i chooses the destination j with a prior probability q_{ij}, $\sum_j q_{ij} = 1$, then the expected flow from i to j is $x_{ij}^0 = a_i q_{ij}$. Clearly $\sum_j x_{ij}^0 = a_i$, $i = \overline{1, m}$, but there may be overestimation $\sum_i x_{ij}^0 > b_j$ or underestimation $\sum_i x_{ij}^0 < b_j$ of the available b_j.

(ii). Calculate relative imbalances $\beta_j^0 = b_j / \sum_i x_{ij}^0$ and $z_{ij}^0 = x_{ij}^0 \beta_j^0$, $i = \overline{1, m}$, $j = \overline{1, n}$.

(iii). Clearly, $\sum_i z_{ij}^0 = b_j$, $j = \overline{1, n}$, but the estimate z_{ij}^0 may overestimate or underestimate the known demand for departures a_i from i. Therefore, calculate $\alpha_i^0 = a_i / \sum_j z_{ij}^0$, $x_{ij}^1 = z_{ij}^0 \alpha_i^0$, and so on.

---

[1] The terms matrix estimation and matrix balancing have also been used in [5] to describe this type of methods. The terms downscaling and upscaling have a rather general meaning. For example, we can speak about a consistent downscaling and upscaling of certain differential equations, random processes, or potential climate changes.

This balancing procedure can be summarized also as the following. We can represent x_{ij}^1 as $x_{ij}^1 = a_i q_{ij}^1$, and $q_{ij}^1 = (q_{ij}\beta_j^0)/(\sum_j q_{ij}\beta_j^0)$, $i = \overline{1,m}$, $j = \overline{1,n}$. Assume $x^k = \{x_{ij}^k\}$ has been calculated. Then find $\beta_j^k = b_j / \sum_i x_{ij}^k$ and calculate $x_{ij}^{k+1} = a_i q_{ij}^{k+1}$, $q_{ij}^{k+1} = (q_{ij}\beta_j^k / \sum_j q_{ij}\beta_j^k)$, $i = \overline{1,m}$, $j = \overline{1,n}$, and so on. In this form the procedure can be viewed as a sequential redistribution of demands a_i from locations $i = \overline{1,m}$ among locations $j = \overline{1,n}$ by using a Bayesian type of rule for updating the prior distribution q_{ij}: $q_{ij}^{k+1} = q_{ij}\beta_j^k / \sum_j q_{ij}\beta_j^k$, $q_{ij}^0 = q_{ij}$.

Initially this method was proposed by the architect Sheleikhovskii (see references in [1]) for estimating passenger flows between districts of a city (including possible new districts). Proof of convergence to the solution maximizing the cross-entropy type function (Remark 1) $\sum_{ij} x_{ij}ln(x_{ij}/q_{ij})$ was given in [1] using extremely lengthy and complex arguments essentially relying on specific mappings associated with the transportation constraints. In Section 3 we propose a similar method for general constraints (2). We apply duality theory, which allows us to significantly simplify and clarify the convergence analysis (Proposition 1). This also opens up an opportunity for various modifications, in particular, to situations with more general constraints and uncertain parameters.

2.3   Estimation of Stochastic Matrices

It is interesting to note that a similar procedure is used in the conventional statistical theory for estimating doubly stochastic matrices (see discussion in [15], [20]). Suppose we can observe transitions of a Markov chain with n states and stochastic matrix $\{P_{ij}\}$. The usual estimate of P_{ij} is $x_{ij} = \alpha_{ij}/a_i$ where α_{ij} is the number of transitions from i to j, which are observed, and $a_i = \sum_j \alpha_{ij}$. This amounts to a normalization of the rows of matrix $\{\alpha_{ij}\}$. If it was known that $\{P_{ij}\}$ is in fact a doubly stochastic matrix, i.e., $\sum_i P_{ij} = 1$, then it was proposed to alternately normalize (as in Section 2.2) the rows and columns of $\{\alpha_{ij}\}$ in the belief that this iterative process would converge to an estimate of $\{P_{ij}\}$. Proof of convergence of this procedure to a doubly stochastic matrix for rather special cases was given in [15]. From the results in [1] follows the convergence for general doubly stochastic matrixes and the optimality of the resulting estimates as the cross-entropy maximizing solution.

3   Sequential Downscaling Methods

Consider the following problem: minimize

$$\sum_{j=1}^{n}\sum_{i=1}^{m} x_{ij}ln(x_{ij}/q_{ij}), \tag{6}$$

subject to constraints (1), (2), where $q_{ij} > 0$, $d_{ij} > 0$ are given, $a_i > 0$, $b_j > 0$, $i = \overline{1,m}$, $j = \overline{1,n}$. As Remark 1 indicates, the minimization of (6) is equivalent to the maximization of a generalized cross-entropy function. Values $x_{ij} = 0$ are also possible when $q_{ij} = 0$ or $d_{ij} = 0$. Without loss of generality, we assume $x_{ij} > 0$, $q_{ij} > 0$, $\sum_{j=1}^{n} q_{ij} = 1$, $i = \overline{1,m}$, and the set of feasible solutions defined by (1), (2) is not empty.

Consider the following sequential procedure.

Step 1: Compute an initial estimate $x_{ij}^0 = a_i q_{ij}$. Clearly, x_{ij}^0 satisfies (1), $\sum_j x_{ij}^0 = a_i$, since $\sum_j q_{ij} = 1$ but, in general, constraints (2) are violated.

Step 2: For given $x^k = x_{ij}^k$, find β_j^{k+1} satisfying equations

$$\sum_{i=1}^{m} d_{ij} x_{ij}^k e^{d_{ij}\beta_j} = b_j, j = \overline{1,n}. \tag{7}$$

The left hand side of this equality is a strictly monotonic function and β_j^{k+1} can be easily calculated.

Step 3: Calculate $z_{ij}^{k+1} = x_{ij}^k e^{d_{ij}\beta_j^{k+1}}$, and

$$\alpha_i^{k+1} = a_i / \sum_j z_{ij}^{k+1}, i = \overline{1,m}, j = \overline{1,n}. \tag{8}$$

Step 4: Update x_{ij}^k to

$$x_{ij}^{k+1} = \alpha_i^{k+1} z_{ij}^{k+1}, i = \overline{1,m}, j = \overline{1,n}. \tag{9}$$

and so on with Steps 2 - 4, until desirable convergence is reached, e.g., constraints (1), (2) are satisfied with a given accuracy.

In summary, this procedure, similar to Sections 2.2, 2.3 involves a sequential updating of a priori probability distribution q_{ij} by using a Bayesian type of rule: $x_{ij}^{k+1} = a_i q_{ij}^{k+1}$, $q_{ij}^{k+1} = q_{ij}\gamma_j^k / \sum_j q_{ij}\gamma_j^k$, $\gamma_j^k = e^{d_{ij}\beta_j^k}$, where values γ_j^k are calculated using observations of imbalances rather than using observations of real random variables.

Proposition 1. The sequence $x^k = \{x_{ij}^k, i = \overline{1,m}, j = \overline{1,n}\}$, $k = 0, 1, ...$, generated by iteration (7)-(9) converges to the solution x^* of constraints (1), (2) minimizing the function (6).

Lemma. There exist such $\alpha_i > 0$, β_j, $i = \overline{1,m}$, $j = \overline{1,n}$, that the optimal solution $x_{ij} = x_{ij}(\alpha,\beta)$ minimizing (6) subject to constraints (1), (2) satisfies the following necessary and sufficient optimality conditions:

$$a_i - \sum_j x_{ij} = 0; i = \overline{1,m};$$
$$b_j - \sum_i d_{ij} x_{ij} = 0; j = \overline{1,n};$$
$$x_{ij} = q_{ij}\alpha_i e^{d_{ij}\beta_j}, i = \overline{1,m}, j = \overline{1,n}.$$

Proof. For a continuous, strictly convex function (6) on a non-empty compact set of an Euclidian space there is a unique optimal solution to the minimization problem. Consider the Lagrangian function:

$$L(x, \lambda, \mu) = \sum_{i,j} x_{ij} ln(x_{ij}/q_{ij}) + \sum_{i=1}^{m} \lambda_i \Big(a_i - \sum_{j=1}^{n} x_{ij}\Big) + \sum_{j=1}^{n} \mu_j \Big(b_j - \sum_{i=1}^{m} d_{ij} x_{ij}\Big)$$

Since the optimal solution is positive, the optimality conditions lead to

$$\frac{\partial L}{\partial x_{ij}} = ln\frac{x_{ij}}{q_{ij}} + 1 - \lambda_i - d_{ij}\mu_j = 0,$$

$i = \overline{1, m}$, $j = \overline{1, n}$, i.e., the optimal solution can be represented analytically as $x_{ij}(\lambda, \mu) = q_{ij} e^{\lambda_i - 1} e^{d_{ij}\mu_j}$, $i = \overline{1, m}$, $j = \overline{1, n}$. The dual problem reads: find Lagrange multipliers (λ_i, μ_j), $i = \overline{1, m}$, $j = \overline{1, n}$, maximizing function

$$\varphi(\lambda, \mu) = min_x L(x, \lambda, \mu) = L(x(\lambda, \mu), \lambda, \mu).$$

From basic results of convex analysis it follows that $\varphi(\lambda, \mu)$ is a strictly concave continuously differentiable function and the optimality condition can be written as

$$\frac{\partial \varphi}{d\lambda_i} = a_i - \sum_{j=1}^{n} x_{ij}(\lambda, \mu) = 0, i = \overline{1, m},$$

$$\frac{\partial \varphi}{d\mu_j} = b_j - \sum_{i=1}^{m} d_{ij} x_{ij}(\lambda, \mu) = 0, j = \overline{1, n}.$$

By using new notations $\alpha_i = e^{(\lambda_i - 1)}$, $\beta_j = \mu_j$, and the same notations $x_{ij}(\alpha, \beta)$ for $x_{ij}(\lambda, \mu)$ with $\lambda_i = \lambda_i(\alpha_i) = ln\alpha_i + 1$, $\mu_j = \beta_j$ we obtain the proof due to the strict monotonicity of $e^{(\lambda_i - 1)}$.

Proof of Proposition 1. It is easy to see that the sequential method (7)-(9) updates variables $\alpha = (\alpha_1, ..., \alpha_m)$, $\beta = (\beta_1, ..., \beta_n)$ and $x = \{x_{ij}\}$ to satisfy the optimality conditions of **Lemma**. Namely, equations (7) require that the gradient of the strictly concave function of the dual problem $\varphi_\mu(\lambda^k, \mu^{k+1}) = 0$, whereas equations (8) require that the gradient $\varphi_\lambda(\lambda^{k+1}, \mu^{k+1}) = 0$, for some λ^k, μ^k, $k = 0, 1, ...$.

Indeed, let us illustrate just a few steps of the method. Solution x_{ij}^0 can be represented as $x_{ij}^0 = \alpha_i^0 q_{ij} e^{d_{ij}\beta_j^0}$, $\alpha_i^0 = a_i$, $\beta_j^0 = 0$. Clearly, that $\sum_j x_{ij}^0 = a_i$, i.e., $\varphi_{\lambda_i}(\lambda^0, \mu^0) = 0$ for $\lambda_i^0 = \lambda_i(\tilde{\alpha}_i^0)$, $\tilde{\alpha}_i^0 = \alpha_i^0$, $\mu_j^0 = \beta_j^0$. At **Step 2** values β_j^1 modify x_{ij}^0 to $y_{ij}^1 = \alpha_i^0 q_{ij} e^{d_{ij}(\beta_j^0 + \beta_j^1)}$, $\sum_i d_{ij} y_{ij}^1 = b_j$, i.e., $\varphi_{\mu_j}(\lambda^0, \mu^1) = 0$, $\mu_j^1 = \beta_j^0 + \beta_j^1$. At **Step 3** values α_i^1 modify y_{ij}^1 to $x_{ij}^1 = \alpha_i^0 \alpha_i^1 q_{ij} e^{d_{ij}(\beta_j^0 + \beta_j^1)}$, $\sum_j x_{ij}^1 = a_i$, i.e., $\varphi_{\lambda_i}(\lambda^1, \mu^1) = 0$, $\lambda_i^1 = \lambda_i(\tilde{\alpha}_i^1)$, $\tilde{\alpha}_i^1 = \alpha_i^0 \alpha_i^1$ and so on.

Therefore, the convergence of vectors λ^k, μ^k and hence $\{x_{ij}^k\}$ to the optimal solutions of the dual and the primal problems follows from the convergence of the cyclic ascent method [19].

Remark 2. It follows from the above that for transportation constraints, i.e., for $d_{ij} = 1$, $i = \overline{1, m}$, $j = \overline{1, n}$, the proposed method is reduced to

Sheleikovskii's method. In this case, it also follows that the optimal solution is represented as $x_{ij}(\alpha, \beta) = q_{ij}\alpha_i\beta_j$, $\alpha_i > 0$, $\beta_j > 0$, $i = \overline{1, m}$, $j = \overline{1, n}$, what is typical for the so-called gravity models [4].

4 Minimax Likelihood and Maximum Entropy

Definitely that besides a cross-entropy maximization there exists a vast variety of optimization principles to single out a solution of equations (3), (4). Let us show that minimization of (5) is a natural generalization of the fundamental maximum likelihood principle of statistical theory to problems involving non-statistical uncertainty. Namely, it can be viewed as minimax likelihood principle.

The standard statistical estimation theory deals with the situation when the information on unknown distribution can be derived from observations of underlying random variables. In such a case, the most natural principle for selecting an estimate from a given sample of observations is the maximum likelihood proposed by Fisher [9]. This principle requires that the estimate has to maximize the probability that a given sample is observed.

A downscaling problem deals with the estimation of often unobservable variables. Yet, the uncertainty can also be characterized or interpreted in probabilistic terms. For example, in the estimation of crop production values defined by equations (3), (4), we can think of values $y_{ij} > 0$, $\sum_{j=1}^{n} y_{ij} = 1$ as the probability (the degree of our belief) that a unit area of pixel i is allocated to crop j. It is easy to see that the maximum entropy principle can be viewed as an extension of the maximum likelihood principle.

Consider a situation similar to problems posed in Section 2. Namely, let us assume that there is an underlying random variable ξ with a finite number of possible values $\xi_1, ..., \xi_r$ and the unknown true probability distribution of ξ is concentrated at these points with associated probabilities $p_1^*, ..., p_r^*$, $Prob[\xi = \xi_j] = p_j^*$.

In the statistical estimation the available information is given by a random sample $\xi^1, ..., \xi^N$ of N independent observations of ξ on $(p_1^*, ..., p_r^*)$. A maximum likelihood estimate of the unknown probabilities $(p_1^*, ..., p_r^*)$ is obtained by maximizing the probability (likelihood) of observing $\xi^1, ..., \xi^N$

$$\prod_{k=1}^{N} Prob[\xi = \xi^k] = \prod_{j=1}^{r} p_j^{v_j} \tag{10}$$

subject to constraints $\sum_{j=1}^{r} p_j = 1$, $p_j > 0$, $j = \overline{1, r}$, where v_j is the number of times the value ξ_j has been observed, $\sum_{j=1}^{r} v_j = N$. Since lny is a monotonously increasing function of y, the maximization of (10) is equivalent to maximization of the log likelihood function $ln \prod_{j=1}^{r} p_j^{v_j} = \sum_{j=1}^{r} v_j lnp_j$ or normalized by the number of observations N, $\sum_{j=1}^{r} v_j = N$, the sample mean function

$$\frac{1}{N} \sum_{j=1}^{n} v_j lnp_j. \tag{11}$$

This is a continuous, strictly concave function on the set of R^n determined by linear constraints. By using the Lagrangian function (or the more general fact of Proposition 2 below) we can derive the well known result (see, e.g., [17]) that the unique solution maximizing (11) is the empirical probability function

$$p_j^N = v_j/N, j = \overline{1,r}. \tag{12}$$

Let us consider this differently. The log likelihood function (11) is the sample mean approximation of the expectation

$$Elnp_\xi = \sum_{j=1}^{r} p_j^* lnp_j, \tag{13}$$

where the unknown probability distribution p_j^* is approximated by the frequencies v_j/N derived from an available sample of observations $\xi^1, ..., \xi^N$. In downscaling problems the available information about the unknown probability distribution p_j^*, $j = \overline{1,r}$ is given not by a sample of observations, but by a number of constraints (3) and (4), i.e., $p^* \in P$, where P is the set of all feasible distributions. If $y = (y_1, ..., y_r) \in P$, then we can consider

$$\sum_{j=1}^{r} y_j lnp_j, \tag{14}$$

as an approximation of the expectation function (13) similar to the sample function (11). The log likelihood function (14) is defined for any feasible probability distribution $y \in P$. The worst-case estimate from P leads to minimization of the function

$$V(y) = \max_{p \in P} \sum_{j=1}^{r} y_j lnp_j. \tag{15}$$

w.r.t. $y \in P$. Therefore, the minimization of (15) w.r.t. $p \in P$ is a counterpart to the minimization of (11)

Proposition 2.

$$\min_{y \in P} \max_{p \in P} \sum_{j=1}^{r} y_j lnp_j = \min_{y \in P} \sum_{j=1}^{r} y_j lny_j. \tag{16}$$

Proof. It follows from analogous to (12) fact: if $y \in P$, then $V(y) = \sum_{j=1}^{r} y_j lny_j$. Indeed, for a given $y = (y_1, ..., y_r) \in P$ and $p \in P$ we have $\sum_{j=1}^{r} y_j lnp_j - \sum_{j=1}^{r} y_j lny_j = \sum_{j=1}^{r} y_j ln\frac{p_j}{y_j} < ln \sum_{j=1}^{r} y_j \frac{p_j}{y_j} = 0.$

Remark 3. In other words, the worst-case estimate leads to the principle of maximizing entropy $- \sum_{j=1}^{r} y_j ln y_j$. In the case of a given prior distribution $q = (q_1, ..., q_r)$, we may require the minimization of the difference between the function (14) for $p \in P$ and $\sum_{j=1}^{r} q_j ln p_j$ for the given prior q from P:

$$\min_{y \in P}[\max_{p \in P} \sum_{j=1}^{r} y_j ln p_j - \sum_{j=1}^{r} y_j ln q_j] = \min_{y \in P} \sum y_j ln \frac{y_j}{q_j}, \tag{17}$$

i.e., the maximization of cross-entropy function $- \sum_{j=1}^{r} y_j ln \frac{y_j}{q_j}$ or the Kullback-Leibler distance between distributions y and q. Clearly, instead of selecting a worst-case distribution $y \in P$ in (15) we can take other distributions, which may lead to different downscaling principles. Since the estimation is usually used to support decision making processes, these more general principles may be specific to different types of problems, i.e., explicitly connected with the goals of a decision making problem.

5   Practical Applications

The proposed method has been applied for downscaling aggregate national and subnational data on crop production and land use (Section 2.1) for all main countries of the world. The downscaling was performed country-by-country. For this, the territory of each country was subdivided into grid cells with cultivation share, each cell with spatial resolution of 5 by 5 min latitude-longitude, i.e., urban areas, infrastructure, and water bodies were excluded from the analysis. To illustrate the dimensionality, the number of grid cells with cultivation in France equaled 9042, in Germany 6510, and in Austria 1165. For larger countries, such as United States and Russia, the number of grid cells with active agricultural use reached approximately 95 thousand, for Brazil 80 thousand and about 75 thousand for China. The data on aggregate country-specific agricultural production was obtained from FAO. The list of crops comprised 28 major crops such as wheat, rice, maize, potato, soybean, pulses, oil crops, coffee, tea, tobacco, cotton, etc. Figure 1 shows spatial distribution of downscaled total crop production value for Europe in terms of international prices (Geary-Khamis (GK) dollars of $2000 - 2001$ per spatial land unit (grid cell).

Calculation of the prior included important spatial information on percentage of cultivated, rainfed and irrigated land in each grid cell derived using satellite images of land cover classes as well as aggregate statistics of arable land used for annual and perennial crops in each country. For example, Figure 2 shows cultivated land share by grid cell. In addition, the calculation of prior included information on multicropping index, i.e., how many harvests may be obtained per year from a piece of land, derived with AEZ methodology [10], [11], crop suitability (including climate, soil, and terrain conditions) and attainable yields in each spatial land unit, as well as information on characteristics of prevailing farming systems and population distribution.

Versions of the algorithm were written in FORTRAN and MATLAB programming languages and performed on PC. They showed fast convergence and, thus, efficient performance in dealing with large and spatially detailed data. Clearly, the time performance of the algorithm depends on the size of a study region, i.e., number of grid cells or land units considered and the number of crops that can be grown in each location. Applications of the algorithms to global studies showed that to attain high precision (10^{-7}) solution time increased roughly linearly with the increase of problem dimensionality. The performance is often remarkably fast, which is explained by the quality of the prior and the corresponding initial approximation. Thus, for Austria, 7 iterations were needed, for Germany and France about 20 to 30 and for China about 60, which indicates that the algorithm can be efficiently used in large real-world downscaling problems.

Fig. 1. Total crop production value, GK dollars per grid cell.

Remark 4. The proposed method can easily be modified to reflect problem-specific peculiarities of constraints (1) and (2). An important special case is the transportation constraints, i.e., $d_{ij} = 1$, $i = \overline{1,m}$, $j = \overline{1,n}$. If coefficients d_{ij} are reasonably well approximated by a product of some parameters δ_i, $i = \overline{1,m}$, σ_j, $j = \overline{1,n}$, for instance $d_{ij} = \delta_i \sigma_j$, $i = \overline{1,m}$, $j = \overline{1,n}$, then (1), (2) can be reduced to the transportation constraints by introducing new variables $y_{ij} = \delta_j x_{ij}$ and substituting b_j by b_j/σ_j and a_i by a_i/δ_i, i.e., simply by rescaling. Another simplifying situation occurs when function $e^{d_{ij}\beta_j}$ is approximated by a function $A_{ij} f_j^{\beta_j}$, $i = \overline{1,m}$, $j = \overline{1,n}$, for some parameters

Fig. 2. Share of cultivated land per grid cell.

$A_{ij} > 0$, $f_j > 0$, $i = \overline{1,m}$, $j = \overline{1,n}$, and β_j varying within the range of plausible solutions of (7).

6   Concluding Remarks

In this paper we analyze numerical downscaling procedures only for situations when aggregate observed information is available and used as constraints on average values. For many practical situations this assumption may be insufficient and the procedures may need to be extended into more rigorous treatment of uncertainty regarding a prior probability q_{ij} and parameters of constraints (1), (2).

For practical applications, the choice of appropriate "priors", their inherent uncertainties and imprecision, are among the major challenges of the downscaling methodology, ultimately determining the success of these procedures.

An important issue for future research, besides the uncertainty of "priors" and other parameters, is concerned with the incorporation of downscaling methods within the overall decision making problems, i.e., similar to the stochastic optimization theory.

Acknowledgments. The authors are grateful to the participants of the IFIP/ IIASA/GAMM workshop on Coping with Uncertainty, December 13-16, 2004, IIASA, and to the anonymous referees for critical suggestions that led to improvements of this paper. In particular, we are thankful to one of them who pointed out an important relevant book ([5]).

References

1. Bregman, L.M. (1967): Proof of the Convergence of Sheleikhovskii's Method for a Problem with Transportation Constraints. Journal of Computational Mathematics and Mathematical Physics 7/1, 191-204 (Zhournal Vychislitel'noi Matematiki, USSR, Leningrad, 1967).
2. Bierkens, M.F.P., Finke, P.A., de Willigen, P. (2000): Upscaling and Downscaling Methods for Environmental Research. Kluwer, Dordrecht, The Netherlands.
3. Borwein, J.M., Lewis, A.S., Nussbaum, R.D. (1994): Entropy Minimization, DAD Problems, and Doubly Stochastic Kernels. Journal of Functional Analysis 123, 264-307.
4. Carruthers, G.A.P. (1956): An Historical Review of the Gravity and Potential Concepts of Human Interaction. Journal of American Institute of Planners 22.
5. Censot, Y., Zenios, S.A. (1997): Parallel Optimization. Oxford University Press.
6. Esopo D.A., Lefkowitz, B. (1963): An Algorithm for Computing Interzonal Transfers Using the Gravity Model. Operation Research 11/6, 901-907.
7. Ermoliev, Y (1976): Stochastic Programming Methods. Nauka, Moscow (In Russian).
8. Ermoliev, Y. and Wets R. (Eds.) (1988): Numerical Techniques of Stochastic Optimization: Computational Mathematics. Springer-Verlag, Berlin, Germany.
9. Fisher, R.A. (1922). On the Mathematical Foundations of Theoretical Statistics. Philosophical Transactions of the Royal Society of London, Series A222, 309-368.
10. Fischer, G., van Velthuizen, H.T., Nachtergaele, F.O., and Medow, S. (2000): Global Agro-Ecological Zones 2000. International Institute for Applied Systems Analysis and Food and Agriculture Organization of the United Nations, Laxenburg, Austria. CD-ROM and web-site http://www.iiasa.ac.at/Research/LUC/GAEZ/index.htm.
11. Fischer, G., H.T. van Velthuizen, M.M. Shah, and F.O. Nachtergaele (2002): Global Agro-ecological Assessment for Agriculture in the 21st Century: Methodology and Results. Research Report RR-02-02. International Institute for Applied Systems Analysis, Laxenburg, Austria.
12. Golan, A., Judge, G., Miller, D. (1996): Maximum Entropy Econometrics: Robust Estimation with Limited Data. Series in Financial Economics and Quantitative Analysis, John Wiley and Sons Ltd, Baffins Lane, Chichester, West Sussex PO19 1UD, England.
13. Kullback, J. (1959): Information Theory and Statistics. John Wiley and Sons, New York.
14. Morgenstern, O. (1963): On the Accuracy of Econmic Observations. Princeton University Press, Princeton, N.J., USA. University Press, Princeton, N.J., USA.
15. Sinkhorn, R. (1964): A Relationship between Arbitrary Positive Matrices and Doubly Stochastic Matrices. Annals of Mathematical Statistics 35/2, 876-879.
16. Wald, A. (1949): Note on the Consistency of the Maximum Likelihood Estimate. Annals of Mathematical Statistics 20, 595-601.

17. Wets, R. (1999): Statistical Estimation from An Optimization Viewpoint. Annals of Operation Research 85, 79-101.
18. Wood, S., Sebastian, K., Scherr, S. (2000): Pilot Analysis of Global Ecosystems: Agroecosystem. A Joint Study by International Food Policy Research Institute and World Resource Institute, Washington D.C.
19. Zangwill, W.I. (1969): Nonlinear Programming, A Unified Approach. Prentice Hall, Englewoods Cliffs, New Jersey.
20. Zenios, A., Zenious, S. (1992): Robust Optimization for Matrix Balancing with Noisy Data. Report 92-01-03, Dept. of Decision Sciences, Wharton School, Univ. of Pennsylvania.

Optimal Control for a Class of Uncertain Systems

F.L.Chernousko

Institute for Problems in Mechanics
of the Russian Academy of Sciences,
Moscow, Russia

Abstract. Linear dynamical control systems subject to uncertain but bounded disturbances are considered. The bounds imposed on the disturbances depend on the control magnitude and grow with the control. This situation is typical for the cases where the disturbance is due to the inaccuracy of the control implementation and often takes place in engineering applications such as transportation, aerospace, and robotic systems.

Under certain assumptions, the minimax control problem is formulated and solved. The explicit expressions for the optimal control (both open-loop and feedback) are obtained that provide the minimax to the given performance index for arbitrary but bounded disturbances. Examples are given.

1   Introduction

Dynamical control systems subject to uncertain disturbances are usually described by ordinary differential equations

$$\dot{x} = f(x, u, v, t). \tag{1}$$

Here, x is the vector of state, t is time, u is the control vector, and v is the vector of disturbances.

The disturbance $v(t)$ in (1) can be due either to external causes, for example, external forces acting upon the system, or to the errors or inaccuracies in the control implementation. The latter case often takes place in engineering applications such as transportation, aerospace, and robotic systems, if the desired program of the control force (for example, thrust) is carried out by engines or actuators with unpredictable errors.

In this case, the bounds on the errors, or the magnitude of the admissible disturbances $v(t)$ in (1), depend on the control applied: these bounds are zero for the zero control and usually grow with the control magnitude.

In this paper, the model of the situation described above is formalized in the framework of linear control systems. The minimax control problem with a general performance index is considered.

The bounds on admissible disturbances are assumed to depend on the control $u(t)$ in such a way that the explicit solution of the minimax control problem is available in an explicit form.

The optimal control, both in open-loop and feedback forms, is obtained that provides the minimax value of the given performance index for arbitrary but bounded disturbances. Examples for the control of a mechanical system are presented.

2   Statement of the Problem

Consider a linear control system subject to the control and several disturbances

$$\dot{x} = A(t)x + B(t)u + \sum_{i=1}^{r} C_i(t)v_i + g(t). \tag{2}$$

Here, t is time, $t \in [t_0, T]$, $x(t)$ is the n-dimensional vector of state, $u(t)$ is the m-dimensional vector of control, $v_i(t)$ is the k_i-dimensional vector of the ith disturbance, and $A(t)$, $B(t)$, $C_i(t)$, and $g(t)$ are given matrix functions of time of dimensions $n \times n$, $n \times m$, $n \times k_i$, and $n \times 1$, respectively, $i = 1, ..., r$.

The disturbances v_i are bounded by the constraints

$$(G_i(t)v_i(t), v_i(t))^{1/2} \le \rho_i(t), \quad i = 1, \dots, r, \tag{3}$$

where the following notation is introduced

$$\rho_i(t) = (K_i(t)u(t), u(t)) + (p_i(t), u(t)) + q_i(t). \tag{4}$$

Here, $G_i(t)$ is a symmetric positive definite $k_i \times k_i$ matrix, $K_i(t)$ is a symmetric nonnegative definite $m \times m$ matrix, $p_i(t)$ is an m-dimensional vector, and $q_i(t)$ is a scalar. All these functions are assumed to be specified for all $i = 1, \dots, r$ and $t \in [t_0, T]$. The symbol $(.,.)$ stands for the scalar product of vectors. The constraints (3) and (4) include, in particular, the case of constant bounds on v_i typical for external disturbances (in this case, G_i is the identity matrix, K_i and p_i are zero, and q_i is a constant) as well as the case of disturbances due to the errors in the control implementation and proportional to the squared control magnitude (in this case, G_i and K_i are the identity matrices whereas p_i and q_i are zero).

Since the left-hand side of inequality (3) is nonnegative, its right-hand side $\rho_i(t)$ should be nonnegative too. To ensure this condition, we will assume that either

$$K_i(t) \equiv 0, \quad p_i(t) \equiv 0, \quad q_i(t) > 0, \tag{5}$$

or the matrix $K_i(t)$ is positive definite $(K_i(t) > 0)$ and the minimum of $\rho_i(t)$ over all u is positive. This minimum is attained at $u = -K_i^{-1}p_i/2$. Substituting this expression into (4), we find that the minimum of ρ_i is

$$q_i - \frac{1}{4}(K_i^{-1}p_i, p_i) \ge 0. \tag{6}$$

Therefore, we suppose that, for each $i = 1, \ldots r$, one of the following two assumptions holds: either (5), or $K_i > 0$ and (6).

At the initial time instant $t = t_0$, the initial condition for system (2) is specified:

$$x(t_0) = x^0, \tag{7}$$

where x^0 is a given n-dimensional vector.

The performance index for system (2) is defined as follows

$$J = \int_{t_0}^{T} \left[(a(t), x(t)) + (D(t)u(t), u(t)) + (b(t), u(t)) \right] dt + F(x(T)). \tag{8}$$

Here, the n-dimensional vector function $a(t)$, the symmetric positive definite $m \times m$ matrix $D(t)$, and the m-dimensional vector function $b(t)$ are specified for $t \in [t_0, T]$, and $F(x)$ is a given smooth scalar function of the state x. The performance index (8) includes a linear integral term proportional to the projection of the state vector onto a given direction, integral terms linear and quadratic in control, and a nonlinear terminal term that can serve as a measure of the deviation of the terminal state from the prescribed position. All functions of time introduced in equations (2), (3), (4), and (8) are supposed to be piecewise continuous in the interval $[t_0, T]$.

We seek for the control $u(t)$ that yields the minimax value to the performance index (8):

$$J^* = \min_u \max_v J, \quad i = 1, \ldots, r, \tag{9}$$

provided that conditions (2)–(4) and (7) hold.

The maximum in (9) is taken over all disturbances $v(t)$ satisfying inequalities (3), and it is assumed that the disturbances "know" the control $u(t)$. The minimum in (9) is taken over all piecewise smooth controls $u(t)$. We will find the optimal open-loop control $u(t)$ that is the "best" reaction to the "worst" disturbances as well as the optimal feedback control $u^*(x, t)$ that depends on the state and time. Our problem can be also considered as the problem of the guaranteed, in the sense of the performance index (8), control of the ensemble of trajectories of system (2) under constraints (3).

3   Transformations

To simplify our problem, let us carry our certain transformations. Denote by $\Phi(t)$ the fundamental matrix os system (2). This matrix satisfies the following initial value problem

$$\dot{\Phi} = A(t)\Phi, \quad \Phi(t_0) = E_n. \tag{10}$$

Here, E_n is the $n \times n$ identity matrix.

Let us make the following change of the state variable:

$$x = \Phi(t)y. \tag{11}$$

Then equation (2) becomes

$$\dot{y} = B_1 u + \sum_{i=1}^{r} C_{i1} v_i + g_1, \tag{12}$$

where the following notation is introduced

$$B_1(t) = \Phi^{-1}(t)B(t), \quad C_{i1}(t) = \Phi^{-1}(t)C_i(t),$$
$$g_i(t) = \Phi(t)g(t), \qquad i = 1, \ldots, r. \tag{13}$$

The initial condition (7) and the performance index (8) take the form

$$y(t_0) = x^0, \tag{14}$$

$$J = \int_{t_0}^{T} [(a_1, y) + (Du, u) + (b, u)]dt + F_1(y(T)). \tag{15}$$

Here, the following denotations are used:

$$a_1(t) = \Phi^T(t)a(t), \qquad F_1(y) = F(\Phi(T)y), \tag{16}$$

where ^{T} denotes the transposed matrix.

Hence, our problem is reduced to the minimax problem (9) for system (12) under the initial condition (14) and constraints (3). The performance index J is defined by (15).

4   Solution of the Problem

Let us find first the maximum of J with respect to v, and then the minimum with respect to u. For both stages, we will apply the maximum principle [1].

The Hamiltonian for the first stage is given by

$$H = (\psi, B_1 u) + \left(\psi, \sum_{i=1}^{r} C_{i1} v_i\right) + (\psi, g_1) + (a_1, y) + (Du, u) + (b, u). \tag{17}$$

Here, $\psi(t)$ is an n-dimensional adjoint vector that satisfies the following equation and initial condition at $t = T$:

$$\dot{\psi} = -a_1(t), \qquad \psi(T) = \frac{\partial F_1(y(T))}{\partial y}. \tag{18}$$

The solution of the initial value problem (18) is

$$\psi(t, y(T)) = \int_t^T a_1(\tau)d\tau + \frac{\partial F_1(y(T))}{\partial y}. \tag{19}$$

Note that the adjoint vector ψ depends on the terminal state $y(T)$ yet unknown.

To find the maximum of the Hamiltonian (17) with respect to v_i, it is necessary to maximize the scalar product

$$(C_{i1}^T \varphi, v_i) \rightarrow \max, \quad i = 1, \dots, r,$$

under the quadratic constraint (3). As a result, we obtain

$$v_i = \lambda_i^{-1} \rho_i G_i^{-1} C_{i1}^T \psi, \quad i = 1, \dots, r. \tag{20}$$

Here, ρ_i is defined by equation (4),

$$\lambda_i = \left(G_{i1}^{-1} C_{i1}^T \psi, C_{i1}^T \psi \right)^{1/2}, \quad i = 1, \dots, r, \tag{21}$$

and we assume that $C_{i1}^T \psi \neq 0$ for $i = 1, \dots, r$.

Substituting v_i from (20) into equation (12), we obtain the equation

$$\dot{y} = B_1 u + \sum_{i=1}^r \lambda_i^{-1} \rho_i C_{i1} G_i^{-1} C_{i1}^T \psi + g_1. \tag{22}$$

Consider now the problem of minimization of the performance index (15) with respect to u for system (22). The Hamiltonian for this optimal control problem is given by

$$H_1 = (\psi_1, B_1 u) + \sum_{i=1}^r \lambda_i^{-1} \rho_i (\psi_1, C_{i1} G_i^{-1} C_{i1}^T \psi) + (\psi_1, g_1) - (a_1, y) - (Du, u) - (b, u). \tag{23}$$

Here, $\psi_1(t)$ is the adjoint vector satisfying the following adjoint equation and initial condition:

$$\dot{\psi}_1 = a_1(t), \quad \psi_1(T) = -\frac{\partial F_1(y(T))}{\partial y}. \tag{24}$$

Comparing equations (18) and (24), we see that

$$\psi_1 = -\psi(t, y(T)), \quad t \in [t_0, T], \tag{25}$$

where $\psi(t, y(T))$ is defined by (19).

Using expressions (25) for ψ_1, (21) for λ_i, and (4) for ρ_i, we transform the Hamiltonian (23) to the form:

$$H_1 = -(B_1^T \psi, u) - \sum_{i=1}^{r} \lambda_i [(K_i u, u) + (p_i, u) + q_i] - (\psi, g_1)$$

$$-(a_1, y) - (Du, u) - (b, u). \qquad (26)$$

The maximum of the Hamiltonian (26) over u is attained at

$$u = -\frac{1}{2} \left(D + \sum_{i=1}^{r} \lambda_i K_i \right)^{-1} \left(B_1^T \psi + \sum_{i=1}^{r} \lambda_i p_i + b \right). \qquad (27)$$

Thus, the open-loop optimal control $u(t)$ is determined by equation (27), where λ_i is defined by (21) and ψ is given by (19). The "worst" disturbances $v_i(t)$ are specified by equation (20) in which ρ_i is defined by (4). Hence, the controls expressed through the given functions of time and the adjoint vector $\psi(t, y(T))$. The "worst" disturbances are expressed through the same functions and the control. To finalize the solution, we need to determine the terminal state $y(T)$ on which the adjoint vector ψ depends.

Let us substitute the control u from (27) into equation (22) and integrate this equation under the initial condition (14). We obtain

$$y(t) = y^0 + \int_{t_0}^{t} L(\psi, \tau) d\tau, \qquad (28)$$

where

$$L(\psi, t) = B_1 u + \sum_{i=1}^{r} \lambda_i^{-1} [(K_i u, u) + (p_i, u) + q_i] C_{i1} G_i^{-1} C_{i1}^T \psi + g_1, \qquad (29)$$

and u is given by equation (27). Setting $t = T$ in (28), we obtain a system of n algebraic equations

$$y(T) = y^0 + \int_{t_0}^{T} L(\psi(t, y(T)), t) dt \qquad (30)$$

for the n-vector $y(T)$. In general, this system of equations is very complicated, and one cannot ensure the existence and uniqueness of its solution $y(T)$. Some particular cases are considered below.

If the unique solution $y(T)$ of system (30) is found, then the adjoint vector ψ, the control u, and the disturbances v_i are successively defined by respective equations (19), (27), and (20). The phase trajectory and the minimax value J^* of the performance index can be found from the respective equations (28) and (15). To return to the original state variables, we are to use the transformation (11).

Thus, we have described the procedure for constructing the optimal open-loop control $u(t)$ for the given initial condition $x(t_0) = x^0$ from (7). This control depending on the initial data can be represented as a function $u = \tilde{u}(t; t_0, x^0)$. To obtain the feedback optimal control, one needs just to let the current state coincide with the initial one. The resulting function

$$u^*(x, t) = \tilde{u}(t; t, x) \tag{31}$$

is a feedback optimal control. Note that if the disturbances differ from the "worst" case, then the feedback control (31) ensures the smaller value of the performance index J compared to its value for the open-loop optimal control.

The results described above are a generalization of results presented in earlier papers [2] and [3], where certain particular cases were considered.

5   Linear-Quadratic Performance Index

Let the function $F(x)$ be linear in x and the performance index (8) have the form

$$J = \int_{t_0}^{T} [(a(t), x(t)) + (D(t)u(t), u(t)) + (b(t), u(t))]dt + (c, x(T)), \tag{32}$$

where c is a given n-vector. Carrying out the transformations described above, we find that equations (18) for the adjoint vector become

$$\dot{\psi} = -a_1(t), \qquad \psi(T) = \Phi^T(T)c. \tag{33}$$

Hence, in case of the performance index (32), the adjoint vector is given by an explicit formula

$$\psi(t) = \int_{t}^{T} a_1(\tau)d\tau + \Phi^T(T)c \tag{34}$$

and does not depend on the vector $y(T)$. Hence, it is not necessary here to solve equations for the vector $y(T)$. This vector, as well as the control $u(t)$ and disturbances $v_i(t)$, are given by the respective equations (30), (27), and (20) as explicit functions of time. The phase trajectory is defined by (28).

Note that in this case the feedback control coincides with the open-loop one. In other words, the optimal control $u(t)$ does not depend on the current state and depends only on time [2].

6   Examples

Consider a simple mechanical system with one degree of freedom subject to the control force u and two different uncertain disturbances v_1 and v_2. The equations of the system are

$$\dot{x}_1 = x_2, \qquad \dot{x}_2 = u + v_1 + v_2. \tag{35}$$

Here, x_1 is the coordinate of the system and x_2 is the velocity. Suppose the disturbance v_1 is due to the errors in the control implementation, whereas v_2 is the external disturbance. The bounds (3) and (4) on the disturbances are specified by

$$|v_1| \le ku^2, \qquad |v_2| \le q, \tag{36}$$

where k and q are given positive constants. The initial conditions (7) for system (35) are

$$x_1(t_0) = x_1^0, \qquad x_2(t_0) = x_2^0. \tag{37}$$

The performance index (8) is taken as follows

$$J = D \int_{t_0}^{T} u^2 dt + F(x(T)), \tag{38}$$

where $D > 0$ is a constant. Four cases 1–4 will be considered in which the function $F(x)$ is given by

$$F^{(1)} = x_1, \quad F^{(2)}(x) = x_2, \quad F^{(3)}(x) = x_1^2/2, \quad F^{(4)}(x) = x_2^2/2. \tag{39}$$

In the notation of Section 3, we have for our examples

$$n = m = 1, \qquad r = 2, \qquad g = a = b = 0,$$

$$A = \left\| \begin{matrix} 0 & 1 \\ 0 & 0 \end{matrix} \right\|, \qquad B = C_1 = C_2 = \left\| \begin{matrix} 0 \\ 1 \end{matrix} \right\|, \tag{40}$$

$$G_1 = G_2 = 1, \quad K_1 = K, \quad q_2 = q,$$

$$K_2 = p_1 = p_2 = q_1 = 0, \quad \rho_1 = Ku^2, \quad \rho_2 = q.$$

Following the procedure of Section 3, we introduce the fundamental matrix Φ of system (35) and the inverse matrix:

$$\Phi(t) = \left\| \begin{matrix} 1 & t - t_0 \\ 0 & 1 \end{matrix} \right\|, \qquad \Phi^{-1}(t) = \left\| \begin{matrix} 1 & t_0 - t \\ 0 & 1 \end{matrix} \right\|. \tag{41}$$

Using equations (40) and (41), we obtain from (13) and (16)

$$B_1 = C_{11} = C_{21} = \left\| \begin{matrix} t_0 - t \\ 1 \end{matrix} \right\|, \qquad g_1 = a_1 = 0. \tag{42}$$

Denote

$$\varphi(t) = (t_0 - t)\psi_1(t) + \psi_2(t), \tag{43}$$

where ψ_1 and ψ_2 are the components of the adjoint vector $\psi(t)$. Substituting formulas (40) and (42) into equations (20), (21), and (27) and using the denotation (43), we have

$$v_1 = Ku^2|\varphi|^{-1}\varphi, \quad v_2 = q|\varphi|^{-1}\varphi, \quad \lambda_1 = \lambda_2 = |\varphi|, \quad u = (D+K|\varphi|)^{-1}(\varphi/2). \tag{44}$$

Using formulas (40) and (42)–(44), we obtain from (29)

$$L(\psi, t) = \left[q - \frac{1}{4K} + \frac{D^2}{4K(D+K|\varphi|)^2}\right] \frac{\varphi}{|\varphi|} \left\|\begin{array}{c} t_0 - t \\ 1 \end{array}\right\|, \tag{45}$$

The adjoint vector $\psi(t)$ is defined by equations (19) and (16). Taking into account equations (41) and (42), we obtain

$$\psi^{(1)} = \left\|\begin{array}{c} 1 \\ T - t_0 \end{array}\right\|, \quad \psi^{(2)} = \left\|\begin{array}{c} 0 \\ 1 \end{array}\right\|, \quad \psi^{(3)} = \eta \left\|\begin{array}{c} 1 \\ T - t_0 \end{array}\right\|,$$

$$\psi^{(4)} = \left\|\begin{array}{c} 0 \\ y_2(T) \end{array}\right\|, \quad \eta = y_1(T) + (T - t_0)y_2(T), \tag{46}$$

where superscripts correspond to cases 1–4 in (39). All functions $\psi^{(i)}$ are constant, $i = 1, 2, 3, 4$.

The function $\varphi(t)$ defined by (43) is equal to

$$\varphi^{(1)} = T - t, \quad \varphi^{(2)} = 1, \quad \varphi^{(3)} = (T - t)\eta, \quad \varphi^{(4)} = y_2(T) \tag{47}$$

for the respective cases 1–4 in (39).

For cases 1 and 2, the function $F(x)$ in (39) is linear. In these cases, according to Section 5, one need not solve equations (30) for the vector $y(T)$. The control $u(t)$, both the open-loop and feedback one, as well as the "worst" disturbances $v_1(t)$ and $v_2(t)$, are determined by equations (44), into which the respective functions $\varphi^{(1)}$ and $\varphi^{(2)}$ should be inserted for the respective cases 1 and 2. The controls for these cases are

$$u^{(1)}(t) = \frac{t - T}{2[D + K(T - t)]}, \quad u^{(2)}(t) = -\frac{1}{2((D + K)}. \tag{48}$$

By substituting (45) and (47) into equation (28) and integrating, we obtain the phase trajectory for case 1 as follows

$$y_1^{(1)}(t) = y_1^0 - \left(q - \frac{1}{4K}\right)\frac{(t - t_0)^2}{2}$$

$$+ \frac{D^2}{4K^3}\left[\ln\frac{D + K(T - t_0)}{D + K(T - t)} - \frac{K(t - t_0)}{D + K(T - t)}\right],$$

$$y_2^{(2)}(t) = y_2^0 + \left(q - \frac{1}{4K}\right)(t - t_0) \tag{49}$$

$$+ \frac{D^2(t - t_0)}{4K[D + K(T - t)][D + K(T - t_0)]}.$$

Similarly, for case 2 we have

$$y_1^{(2)} = y_1^0 - \kappa_2(t - t_0)^2/2, \quad y_2^{(2)} = y_2^0 + \kappa_2(t - t_0),$$
$$\kappa_2 = q - \frac{1}{4K} + \frac{D^2}{4K(D + K)^2}. \tag{50}$$

To determine the controls and disturbances for cases 3 and 4, we substitute expressions (47) into equations (44). We obtain

$$u^{(3)} = \frac{\eta(t - T)}{2[D + K|\eta|(T - t)]}, \quad u^{(4)} = -\frac{y_2(T)}{2(D + K|y_2(T)|)}, \tag{51}$$

where η is defined in (46). The control $u^{(4)}$ in case 4 is constant.

Phase trajectories $y^{(3)}(t)$ and $y^{(4)}(t)$ for cases 3 and 4 are obtained similarly to (49) and (50). As a result, we have

$$y_1^{(3)}(t) = y_1^0 - \left(q - \frac{1}{4K}\right)\frac{\eta(t - t_0)^2}{2|\eta|}$$
$$+ \frac{D^2}{4K^3\eta|\eta|}\left[\ln\frac{D + K|\eta|(T - t_0)}{D + K|\eta|(T - t)} - \frac{K|\eta|(t - t_0)}{D + K|\eta|(T - t)}\right],$$
$$y_2^{(3)}(t) = y_2^0 + \left(q - \frac{1}{4K}\right)\frac{\eta(t - t_0)}{|\eta|}$$
$$+ \frac{D^2\eta(t - t_0)}{4K|\eta|[D + K|\eta|(T - t)][D + K|\eta|(T - t_0)]},$$
$$y_1^{(4)}(t) = y_1^0 - \kappa_4(t - t_0)^2/2, \quad y_2^{(4)}(t) = y_2^0 + \kappa_4(t - t_0), \tag{52}$$
$$\kappa_4 = \left\{q - \frac{1}{4K} + \frac{D^2}{4K[D + K|y_2(T)|]^2}\right\}\frac{y_2(T)}{|y_2(T)|}.$$

Expressions (51) and (52) depend on the unknowns η and $y_2(T)$ for the respective cases 3 and 4. To find these unknowns, we are to solve equations (30) for the vector $y(T)$. It occurs, however, that two equations (30) can be reduced to one equation in both cases 3 and 4.

In case 3, calculating η according to formulas (46) and (52), we obtain

$$\eta = y_1^0 + (T - t_0)y_2^0 + \left(q - \frac{1}{4K}\right)\frac{\eta(T - t_0)^2}{2|\eta|}$$
$$+ \frac{D^2}{4K^3\eta|\eta|}\ln\left[1 + \frac{K|\eta|(T - t_0)}{D}\right] - \frac{D^2(T - t_0)}{4K^2\eta[D + K|\eta|(T - t_0)]}. \tag{53}$$

Denote

$$Y = y_1^0 + (T - t_0)y_2^0, \quad z = KD^{-1}|\eta|(T - t_0). \tag{54}$$

Then equation (53) can be reduced to the following one:

$$\text{sign}\eta \left[\frac{Dz}{K(T-t_0)} + \frac{(T-t_0)^2}{4K}\Psi(z) - \frac{q(T-T_o)^2}{2} \right] = Y, \qquad (55)$$

where

$$\Psi(z) = \frac{2+z+z^2}{2z(1+z)} - \frac{\ln(1+z)}{z^2}. \qquad (56)$$

The function $\Psi(z)$ defined by (56) is positive and grows monotonically from 0 to 0.5 as z changes from 0 to ∞, see Fig. 1. Hence, the function

$$w(z) = \frac{Dz}{K(T-t_0)} + \frac{(T-t_0)^2}{4K}\Psi(z) \qquad (57)$$

is also positive and monotone; it changes from 0 to ∞ as z grows from 0 to ∞. Equation (55) can be rewritten as follows:

$$\text{sign}\eta[w(z) - q(T-t_0)^2/2] = Y. \qquad (58)$$

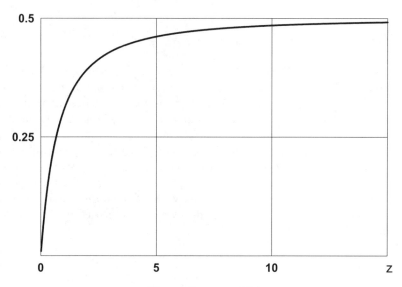

Fig. 1. Function $\Psi(z)$

A priori two possibilities should be considered: either $\text{sign}\eta = \text{sign}Y$, or $\text{sign}\eta = -\text{sign}Y$. It follows from (20), (42), and (46) that for case 3 the signs of the disturbances v_1 and v_2 coincide with the sign of η. Hence, the second possibility implies that the signs of v_1 and v_2 are opposite to the sign of $Y = y_1^0 + (T - t_0)y_2^0$. The latter quantity, according to (11) and (41), is equal to the terminal value $x_1(T)$ in the absence of disturbances and control.

Therefore, the second possibility means that the disturbances act in such a way as to decrease $x_1^2(T)$. Thus, the second possibility does not correspond to the "worst" disturbances, and, hence, to the minimax of the performance index (38) in case 3. Hence, only the first possibility corresponds to the minimax of J, and it follows from (58) that

$$\text{sign}\eta = \text{sign}Y, \qquad w(z) = |Y| + q(T - t_0)^2/2. \tag{59}$$

The first equality (59) defines the sign of η, whereas the second one is an equation for z. Since $w(z)$ is a monotone function changing from 0 to ∞, there exists a unique solution for z, which, according to the second equality (54), defines the absolute value of η. Thus, equations (59) define a unique solution for η in case 3.

In case 4, we obtain from (52)

$$\text{sign}\zeta \left[|\zeta| - q(T - t_0) + \frac{|\zeta|(2D + K|\zeta|)(T - t_0)}{4(D + K|\zeta|)^2} \right] = y_2^0, \tag{60}$$

where $\zeta = y_2(T)$. Similarly to case 3, it can be shown that

$$\text{sign}\zeta = \text{sign}y_2^0, \quad |\zeta| + \frac{|\zeta|(2D + K|\zeta|)(T - t_0)}{4(D + K|\zeta|)^2} = |y_2^0| + q(T - t_0). \tag{61}$$

The first equality (61) defines the sign of $\zeta = y_2(T)$, and the second one is an equation for $|\zeta|$. The left-hand side of this equation is a monotone function of $|\zeta|$ that changes from 0 to ∞ as $|\zeta|$ grows from 0 to ∞. Hence, there exists a unique solution $|\zeta|$ of this equation. Therefore, $\zeta = y_2(T)$ is defined uniquely by (61).

For all cases 1–4, the optimal controls and phase trajectories are defined by equations (48)–(52). The quantities η and ζ for the respective cases 3 and 4 are determined uniquely. The "worst" disturbances can be easily obtained by means of equations (20). The minimax value of the performance index J^* and the feedback controls can be found following the procedure described in Section 4.

7   Conclusions

Optimal control is obtained for a class of linear dynamical systems subject to unknown but bounded disturbances. Under certain assumptions about the performance index and constraints imposed on the disturbances, the solution of the corresponding minimax problem is found in an explicit form. The magnitude of possible disturbances is supposed to depend on the magnitude of the control and grow with the latter. This model of disturbances can describe errors and inaccuracies in the control implementation that occur in engineering applications.

Acknowledgments. The work was supported by the Russian Foundation for Basic Research (Project 05–01–00647) and the Grant for Russian Scientific Schools (NSh 1627.2003.1).

References

1. Pontryagin, L.S., Boltyanski, V.G., Gamkrelidze, R.V., and Mishchenko, E.F. (1962): Mathematical Theory of Optimal Processes. Wiley-Interscience, New York
2. Chernousko, F.L. (2002): Minimax Control for One Class of Systems Under Disturbances. Doklady Mathematics 65, 2, 310–313
3. Chernousko, F.L. (2004): Optimal Control for a Class of Systems Subjected to Disturbances. Journal of Applied Mathematics and Mechanics 68, 4, 503–510.

Uncertainties in Medical Processes Control

A.G.Nakonechny[1] and V.P.Marzeniuk[2]

[1] Department of System Analysis and Decision Making Theory, Kiev National University, Kiev 01127, UKRAINE
[2] Department of Medical Informatics, Ternopil Medical University, Ternopil 46001, UKRAINE

Abstract. Models describing diseases and pathologic processes in particular are considered. There are presented basic uncertainties arising in such systems. There is shown why is it so necessary to take into account these uncertainties.

Keywords: Disease, Pathologic Process, Uncertainty, Nonlinear Differential Equations

1   Introduction

Here we would like to present our results in field of application of system analysis methods to problem of clinical medicine. We emphasize effects of uncertainty that should be taken into account in such complex systems. It will be shown that even considering deerministic models of such nonlinear systems we see different qualitative behavior closely dealt with parameters values. Let's start from origin of such a problem. Nowadays there are obtained a lot of models describing physiological indexes of human body at different diseases and treatment schemes. Primarily they are based on regression analysis. More complex ones use neural networks and evolutionary programming. The most significant attempts to construct mathematical models at different levels of hierarchy of human organism were made by John Murray [3], Keener and Sneyd [2], G.I.Marchuk [1], Mackey and Glass (they investigated nonlinear phenomina applying dynamic systems and introduced notion of dynamic diseases). Without considering uncertainty all these models can be applied for patients from determined groups (primarily for given age and a lot of another restrictions).

As for projects stimulating given research we would like to note the following. During the last years our Departments are fulfilling investigations initiated by Healthcare Ministry of the Ukraine in order to develop and use general system analysis algorithm to study different diseases [4] - [9]. Namely, in fields of oncology (melanoma, leukemia), infectious diseases (toxic colitis), therapy (bone tissue diseases). Naturally there arises a problem to develop a general model for disease. It is incorrect to state that we managed offering unique universal algorithm to construct disease general model at whole. More correct is to say this approach can be used for diseases of different nature. We believe this approach can be extended to processes in sociology and demography as well as for economy and finance branches tasks. A lot of them have the same nature as human diseases. Let's pay attention on special medical terminology necessary (as small as possible). First of all, the most recognized definition of disease states that disease is a set of pathologic processes weakening vitality and activity of a human organism. Here pathologic process is a set of pathologic (that is not normal) and protectoral reactions within human organism. That is, the most significant is modeling pathologic process.

Based on these reasonings we offered general model for pathologic process including three counterparts

(i) the *reason* or cause of disease (it may be some external factor (like bacteria, chemicals) or own modified cells (tumor cells);

(ii) *immune system* supports organism with help of specific antibodies (sort of predators) and plasmatic cells (their ancastors);

(iii) *normal cells*, tissues and organs (it is necessary to consider them to satisfy to some constraints of toxicity).

For these researches we used our own software - Software Environment for Medical System Researches (SEMSR). Demo-version is located on http: // www.tdma.edu.te.ua /data/structure /med-inf/medicalsystemresearches/ medicalscientificinvestigations.htm. There is developed conceptual model of software environment of system medical investigations support. Implementing it there is offered model of data structure in branch of system medical investigations and invented in terms of XML- technology. There is developed interface which is Web-integrated, user-oriented and adjustable. There are implemented mathematical methods of system analysis of pathologic processes in form of Java-classes hierarchy. There are developed software tools to execute system medical investigations, to prepare results obtained for presentation in Internet and visualization.

2   Generalized Pathologic Process

Immunity is the system of supervision the basic function of which is management of cells proliferation and death of mutated cells. Advantages of immunology confirmed idea of F.Bernet offered firstly in 1959 that anti microbe action is only partial appearance of immunity. Thus infectious immunology became the base of origin of new branch of scientific knowledge, namely, noninfectious immunology. One of the important directions of it is study of tumor immunity. Such immunity depends on tumor reason (virus, chemicals, random tumors). Immunity is specific as for viruses causing tumor (DNA- or RNA- including viruses). It arises in a few days or even hours after viruses injection and is continued during months. Immunity for tumors caused by chemicals is less than for viruses and the least is immunity to the cells of random tumors. Model considered is based on the following ideas on tumor immunity.

Immune system generates immune response (cell-like, due to cytotoxic T- lymphocytes and due to antibodies, e.g., specific IgG and IgM). Immune reactions are induced by specific tumor antigen which can be found in the different parts of tumor cell (primarily on surface).

In the following general model we do the following assumptions.

1. Populations of tumor cells, antibodies and plasmatic cells are inhomogeneous.

2. Change of number of population of tumor cells is due to generalized Gompertzian dynamic laws. That is, we have rapid growth at small population sizes and slow one when approaching carrying capacity.

3. Immune reaction is induced by tumor antigen and specific antibodies.

4. Concentration of tumor antigens at instant is proportional to number of tumor cells of corresponding pool.

5. Tumor cells exert negative influence on increase of antibodies population.

6. Immune protection potential and treatment toxicity is measured by concentration of bone marrow cells measured by bone mineral density.

7. Change of bone mineral density is due to logistic dynamic.

To get mathematical model we use well-known population dynamics techniques. All mathematical and biological steps used to obtaine equations are described in [7]. Thus we get the following system of differential equations

$$
\begin{aligned}
\frac{dL_{P_i}(t)}{dt} &= \left[\left\{ 1 - \alpha_i - \sum_{s=1}^{M} (\mu_{P_i,P_s} + \mu_{P_i,C_s}) \right\} L_{P_i}(t) + \sum_{s=1}^{M} \mu_{P_s,P_i} L_{P_s}(t) \right] G_i + \\
&+ \beta_i L_{C_i}(t) - \delta_{P_i} L_{P_i}(t) - \sum_{k=1}^{m} \left(\kappa_{P_i,k} + \sum_{s=1}^{M} \mu_{P_i,P_s,k} \right) c_k(t) L_{P_i}(t) + \\
&+ \sum_{k=1}^{m} \left(\sum_{s=1}^{M} \mu_{P_s,P_i,k} c_k(t) L_{P_s}(t) + \gamma_{C_i,P_i,k} c_k(t) L_{C_i}(t) \right), \\
G_i &= \log \frac{\Theta_i}{\sum_{s=1}^{M} L_{P_i} + \sum_{s=1}^{M} L_{C_i}}, i = \overline{i,M},
\end{aligned}
\tag{1}
$$

$$
\begin{aligned}
\frac{dL_{C_i}(t)}{dt} &= \left[\alpha_i L_{P_i}(t) + \sum_{s=1}^{M} \mu_{P_s,C_i} L_{P_s}(t) \right] G_i - \\
&\beta_i L_{C_i}(t) - \delta_{C_i} L_{C_i}(t) - \sum_{k=1}^{m} (k_{C_i,k} + \gamma_{C_i,P_i,k}) c_k(t) L_{C_i}(t), i = \overline{1,M},
\end{aligned}
\tag{2}
$$

$$
\begin{aligned}
\frac{dC_{P_i}(t)}{dt} &= \xi(m)\alpha_{P_i} L_{P_i}(t - \tau_{C_{P_i}}) F_{P_i}(t - \tau_{C_{P_i}}) - \mu_{C_{P_i}}(C_{P_i} - C_{P_i,0}) + \\
&+ b_{C_{P_i}} \rho(t) - \sum_{k=1}^{m} \beta_{C_{P_i},k} c_k(t) C_{P_i}(t), i = \overline{1,M},
\end{aligned}
\tag{3}
$$

$$
\begin{aligned}
\frac{dC_{C_i}(t)}{dt} &= \xi(m)\alpha_{C_i} L_i(t - \tau_{C_{C_i}}) F_{P_i}(t - \tau_{C_{C_i}}) - \mu_{C_{C_i}}(C_{C_i} - C_{C_i,0}) + \\
&+ b_{C_{C_i}} \rho(t) - \sum_{k=1}^{m} \beta_{C_i,k} c_k(t) C_i(t), i = \overline{1,M},
\end{aligned}
\tag{4}
$$

$$\frac{dF_{P_i}(t)}{dt} = b_{f_{P_i}} C_{P_i} - (\mu_{f_{P_i}} + \eta_{P_i} \gamma_{L_{P_i}} L_{P_i}(t)) F_{P_i}(t), i = \overline{1, M}, \qquad (5)$$

$$\frac{dF_{C_i}(t)}{dt} = b_{f_{C_i}} C_{C_i} - (\mu_{f_{C_i}} + \eta_{C_i} \gamma_{L_{C_i}} L_{C_i}(t)) F_{C_i}(t), i = \overline{1, M}, \qquad (6)$$

$$\frac{dm(t)}{dt} = \sum_{i=1}^{M} \sigma_{P_i} L_{P_i}(t) + \sum_{i=1}^{M} \sigma_{C_i} L_{C_i}(t) - \mu_m m(t), \qquad (7)$$

$$\frac{d\rho(t)}{dt} = b_\rho \rho(t)(\overline{\rho} - \rho(t)) - \\ - \sum_{i=1}^{M} d_{P_i} C_{P_i} - \sum_{i=1}^{M} d_{C_i} C_{C_i} - \sum_{k=1}^{m} \beta_{\rho,k} c_k(t) \rho(t). \qquad (8)$$

There are used the following denotions for tumor cells subpopulations: $C_{P_i}(t)$ and $C_{C_i}(t)$, $\overline{1, M}$ are concentrations of plasmatic cells producing antibodies specific as for P_i and C_i respectively (we designate their concentrations through $F_{P_i}(t)$ and F_{C_i}; $m(t)$ is level of organ damage, $\rho(t)$ is bone mineral density.

Physical meaning of these equations is the following. Equations (1) and (2) are equations of generalized Gompertzian dynamics describing growth of proliferating and clonogenic tumor cells. Equations (3) and (4) describe growth of populations of corresponding antibodies. Equation (7) describes change of level of organ damage; (8) is for change of bone mineral density.

Below we explain model coefficients which are assumed to be deterministic ones. α_{P_i}, α_{C_i} are coefficients indicating chances of meeting antigen-antibody; $\mu_{C_{P_i}}$, $\mu_{C_{C_i}}$ are coefficients inverse to lifetime of plasmatic cells; $b_{f_{P_i}}$, $b_{f_{C_i}}$ are rates if production of antibodies by one plasmatic cell; $\mu_{f_{P_i}}$, $\mu_{f_{C_i}}$ are coefficients inversely proportional to antibodies decay time; $\eta_{f_{P_i}}$, $\eta_{f_{C_i}}$ are numbers of specific antibodies required for neutralization of one antigen; σ_{P_i}, σ_{C_i} are coefficients indicating rates of cells death because of damaging action of antigen; μ_m is coefficient indicating regeneration rate of organ damaged; $\tau_{C_{P_i}}$, $\tau_{C_{C_i}}$ are delays (time required for a maturation of plasmatic cells cascade); $b_{C_{P_i}}$, $b_{C_{C_i}}$ are production rates of plasmatic cels per unit of bone density; $\beta_{\rho,k}$ is coefficient of decrease of bone mineral density due to toxic action of k-th cytotoxic agent; $\xi(m)$ is continuous nonincreasing function ($0 \le \xi(m) \le 1$) characterizing deviations of normal functioning immune system because of damage of target-organ.

Parameters listed above are positive and specific both as for organ and concrete organism.

Here, $L_{P_i}(t)$, $L_{C_i}(t)$, $C_{P_i}(t)$, $C_{C_i}(t)$, $F_{P_i}(t)$, $F_{C_i}(t)$, $m(t)$, $\rho(t) \in C^1[t_0, \infty)$, $i = \overline{1, M}$ and $c_k(t)$, $k = \overline{1, m}$ are piece-wise functions with values $0 \le c(t) \le 1$ (we can assume it after normalization).

There are given continuous initial conditions at $t \in [t_0 - \tau^*, t_0]$, $\tau^* = max_{i=\overline{1,M}} \left\{ \tau_{C_{P_i}}, \tau_{C_{C_i}} \right\}$:

$$L_{P_i}(t) = L_{P_{i,0}}(t), L_{C_i}(t) = L_{C_{i,0}}(t), C_{P_i}(t) = C_{P_{i,0}}(t), C_{C_i}(t) = C_{C_{i,0}}(t),$$
$$F_{P_i}(t) = F_{P_{i,0}}(t), F_{C_i}(t) = F_{C_{i,0}}(t), m(t) = m_0(t), \rho(t) = \rho_0(t).$$
$$(9)$$

In the work [9] there were developed mathematical methods of system analysis of models for which (1)-(9) is generalization. They include parameters identification and state estimation, stability problems, controllability and optimal control construction, nonlinear dynamics. We refer to this model as for generalized model of pathogenic process.

3   Simplified Model

There is shown one of the simplest model developed. We do the following assumption. Populations of tumor cells, antibodies, and plasmatic cells are homogenous. That is, we dont consider tumor and immune system cells subpopulations.

Thus, in the absence of cytotoxic agent we get the following system of differential equations:

$$\frac{dL(t)}{dt} = \alpha_L L(t) ln \frac{\theta_L}{L(t)} - \gamma_L F(t) L(t), \tag{10}$$

$$\frac{dC(t)}{dt} = \xi(m)\alpha L(t - \tau)F(t - \tau) - \mu_C(C - C_0) + b_C \rho(t), \tag{11}$$

$$\frac{dF(t)}{dt} = b_f C - (\mu_f + \eta\gamma_L L(t))F(t), \tag{12}$$

$$\frac{dm(t)}{dt} = \sigma L(t) - \mu_m m(t), \tag{13}$$

$$\frac{d\rho(t)}{dt} = b_\rho \rho(t)(\overline{\rho} - \rho(t)), \tag{14}$$

Where $L(t)$ is a number of tumor cells, $C(t)$ is concentration of plasmatic cells, $F(t)$ is concentration of antibodies, $m(t)$ is level of organ damage, $\rho(t)$ is bone mineral density. There are given initial conditions at $t \in [t_0, -\tau, t_0]$:

$$L(t) = L_0(t), C(t) = C_0(t), m(t) = m_0(t), \rho(t) = \rho_0(t). \tag{15}$$

Furthermore, when analyzing effect of uncertainty of parameters we will consider model of simplified pathologic process at the following parameter values:

$$\alpha_L = 0.0396, \gamma_L = 4 * 10^{-3}, \theta_L = 0.001,$$

$$\alpha = 10^4, \mu_c = 0.5, \rho = 0.17, \mu_f = 0.17, \eta = 10, \mu_m = 0.12,$$

$$\xi(m) = \begin{cases} 1, m \le 0.1, \\ (1-m)/(10/9), 0.1 \le m \le 1. \end{cases}$$

At $t \in [-\tau, 0]$ there hold initial conditions

$$L(t) = max(0, x + 10^{-6}), C(t) = 1, F(t) = 1, m(t) = 0.$$

These values of parameters correspond to G.I. Marchuk model [1] considering immune response with help of specific IgG antibodies.

From mathematical viewpoint model includes nonlinear differential equations with delays in state. The first equation describes so-called Gompertzian dynamics. Model mentioned here was considered at some control, that is, treatment (including therapeutic and surgery interventions). Note, control was considered in a class of generalized functions. It is essential, for example, because of short-term action of drug injection.

4   Uncertainties

Uncertainties in such models may be parametric. Some of the parameters may be unknown functions. As for uncertainty in control it is necessary to take into account all possible scenarios. Note, the purpose of this article is not to present methods to identify these uncertainties. For these purpose we need to present powerful and deep mathematical apparatus of adjoint systems, sensitivity functions and minimax aposteriorial estimation. Here we would like to answer two questions

(i) why is it so important to take into account uncertainties?

(ii) the basic uncertainties in models of diseases.

When answering the first question we should say that as it will be shown below even mathematical solutions of equations (10)-(15) have different qualitative behavior. In practice we can observe different forms of disease (subclinical, acute, chronic, lethal). Search of treatment scheme is dependent on such forms.

In our researches we investigated uncertainties in the following issues: maturation time for plasmatic cells τ, influence of antigen on target-organ damage rate σ, relation between target-organ damage rate and immune response $\xi(m)$, therapy scheme (polychemiotherapy, radiotherapy), surgery interventions. Note, the three last ones are non- parametric. They depend on unknown function like controller.

Let's consider uncertainty in value of maturation time for plasmatic cells. Consider behaviour of system if $\sigma = 200$ and we have uncertainties in the value of maturation time $\tau \ge 0$. In case $0 \le \tau < \tau_1$, where $\tau_1 \approx 4.1$ pathologic process has chronic form resulting to some stationary state of disease (it correspond to attractor - stable focus). At $\tau = \tau_1$ pathologic process has chronic form resulting to some dynamic state of disease (mathematically this phenomenon corresponds to Hopf bifurcation - we get limit cycle). Increasing τ we get chronic form with lethal results (it corresponds to unstable limit cycle). At $5.5 < \tau < 6.9$ form of pathologic process is unpredictible. This phenomenon is called in chaos theory as strange attractor. We have some irregular oscillations. Note, their behavior is similar for tumor cells, organ damage level, plasmatic cells and antibodies. At $7 < \tau \leq 16.31648422$ we return to chronic form with some periodic state, which at $\tau \approx 16.3165$ transforms to stationary one.

Influence of pathogenic factor on organ damage in simplified pathologic process model is described by coefficient σ. Consider behavior of model if $\tau = 5$ and we have uncertainty in value of σ. If $0 \leq \sigma < \sigma_1$, where $\sigma_1 \approx 100$ we have chronic form tending to some cyclic state (i.e., stable limit cycle). At $\sigma = \sigma_1$ it transforms to some lethal form (unstable limit cycle). At $\sigma = \sigma_2 \approx 150$ chronic form of pathologic process tends to stationary state (stable node). At $\sigma = \sigma_3 \approx 312.5$ we have unpredictable behavior of pathologic process (strange attractor). Mathematical reason is period doubling bifurcation.

As an example of uncertainty in the treatment scheme let us consider one concrete disease. Melanoma is tumor for which treatment there are applied gamma-therapy, X- ray therapy, polychemotherapy and surgery interventions. To simulate it we use the following model

$$\frac{dP(t)}{dt} = \alpha_L P(t) \ln \frac{\theta_L}{P(t)} - \delta(t - t_o) \kappa_o(P(t)) P(t)$$
$$- \sum_{i=1}^{n} \delta(t - t_i^\gamma) \rho(f_i^\gamma, P(t)) P(t) - \sum_{i=1}^{m} \delta(t - t_i^x) \rho(f_i^x, P(t)) P(t)$$
$$- \sum_{i=1}^{k} \delta(t - t_i^c) \rho_c(P(t)) P(t),$$

$$P(0) = 1.$$

Here $P(t)$ is number of proliferating cells at instant t, α_L is growth rate of melanoma (days-1); θ_L is maximal number of proliferating cells which organism can carry (carrying capacity); t_o is moment of surgery intervention (day); $\kappa_o(P(t))$ is probability of sterilization as a result of surgery intervention; t_i^γ is moment of gamma-therapy (day); f_i^γ is one dose of gamma-therapy (Gy); t_i^x is instant of X-ray therapy (day); f_i^x is one dose of X-ray therapy (Gy); $\rho(f, P)$ is probability of sterilization as a result of radiation of dose f (Gy) of P proliferating cells of melanoma which is calculated due to PS $= \exp\{-P \exp\{-0.255 f\}\}$; t_i^c is instant of polychemotherapy (day); $\rho_c(P)$ is probability of sterilization as a result of polychemotherapy at presence of P proliferating melanoma cells. Even for this model considering unique melanoma cells population we deal with a lot of parameters that depend on particularities of concrete patient. Analyzing solution of such a model we concluded essential influence of these unknown parameters on form of trajectory.

5   Conclusions

So, even without considering probabilistic nature of the most of quantities and parameters we saw the complex qualitative behavior of diseases models depending on parameters and controllers. At different values of these quantities we observed subclinical, acute, chronic or lethal forms of pathologic processes.

Taking into account complexity of mathematical equations (nonlinear systems with delays) requires appearance of new powerfull methods of exact parameter identification and qualitative analysis.

From viewpoint of theoretical medicine uncertainties arising in models of diseases require to develop treatment schemes that are effective, take into account toxicity constraints, enable life quality, cost benefit.

In future works our idea is to compare behavior of pathologic processes using both deterministic and stochastic models and to extend such models to demographic processes and finance branch.

References

1. Mathematical modelling in immunology and medicine. - Proc. of the IFIP TC-7 Working Conf., Moscow, USSR, 5-11 July 1982, Ed. by G.I.Marchuk, L.N.Belykh. Amsterdam, New York, Oxford: North-Holland, 1983.
2. Keener, J., Sneyd, J. Mathematical Physiology. New York: Springer Verlag,1998
3. Murray J.M. Mathematical Biology. New York: Springer-Verlag, 1989
4. V.P.Martsenyuk, On the Problem of Chemotherapy Scheme Search Based on Control Theory. Vol. 35/4 (2003) - Journal of Automation and Information Sciences
5. V.P.Martsenyuk, On Hopf Bifurcation and Periodic Solutions in G.I.Marchuk Model of Immune Protection - Vol. 35/8 (2003) - Journal of Automation and Information Sciences
6. V.P. Marzeniuk, Taking Into Account Delay in the Problem of Immune Protection of Organism, Nonlinear Analysis: Real World Applications, Vol 2/4 (2001), pp. 483-496.
7. A. G. Nakonechnyi, V. P. Martsenyuk, Controllability Problems for Differential Gompertzian Dynamic Equations, Cybernetics and Systems Analysis 40 (2): 252-259, 2004
8. V. P. Martsenyuk, On Stability of Immune Protection Model with Regard for Damage of Target Organ: The Degenerate Liapunov Functionals Method, Cybernetics and Systems Analysis 40 (1): 126-136, 2004
9. V.P.Marzeniuk, Qualitative analysis of human cells dynamics: stability, periodicity, bifurcations, control problems, Advances in Mathematics Researches, vol.5, 2003. - New York: Nova Science Publishers. pp. 137-200.

Part IV

Applications of Stochastic Optimization

Impacts of Uncertainty and Increasing Returns on Sustainable Energy Development and Climate Change: A Stochastic Optimization Approach

A. Gritsevskyi and H.-H. Rogner

International Atomic Energy Agency, Wagramerstrasse 5, A1400 Vienna, Austria

Abstract. In this article we discuss a stochastic optimization model used for evaluation of long-term energy development. The model includes the following features:

1. Increasing returns to scale for the costs of new technologies with uncertain learning rates;
2. Uncertain costs of all technologies and cost/quantities for energy sources, both renewable and depletable;
3. Uncertainties in future energy demands and their volatilities;
4. Uncertain environmental constrains;
5. Clusters of linked technologies that induced technological advances.

In particular, this allows us to identify robust dynamic technology portfolios, which supply (in a sense) potential energy demand, while minimizing adjusted to risks expected costs together with investment and environmental risks. Formally, the discussed problem involves a non-convex, large-scale stochastic optimization model requiring special global optimization technique which takes advantage of the specific structure of the problem.

This article primarily concentrated with main motivations, critical importance of non-convexity (increasing returns) and explicit treatment of uncertainty by using stochastic optimization approach.

1   Introduction

There are deep uncertainties and controversies regarding feasible transitions to sustainable energy systems. Let us outline some of them determining main elements of our model.

Globally, energy resources are plentiful and are unlikely to constrain sustainable development even beyond the 21st century. The fossil resource base is at least 600 times current the fossil fuel use, or 16 times the cumulative fossil fuel consumption between 1860 and 2000. While the availability and costs of fossil fuels are unlikely to impede sustainable development, current practices for their use and waste disposal are not sustainable. Thus, the uncertainty related to the economic and environmental performance of fossil, nuclear, and renewable conversion technologies - from resource extraction to waste disposal - will determine the extent to which an energy resource can be considered sustainable. In particular, a transition to sustainable energy systems that continue to rely predominantly on fossil fuels will depend on the development and commercialization of fossil technologies that do not close their fuel cycle through the atmosphere. Alternatively, the transition will likely require determined policies to move away from fossil fuels within a rather uncertain long time horizon.

Fundamental changes in global energy systems tend to occur slowly. The replacement of traditional energy sources - such as the substitution of coal for fuelwood with the advent of steam, steel, and railways - took almost all of the nineteenth century. The subsequent replacement of coal with oil and gas and associated technologies lasted the better part of the twentieth century. In contrast to these very slow processes of change in the global energy system, for some parts of the energy system change can be more dynamic - especially in the evolution of end-use technologies. However, the fact that fundamental changes occur over many decades rather than a few years means that technological changes that have inherently shorter time horizons need to be consistent with the overall slower processes of change in the energy system. Besides the horizons, the directions of these future transitions are deeply uncertain. Future energy systems could rely on renewable energy sources, on clean coal, on less carbon-intensive fossils such as natural gas, or on nuclear power as well as on their combination that may change over time.

The transition strategies are path dependent, and the choice of a robust long term path requires an appropriate dynamic model. Our analysis shows that endogenous increasing returns and uncertainty will have the greatest impact on the emerging energy system structures during next few decades. Under deep uncertainty, the near-term investment decisions in new technologies which take into account long-term perspectives and risks becomes the most important in determining the direction of long-term development. The analysis also shows that fundamentally different future energy system structures might be reachable with similar overall costs, so future energy systems with low carbon dioxide emissions, and possible those that meet other environmental constraints, need not be associated with costs higher than those of systems with high emissions and not sustainable.

The essential feature of our model is also the following. There are multiple factors that lead to cost reductions. Investments defining transition paths may have uncertain outcomes and are subject to high risk, but at the same time, they may turn out to be highly profitable and be subject to increasing returns to adoption. Actual costs reductions will not only depend on technical potentials but also on actually achieved performance and the diffusion rates which are realized on potential markets. These elements are not only uncertain but also related. Faster and deeper cost reductions may significantly accelerate early market adoption of a technology and, subsequently, may lead to even more dramatic technological improvements.

Accordingly, our model has the structure of a dynamic random network with flows where nodes correspond to positions of different technologies at various time moments. Arrows describe feasible transitions.

Induced technological change is modeled by so-called stochastic learning curves. Technologies are improved with cumulative experience as expressed by the scale of their application. The learning curves indicate how costs and uncertainty decline with increasing scale of application (see Section 3). They reflect the fact that the process of technological learning is uncertain even as cumulative experience increases.

Technologies also are related to each other forming certain clusters of technologies. For example, jet engines and gas turbines for electricity generation are related technologies and the latter were initially derived from the former. These relationships among technologies are frequent and imply that improvement in some of the technologies may be transferable to other related technologies, i.e., they lead to so-called (positive) spillover effects discussed in Section 3.4.

Despite the fundamental importance of technological learning under increasing returns and uncertainty, modeling of these processes has received inadequate attention in the literature. Several reasons may explain the apparent lack of systematic approaches. Among them, the complexity of appropriate modeling approaches is perhaps the most critical. Increasing returns to scale lead to nonconvexities. In conjunction with the treatment of uncertainties, adequate modeling becomes methodologically and computationally very challenging. It requires the development of specific global nonsmooth stochastic optimization techniques, which are only now under development (see, e.g., [4], [5], [16] and [30]). The aim of this article is to discuss main motivations for explicit treatment of uncertainties, critical importance of non-convexities (increasing returns) as well as global stochastic optimization methods. We also are presenting some selected important counterintuitive conclusions that could be learned from our study. We only outlined the implemented random global search method that was used since its formal discussion probably requires a separate paper. In particular, this method utilizes some rather specific features of the numerical model that itself requires a rather detailed and lengthy description of equations and data sets. Section 2 presents the stochastic optimization model. Section 3 discusses details of its structure. Sections 4, 5 outline the implementation and summarize some numerical results.

2   Modeling Approach and a Motivating Example

Any realistic policy in the presence of uncertainties bears a risk. Explicit introduction of these uncertainties and related risks in our model creates a driving force for the development of new technologies necessary for making the energy system flexible enough against possible instabilities and surprises.

The proposed modeling approach relies on so-called two-stage dynamic stochastic optimization model which seems to be rather appropriate for problems with large number of decision variables and uncertain parameters. Conventional approaches of the control theory relying on standard dynamic programming equation are applicable only in the case of a small number of variables. This is also true for multistage stochastic optimization models. Although the large-scale optimization techniques are used in such case instead of the recurrent equations, the actual size of solvable problems with realistic number of scenarios is small again.

2.1   Stochastic Optimization Model

Our two-stage dynamic stochastic optimization model is based on the idea of representing energy systems development as a dynamic network where flows from one energy form to another correspond to energy technologies such as electricity generation from coal or gas power plants. Figure 1 illustrates the assumed reference energy system as composed of about one hundred different technologies. Five different stages of energy flows are shown – energy extraction from energy resources, primary energy conversion into secondary energy forms, transport and distribution of energy to the point of end use that results in the delivery of final energy, and finally the conversion at the point of end use into useful energy forms that fulfill the specified demands (as discussed above). All possible connections between the individual energy technologies are also specified by arrows. Various demands for useful energy are shown for different sectors of the economy. Each technology in the system is characterized by its associated costs, unit size, efficiency, lifetime, emissions, etc.

In addition to various balance constraints, there are limitations imposed by the resource availability as a function of (uncertain) costs. The overall objective is to minimize the total expected discounted energy system costs adjusted to risks to fulfill various demands by the utilization of technologies and resources.

Formally, the problem is formulated as the minimization of function

$$\sum_{t=0}^{T} d^t \mathrm{E}[\langle C^t(x|_0^t, \omega), x^t \rangle + \sum_{t=0}^{T} \langle a^t(\omega), y^t(\omega) \rangle \\ + \rho(\langle C^t(x|_0^t, \omega) - \mathrm{E}C^t(x|_0^t, \omega), x^t \rangle)] \tag{1}$$

subject to constraints of the following type:

$$\sum_{k=0}^{t} A_k(\omega) x^k + B_t(\omega) y^t(\omega) = b^t(\omega), t = 0, 1, \ldots, T, \tag{2}$$

$$x^t \in X_t \subseteq R^n, y^t \in Y_t \subseteq R^m \tag{3}$$

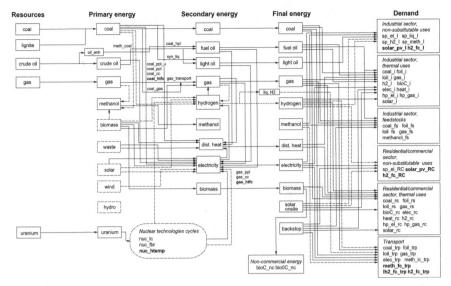

Fig. 1. Schematic diagram of the reference energy system showing some hundred individual technologies. *Source: Gritsevskyi and Nakićenović et al., 2002*

where $x = x|_0^T$, ω defines random (uncertain) variables and d^t is a discount factor at the time t; $x|_0^t = (x^0, x^1, \ldots, x^t)$; $C^t(x|_0^t, \omega)$ are stochastic costs a given technology path $x|_0^t$; matrices $A_t(\omega)$, $B_t(\omega)$ and vectors $b^t(\omega)$ reflect uncertain quantity-to-cost relations for resources, establish link between technology activities and energy demands, characterize different system or environmental constrains; ω denotes uncertain parameters of the model; $\rho(y)$ is a piece–wise linear risk function, e.g., $\rho(y) = \max\{0, y\}$ or as in Markovitz, 1959. This model is a specific case of the dynamic two-stage STO models (see [6]) with the first stage decision vector x, and the second stage $y(\omega) = y_t(\omega)$, $t = 0, 1, \ldots, T$. To model the increasing returns, marginal costs $C^t(\cdot, \omega)$ are represented as decreasing piece-wise constant or piece-wise linear function, therefore expected value $E\left[\langle C^t(x_0^t, \omega), x^t\rangle\right]$ is in general a nonconvex non-smooth vector-function. This function can also be represented (in the case of discrete distributions of ω) by mixed integer linear equation which significantly destroy specific structure of constraints (2)-(3) that is utilized in the developed random search algorithm outlined in Section 4. In details, the structure of constraints and the objective function is discussed in Section 3.

2.2 A Motivating Example

The energy systems we are dealing with are rather complex, so in order to illustrate the importance of stochastic approach let us consider an almost trivial example with only two technologies involved. Figures 2-3 illustrate effects of increasing returns, uncertainty and short-term vs. long-term modeling dilemma. The learning curve (linear function in log-log scale, Figure 2) indicates the change of marginal production costs with the size of adopted (by market) technology and operational experience gained.

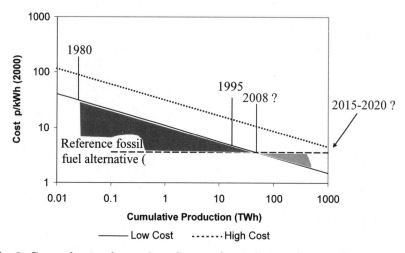

Fig. 2. Cost reduction for on-shore (low cost) vs. off-shore (high cost) wind energy technologies in the EU, production cost vs. cumulative production. *Source: derived from IEA Study of Learning Curves, IEA 2000*

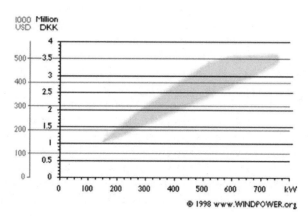

Fig. 3. What does a wind turbine cost? The price banana. *Source: www.windpower.org, 1998*

Figure 2 shows the coexistence of cheap (CCGT, Combined Cycle Gas Turbine) and expensive (wind) technologies. The wind technology slowly takes-off penetrating the market despite higher overall costs. The cost-effective solutions of standard deterministic modeling framework would take only the cheapest technology. The coexistence of these technologies, as in Figure 2, can be justified only within an appropriate time horizon: even if we know in 1980 the cost reduction trends for the wind technology for the coming 20 years precisely, the additional 30 years may not be enough to justify its benefits.

Uncertainties significantly affect time-scales. For example, a slight change in the slope (cost reduction rate) even for best case technology may result in considerable shifts of the breaking point when technology becomes cheaper. This also significantly depends on scenarios for CCGT cost development (say, due to future anticipated increase in gas price and more expensive gas infrastructure to be built) and interdependencies (the portfolio of feasible technologies sharing the same infrastructure or resource).

Stochastic models enable us to justify in a rather natural way the coexistence of cheap and expensive technologies and, hence to justify S-shape characters of technological development and necessity of upfront investments in new technologies and RD&D. For example, the explicit introduction of risk related to the overestimation of costs as in 1 results in coexistence of technologies which are more expensive with respect to average costs.

3   Model Structure

Deterministic IIASA's energy model MESSAGE has been used in stylized form in a number of energy modeling approaches to capture elements of endogenous technological change. This led to a number of important insights (Messner (1995, 1997), Grübler and Messner, 1998, Nakićenović, 1996 and 1997). However, the major drawback was its deterministic character (perfect foresight) so that early investments in new, costly technologies were always rewarded with increasing returns. Yet, it is clear from the illustrative example of Section 2 that increasing returns within a given time interval are possible on average but with a considerable degree of uncertainty. Deterministic models result in unnaturally degenerated sets of technologies without coexistence of technologies even closed with respect to costs.

The basis for introducing uncertainties in the distributions of future costs was the IIASA technology inventory that now contains information on some 1600 energy technologies, on their costs, technical and environmental characteristics (Messner and Strubegger, 1991). Figure 4 shows that the distributions of the investment costs are not symmetric and that they have very pronounced tails with both "pessimistic" and "optimistic" views on future costs per unit capacity. Such cost distributions were introduced explicitly in a simple, stochastic version of MESSAGE and have led to "hedging" against this uncertainty (Golodnikov et al., 1995; Messner et al., 1996). Stochastic versions with increasing returns (Grübler and Gritsevskyi, 1998, Gritsevskyi and Ermoliev, 1999) demonstrated the need for coexistence of technologies

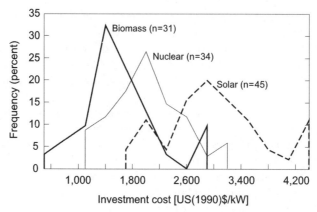

Fig. 4. Range of Future Investment Cost Distributions from the IIASA Technology Inventory for Biomass, Nuclear, and Solar Electricity-Generation Technologies, in US 1990 $ per kilowatt (kW). *Source: Nakićenović et al., 1998*

even if some of them are not rewarding within a given time horizon. Example 1 illustrates that due to the coexistence some of these technologies may become superior in a longer time interval.

Another important outcome was the so-called S-shaped patterns of technological diffusion which were exhibited by optimal dynamic portfolios of technologies without any explicit technology inducement mechanisms other than uncertain increasing returns. The disadvantage of the proposed method was its computational complexity making this method basically infeasible for application with more than few technologies.

The basic approach of Grübler and Gritsevskyi (1998) was retained in the new method (see Gritsevskyi and Nakićenović 2002) enabling to include a hundred technologies by applying parallel computing techniques. Later it was further extended to a multi-actor, multiregion model with uncertain increasing returns (Grübler and Gritsevskyi 2002). A distinguishing feature of the model is that the social planner is replaced by a set of actors, each of which optimizes its own part of a global system while remaining interdependent via negotiated energy and technology trade flows. The newly developed method allowed the introduction of more sophisticated environmental uncertain constrains compared to Grübler and Gritsevskyi (1998). The model which is discussed in this paper containers a large number of variables (from just a few dozen of thousands in relatively simple runs and up to some hundred thousands in case when we used Cray T3E supercomputer). A full detailed description of the specific equations and data sets is lengthy, so only the short outlines of the main blocks of the model are given. A key idea of the approach relies on the network structure of the model and specifics of numerical versions of constraints (2).

3.1   Demands

Future energy demand entering in vector b^t of constraints (2) is highly uncertain. At the same time it is one of the key factors that determine at large extend the structure of future energy systems and environmental impacts. We assume that future demand growth is uncertain, demand randomly fluctuates (could not be perfectly forecast even if growth is know) and, last, quality matters. It is assumed that the regional energy demand is characterized by their energy form, e.g., solids, liquids, and grids. Within each energy form, different combinations of primary resources and conversion technologies can satisfy the demand. In other words, we assume perfect substitutability within each energy form. For instance, the demand for high-quality energy carriers (grids) can be satisfied via electricity generated from coal or from PVs or wind turbines, or alternatively by natural gas. This is reflected by the specific structure of matrix A_t schematically indicated in Figure 1. We also assume asymmetric (and partial) substitutability between energy forms; that is, substitutability is only possible in the direction from high to low exergetic energy forms. Electricity (grids), for instance, can replace coal (solids) as a final energy carrier, but the reverse is not true. All these features are an important improvement of the conventional energy models which have focused only on quantities and prices, largely ignoring energy quality. Quantitatively, the demand is parameterized similar to the peer reviewed long-term energy scenario study reported in IPCC (2000).

3.2   Technologies

The main issue of the model is to capture the different directions of possible future technological change resulting from various technological replacements and incremental improvements that may occur during this century. The basic assumption is that endogenous learning is a function of cumulative experience, measured by cumulative installed capacity, and that this process is uncertain. Although there are many other indicators of technological learning, we chose the cumulative installed capacity because it is relatively easy to measure.

3.3   Uncertain Increasing Returns and Costs

Time horizons of a century or longer are frequently adopted in energy studies. Modeling energy systems developments over such long time horizons imposes a number of methodological challenges. Over longer horizons, technological change becomes fluid and fundamental changes in the energy system are possible. The increasing returns to scale basically mean that technologies are improved with cumulative experience, as expressed by the scale of their application. Accordingly, costs per unit of capacity or output and uncertainty are assumed to decline with increasing scale of application.

Increasing returns, on the other hand, lead to nonconvexities and disequilibrium tendencies by providing positive feedbacks. After (generally large) initial investments in RD&D and early market introduction, the incremental costs of further applications become cheaper and cheaper per unit capacity (or as assumed here, per unit output). Thus, the more widely adopted a technology becomes, the cheaper it becomes (with lower uncertainties, leading to lower risks to adoption). There are many manifestation of this basic principle. One of the better known is the concept of "lock-in": as a technology becomes more widely adopted it tends to increasingly eliminate other possibilities. Related to this concept frequently used in empirical analysis is the so-called technological learning or experience curve: the more experience that is gained with a particular technology, the greater the improvements in efficiency, costs, and other important technology characteristics.

The learning rates are uncertain and are captured by probability distribution functions. We assume that the generic cost reduction function has the following form:

$$CI_t = (2^{-\beta})^{ND_t} \tag{4}$$

where CI_t is the *cost reduction* index, or the ratio between technology unit costs (or more precisely, the annual levelized costs) at time t and initial cost in the base year which depends on already accumulated output by that time; ND_t is the number of doublings of cumulative output achieved by time t compared to the initial output; and β is the random *progress ratio* that indicates the cost reduction rate per doubling of output. It is important to note that the suggested algorithmic approach is not limited to the type of distribution, and, in fact, that it does not require any prior knowledge about the type of the distribution function.

Figure 5 illustrates the uncertain learning index as a function of each doubling (of cumulative output). The expected value for the cost reductions rate is 20% per doubling in the example shown. The numbers between the isolines indicate the probability ranges of occurrence of different learning rates. For example, there is a 50% chance that the cost reductions rate falls between 14 and 25% per each doubling. Let us note that there is a small chance of 5% that the cost reductions would range from very small to actual cost increase and that there is a very small probability of 0.1% that there would be significant cost *increase* per each doubling. This indicates a real possibility of negative learning or "induced forgetting" rather than learning, which illustrates the true risk of investing in new technologies. There is a high chance that technology would improve with accumulated experience, but there is also a small chance that it would be failure and even a smaller chance of a genuine disaster. In the model we extend the application of uncertain learning to many new technologies ranging from wind and photovoltaic to fuel cells and nuclear energy.

It is commonly assumed that traditional, "mature" technologies do not benefit from learning (another interpretation is that cost reductions as the

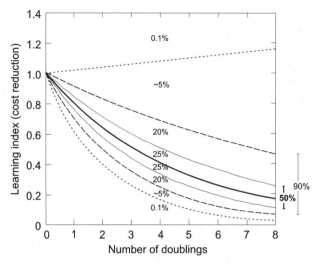

Fig. 5. Uncertain cost reductions represented by the learning index as a function of number of doublings (of cumulative output). The expected value of β (the mean) learning index (rate) corresponds to 20% cost reduction per each doubling (of cumulative output). The numbers between the isolines of different learning indices indicate probability ranges. There is a small probability of no learning at all between any given doubling. *Source: Gritsevskyi and Nakićenović et al., 2002*

result of learning are insignificant compared to other uncertainties that affect costs). We assume that cost distributions of traditional technologies are technologies are *static* over time. For new technologies, due to possible cost reductions to learning (as described above), the costs are defined by conditional probabilities. We assume that all initial cost distribution are *log-normal* with different mean and variance based on the empirical analysis of technological characteristics with the IIASA technology inventory (see Strubegger and Reitgruber, 1995).

More precisely, the cost distribution function for a new technology at any given moment of time t, under the condition that N doublings of cumulative output have been achieved and that the realized value for random learning rate β is equal b, is defined by the following expression:

$$F_t(\zeta | ND_t = N, \beta = b) = F_0(m_t, s_t),$$
$$m_t = m_0(2^{-b})^N, s_t = Km_t,$$

where $F_0(\cdot, \cdot)$ is the initial lognormal distribution function with parameters m_0 and s_0; K defines the spread of the distribution.

In addition, the mean value and the variance of these cost distributions is assumed to decrease with increasing application of new technologies according to the generic cost reduction ratio (4) with *normally* distributed progress ratio. This means that the process of technological learning is uncertain even as cumulative experience increases. The uncertainty of new technologies is characterized by joint probability and scenarios.

The uncertainty associated with magnitudes and costs of energy reserves, resources, and renewable potential and their extraction and production costs is also considered. Following the estimates by Rogner (1997), Nakićenović *et al.* (1995) and others, we assume a very large global fossil resource base corresponding to some 5,000 Gtoe and accordingly large renewable potentials. We also assume that the energy extraction and productions costs are uncertain varying by a factor of more than five. Following the approach proposed by Rogner (1997), we derived aggregate, global, upward-sloping supply curves with uncertain costs.

All distributions for uncertain factors are based on actual technological assessments and supported by empirical data. Nevertheless, they are ultimately subjective views on potentials of the new energy technologies. Therefore, the explicit introduction of different uncertainties and analysis of robust policies with respect to all of them is the main idea of our approach to cope with uncertainties.

3.4   Technological Clusters and Spillover Effect

Technologies are related to one another. For example, we consider the different applications of fuel cells, such as for stationary electricity generation and for vehicle propulsion. We also consider fuel cells that have the same end-use application but different fuels, for example, hydrogen and methanol mobile fuel cells. These fuel cells are different but they are related in the technological sense, so that improvements in one technology may lead to improvements in the other.

In the model, we explicitly consider the possibility of such spillover effects among energy technologies. We simply assume that there are basically two explicit types of spillover effects. One is indirect through the connections among energy technologies within the energy system. For example, cheaper gas turbines mean cheaper electricity, so this could favor electricity end-use technologies for providing a particular energy service compared with other alternatives. The other effect is more direct. Some technologies are related through their proximity from the technological point of view, as was suggested by the example of hydrogen and methanol mobile fuel cells. We explicitly define clusters of technologies, where learning in one technology may spill over into another technology. The spillover effects are assumed to be strong within clusters and weaker across clusters.

Figure 6 gives a schematic diagram of the ten technology clusters and indicates how they are related to each other with respect to the assumed learning spillover effects within the structure of the energy system. Two of the technology clusters are characterized by generally large "unit size" compared to other technologies – nuclear high temperature reactors (HTRs) and infrastructure clusters. Consequently, very large cumulative output is required

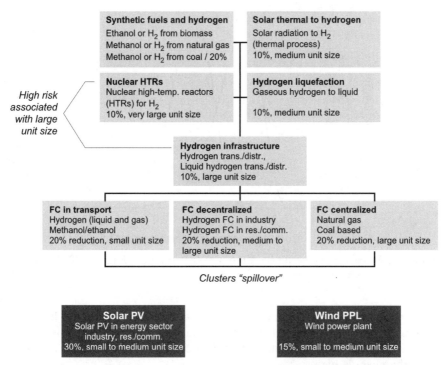

Fig. 6. Schematic diagram of the ten technology clusters and their relationships to each other with respect to the assumed learning spillover effects within the structure of the energy system. Technologies in each cluster are listed together with their assumed expected mean learning rates. *Source: Gritsevskyi and Nakićenović et al., 2002*

for achieving a doubling compared to other clusters. The expected learning rates are indicated for each cluster based on expert opinions. The modular (smaller "unit size") technologies have generally higher mean learning compared to other technologies. The highest mean learning rate is indicated for the photovoltaic cluster. The lowest are shared by the solar-thermal (hydrogen), nuclear (HTR) and infrastructures clusters.

Figure 7 illustrates the spillover effects within one cluster of technologies. There are two density functions of technology costs in 2030 for decentralized fuel cells. The density function with the lower overall costs is for the case of spillover effects within the technology cluster; that with the higher overall costs is for the case without spillover effects. The costs are given in US 1990 $ cents per kilowatt hour (kWh) of electricity generation without the fuel costs. Both the expected costs and their variance are substantially higher without the spillover effects. Thus, the costs, as well as the uncertainty, are expected to be lower with spillover effects. Therefore, the probability of lower costs is overall much higher with spillovers. However, the heavy tail of the density

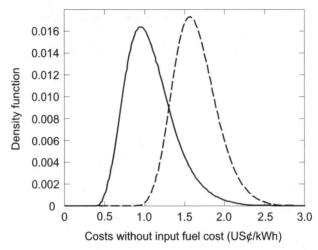

Fig. 7. Spillover Effects within the Cluster of Decentralized Fuel Cell Technologies.
Source: Gritsevskyi and Nakićenović et al., 2002

distribution is more pronounced in the case of spillover effects. This is an interesting feature: the expected costs are generally lower with spillovers, but at the same time the possibility of realizations of very high costs compared with the mean is higher.

3.5   Environmental Constraints and Resources

Resource availability and environmental constraints are treated as uncertain. Resource availability and cost assumptions were derived from Nakićenović et al. (1998) and IPCC (2000). Corresponding resource extraction profiles and costs were based on Rogner (1997). These assumptions ensure convex relationships between the quantity of resources available and their costs and in some sense is simplification of actual situation.

Future environmental regulations and related constrains on energy systems might emerge, influencing technology choices. Such constraints could take the form of either hard quantitative limits or emissions taxes. For the model calculations outlined in the following section, we follow a probabilistic scenario similar to those described in Grübler and Gritsevskyi (1998) and Grübler (1998) for a carbon tax. First, we assume a cumulative probability distribution of the occurrence of the emissions tax over the entire time horizon. Starting near zero in 2000, the starting year of our simulations, the cumulative probability distribution rises over time. The distribution function assumed reflects only a 50 percent chance that a carbon tax is ever implemented. The probability of the tax being introduced rises to 25 percent by 2030, reaching 50 percent toward the end of the model's time horizon. For the magnitude of the tax, we assume a distribution with a very small probability of a high carbon tax level, as represented by a Weibull distribution around a mean value of US $ 75 per ton carbon (C), with a 99 percent probability that the tax will not exceed US $ 150 per ton C if it is implemented at all.

4   Implementation

We use the sample mean approximation to the original problem:

$$
\begin{aligned}
&\min \sum_{t=0}^{T} d^t \mathrm{E}[\langle C^t(x|_0^t, \omega), x^t\rangle + \sum_{t=0}^{T} \langle a^t(\omega), y^t(\omega)\rangle + \\
&+\rho(\langle C^t(x|_0^t, \omega) - \mathrm{E}C^t(x|_0^t, \omega), x^t\rangle)] \approx \\
&\min \frac{1}{N} \sum_{t=0}^{T} \sum_{s=1}^{N} d^t[\langle C^t(x|_0^t, \omega^s), x^t\rangle + \sum_{t=0}^{T} \langle a^t(\omega_s), y^t(\omega_s)\rangle + \\
&+\rho(\langle C^t(x|_0^t, \omega^s) - \frac{1}{N} \sum_{s=1}^{N} C^t(x|_0^t, \omega^s), x^t\rangle)]
\end{aligned}
\tag{5}
$$

subject to constraints (2), (3) for ω from a given set of scenarios, where N is the number of scenarios for exogenous parameters.

The consistency of implemented approximation technique is well known (see [2], [6], [18] and [20]). In our application, we analyzed 520 alternative technology dynamics and have drawn some 250 scenarios for each of them, resulting in a run of about 130 thousand scenarios.

The main difficulties in solving problem (5) is related to the link between cost components of C^t and solution vectors $x|_0^t$. Fortunately, in our case it is only few dozens of them that follow non-convex relation described by equation (4) and attributed to increasing returns to scale effect. Major portion of components C^t is simply constant and those with convex relations could be addressed by applying conventional linearisation technique. Due to the specific structure of our problem each of those component of C^t that are linked to solution vectors $x|_0^t$ in accordance to equation (4) is fully defined by single parameter or so-called "number of doublings" in corresponding total technology output (see Figure 5). The set of parameters p^t which determine the components of C^t with increasing returns relations could be written as

$$
p^t = \sum_{k=0}^{t} \langle D_k, x^k\rangle
\tag{6}
$$

with extremely sparse vector D_t. This means that by choosing some fixed parameters $\{p^t | t = 1, \ldots, T\}$ the problem defined by (5) could be treated as following linear optimization problem

$$
\begin{aligned}
&\mathbf{L}(p) = \min \frac{1}{N} \sum_{t=0}^{T} \sum_{s=1}^{N} d^t[\langle C^t(p^t, \omega^s), x^t\rangle + \sum_{t=0}^{T} \langle a^t(\omega_s), y^t(\omega_s)\rangle + \\
&+\rho(\langle C^t(p^t, \omega^s) - \frac{1}{N} \sum_{s=1}^{N} C^t(p^t, \omega^s), x^t\rangle)]
\end{aligned}
\tag{7}
$$

subject to constraints:

$$\sum_{k=0}^{t} A_k(\omega^s)x^k + B_t(\omega^s)y^t(\omega^s) = b^t(\omega^s),$$
$$\sum_{k=0}^{t} \langle D_k, x^k \rangle = p^t, t = 0, 1, \ldots, T, s = 1, \ldots, N, \tag{8}$$

$$x^t \in X_t \subseteq R^n, y^t \in Y_t \subseteq R^m \tag{9}$$

where $p = (p_0, \ldots, p_T)$.

Physical meaning of parameters p is such that each component of vector $p^t = (p_1^t, \ldots, p_l^t)$ $p_j^t \geq 0$ and for each $p = \{p^t | t = 1, \ldots, T\} \in P$ such that (8) is feasible, $p_j^{t-1} \leq p_j^t, j = 1, \ldots, M, t = 1, \ldots, T$

The original stochastic optimization is now approximated by a sequence of large-scale linear optimization problems in combination with a global random search method for vector p minimizing the function $\mathbf{L}(p)$. In other words, the main idea for solving the original global stochastic optimization problem is based on the representation of the model in the form of a two-level nested structure.

The global optimization part, which defines technological dynamics with respect to new unit installations for technologies with increasing returns to scale, is an implementation of an adaptive (controlled) global optimization random search algorithm specifically "tailored" to network flows optimization problems (see Horst and Pardalos, 1995, Pinter, 1996, Neumaier, 2003). The basic idea is that we start from some (e.g., uniform) distribution on the set P of the vectors p, sample points $p_1, \ldots, p_m$ from P in accordance to that distribution and obtained $\mathbf{L}(p_1), \ldots, \mathbf{L}(p_m)$. Based on these values the distribution is updated using some additional information regarding the model, in particular, the specifics of its network structure and numerical data sets. This allows us to postulate some plausible random updating rules (principles) in a way that so far best obtained values get higher probability. The random search method repeats the drawing and probing parts again. We performed the probing part asynchronously rather then sequentially updating information as soon as we got it and stopping processes that could not return values close enough to the "best obtained so far" allowing us to achieve interlinear speedup.

The algorithm for the inner linear optimization problem is the interior-point method for linear optimization. One of the big advantages of the adaptive random search algorithm is that it refines the approximated solution at the time when information is available. This allowed us to devise a "parallel" adaptation of this technique. The inner linear optimization problem is rather large with number of variables in a range of 10,000 to 50,000.

The algorithm that we used for global search also calculated for each step and lower monotonically increasing bound for $\mathbf{L}(p)$. The difference between the best obtained value and the currently obtained lower bound was used as stopping rule. As a final step we launched a fixed number of gradient free local search algorithms in areas with potentials to get a solution better than obtained by global search. In all our cases it turned out to be rather small set of isolated "spots" which probably reflect very special structure of the original problem and most probably will not be a case for more general formulation.

Initially, the original problem implementation was done on a CRAY T3E-900 supercomputer at the National Energy Research Scientific Computing Center (NERSC), USA (see http://old-www.nersc.gov/research/annrep98/ gritsevskii.html). At a later stage the problem was ported to IIASA computer network environment.

The PCx and pPCx solvers of large-scale linear optimization subproblems used in this study were provided by the Argonne National Laboratory (Wright 1996a, 1996b; Czyzyk et al. 1997). These solvers are written in C code, modified to increase computational efficiency for our specific problem formulation and to link the solvers directly to the global optimization part of the structure.

5   Concluding Observations

Numerical results show that fundamentally different technological dynamics produce a wide range of different emergent energy systems, approximately equivalent with respect to optimality criteria. Thus, one result of the analysis is that different energy system structures emerge with similar overall costs; in other words, there is a large diversity across alternative path dependent energy systems.

The actual energy requirements for a given provision of energy services can range from very high to extremely low compared with current standards. Therefore, the future environmental impacts of energy systems will vary accordingly. For example, CO_2 emissions range from 10 times the current levels to virtually no net emissions by 2100 for scenarios in the literature. Figure 8 shows the range of future CO_2 emissions derived for the set of 520 technology dynamics (130,000 scenarios) versus the set of 53 optimal dynamics (about 13,000 scenarios).

Another result of the analysis is that the endogenous technology learning with increasing returns, uncertainty and spillover effects will have the greatest impact on the emerging energy system structures during the first few decades of the twenty-first century. Over these intermediate periods of time, these mechanisms endogenize future lock-in effects and increasing returns to adoption.

In the very long run, however, none of these effects is of great importance. The reason is that over such long periods many doublings of capacity of all technologies with inherent learning occur, so that little relative cost advantage results from large investments in only a few technologies and clusters.

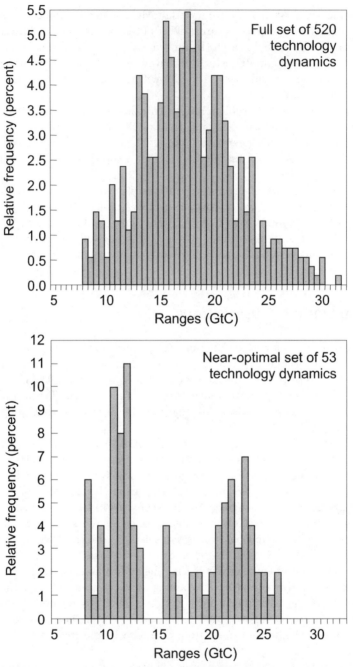

Fig. 8. Global Carbon Dioxide Emissions Ranges for (Top) the Full Set of 130,000 Scenarios with Endogenous Technological Change (Comprising 520 Different Technology Dynamics) versus (Bottom) the Ranges of about 13,000 Optimal Scenarios from 53 Different Technology Dynamics. All scenarios share a given useful energy trajectory; emissions ranges in gigatons of carbon (GtC). *Source: Gritsevskyi and Nakićenović et al., 2002*

Therefore, the main finding is that the near-term investment decisions in new technologies are more important in determining the direction of long-term development of the energy system than are decisions that are made later. In other words, the most dynamic phase in the development of future energy systems will occur during the next few decades, but robust justification of these developments requires truly long-term modeling approach. It is during this period that there is a high degree of freedom of choice across future technologies, and many of these choices lead to high spillover learning effects for related technologies.

One policy implication that can be made based on the emerging dynamics and different directions of energy systems development in this analysis is that future research, development, and demonstration (RD&D) efforts and investments in new technologies should be distributed across related technologies rather than directed at only one technology from the cluster, even if that technology appears to be a winner. Another implication is that it is better not to spread RD&D efforts and technology investments across a large portfolio of future technologies. Rather, it is better to focus on (related) technologies that might form technology clusters. Finally, the results imply that fundamentally different future energy system structures might be reachable with similar overall costs. Thus, future energy systems with low carbon dioxide (CO_2) emissions need not be associated with costs higher than those of systems with high emissions.

Acknowledgments. We would like to thank Yuri Ermoliev, Gordon J. MacDonald, Arnulf Grübler, Nebojša Nakićenović, Sabine Messner, and Manfred Strubegger, all from IIASA, for their help and advice. Initial tests of methodology was performed jointly with Yuri Ermoliev, Sabine Messner and Arnulf Grübler with the conceptual idea clearly attributed to Yuri Ermoliev. Nebojša Nakićenović, Sabine Messner and Gordon J. MacDonald worked with one of the authors on the grant from the National Energy Research Scientific Computing Center at Lawrence Berkeley National Laboratory that is funded by the US Department of Energy. This grant allowed more advanced problem implementation on a Gray T3E-900 supercomputer, and we are grateful for the financial support provided by the US Department of Energy. And finally, a multiregion, multi-actor model of uncertain increasing returns on technological innovation taking into account experience gained from supercomputer runs was suggested and analyzed jointly with Arnulf Grübler.

We would also like to thank colleagues from other institutions who have provided software and support for our research, including Michael Wagner from the Argonne National Laboratory, Steve Wright of Cornell University, and Francesca Verdier from the National Energy Research Scientific Computing Center at Lawrence Berkeley National Laboratory for their assistance.

We would like to thank participants of the IFIP/IIASA/GAMM workshop on Coping with Uncertainty for important and constructive discussions. We would also like to thank our referees for comments which led us to considerable improvements of this paper.

References

1. Bankes, S.C. (1993) Exploratory Modeling and Policy Analysis, RAND/RP-211, Santa Monica, CA, USA.
2. Birge, J., Louveaux, F. (1997) Introduction to Stochastic Programming. Springer-Verlag, New York, USA.
3. Czyzyk, J., Mehrotra, S., Wagner, M., Wright, S. (1997) PCx User Guide (Version 1.1), Technical Report, OTC 96/01, http://www-unix.mcs.anl.gov/otc/Tools/PCx/doc/PCx-user.ps.
4. Ermoliev, Y., Norkin, V. (1997) On Nonsmooth Problems of Stochastic Systems Optimization, *European Journal of Operational Research*, 101:230-244.
5. Ermoliev, Y., Norkin, V. (1998) Monte Carlo Optimization and Path Dependent Nonstationary Laws of Large Numbers, IR-98-009, International Institute for Applied Systems Analysis, Laxenburg, Austria.
6. Ermoliev, Y., Wets, R.J.-B. (1988) Numerical Techniques for Stochastic Optimization, Springer-Verlag, Berlin, Germany.
7. Foray, D., Grübler, A. (1990) Morphological analysis, diffusion and lockout of technologies: Ferrous casting in France and Germany, *Research Policy*,19(6):535–550.
8. Golodnikov, A., Gritsevskyi, A., Messner, S. (1995) A Stochastic Version of the Dynamic Linear Programming Model MESSAGE III, WP-95-094, International Institute for Applied Systems Analysis, Laxenburg, Austria.
9. Grübler, A., Gritsevskyi, A. (1998) A model of endogenized technological change through uncertain returns on learning, http://www.iiasa.ac.at/Research/TNT/WEB/Publications/
10. Gritsevskyi, A., Ermoliev, Y. (1999) An Energy Model Incorporating Technological Uncertainty, Increasing Returns and Economic and Environmental Risks. *Proceedings of International Association for Energy Economics 1999 European Energy Conference "Technological progress and the energy challenges"*, Paris, France.
11. Gritsevskyi, A., Nakićenović, N. (2002) Modeling Uncertainty of Induced Technological Change, pp.251–279, in A. Grübler, N. Nakićenović and W.D. Nordhaus, eds, *Technological Change and the Environment*, Resources for the Future and International Institute for Applied Systems Analysis, published by Resources for the Future, Washington, United States of America.
12. Grübler, A., Gritsevskyi, A. (2002) A Model of Endogenized Technological Change through Uncertain Returns on Innovation, pp.280–319, in A. Grübler, N. Nakićenović and W.D. Nordhaus, eds, *Technological Change and the Environment*, Resources for the Future and International Institute for Applied Systems Analysis, published by Resources for the Future, Washington, United States of America.
13. Grübler, A. (1998) *Technology and Global Change*, Cambridge University Press, Cambridge, UK.
14. Grübler, A., Messner, S. (1996) Technological Uncertainty, in N. Nakićenović, W.D. Nordhaus, R. Richels, and F.L. Toth, eds, *Climate Change: Integrating*

Science, Economics, and Policy, CP-96-001, International Institute for Applied Systems Analysis, Laxenburg, Austria.

15. Grübler, A., Messner, S. (1998) Technological change and the timing of mitigation measures, *Energy Economics,* **20**:495–512.

16. Horst, R., Pardalos, P.M., eds (1995) *Handbook of Global Optimization,* Kluwer, Dordrecht, Netherlands.

17. IPCC (Intergovernmental Panel on Climate Change) (2000) *Emissions Scenarios,* Cambridge University Press, Cambridge, UK.

18. Kall, P., Ruszczynski A. and Frauendorfer K. (1987) Approximation techniques in stochastic programming, in Ermoliev Y.M. and Wets R.(eds.) *Numerical Methods for Stochastic Optimization* , Springer-Verlag.

19. Lempert, R.J., Schlesinger, M.E., Bankes, S.C. (1996) When we don't know the costs or the benefits: Adaptive strategies for abating climate change, *Climatic Change,* **33**(2):235–274.

20. Marti K. (2005) Stochastic Optimization Methods, Springer, Berlin, Germany.

21. Markowitz, H. (1959) *Portfolio Selection,* Wiley, New York, NY, USA.

22. Messner, S. (1995) Endogenized Technological Learning in an Energy Systems Model, WP-95-114, International Institute for Applied Systems Analysis, Laxenburg, Austria.

23. Messner, S. (1997) Endogenized technological learning in an energy systems model, *Journal of Evolutionary Economics,* **7**(3):291–313.

24. Messner, S., and Strubegger, M. (1991) Part A: User's Guide to CO2DB: The IIASA CO2 Technology Data Bank–Version 1.0, WP-91-031, International Institute for Applied Systems Analysis, Laxenburg, Austria.

25. Messner, S., Golodnikov, A., Gritsevskyi, A. (1996) A stochastic version of the dynamic linear programming model MESSAGE III, *Energy,* **21**(9):775–784.

26. Nakićenović, N. (1996) Technological change and learning, in N. Nakićenović, W.D. Nordhaus, R. Richels, and F.L. Toth, eds, *Climate Change: Integrating Science, Economics, and Policy,* CP-96-001, International Institute for Applied Systems Analysis, Laxenburg, Austria.

27. Nakićenović, N. (1997) Technological Change as a Learning Process, paperpresented at the Technological Meeting '97, International Institute for Applied Systems Analysis, Laxenburg, Austria.

28. Nakićenović, N., Grübler, A., Ishitani, H., Johansson, T., Marland, G., Moreira, J.R., Rogner, H.-H. (1996) Energy primer, in *Climate Change 1995: Impacts, Adaptations and Mitigation of Climate Change: Scientific-Technical Analysis,* Contribution of Working Group II to the Second Assessment Report of the IPCC, Cambridge University Press, Cambridge, UK, pp. 75–92.

29. Nakićenović, N., Grübler, A., McDonald, A., eds. (1998) *Global Energy Perspectives,* Cambridge University Press, Cambridge, UK.

30. Norkin V. I., Pflug G. C., Ruszczynski A. (1996) A Branch and Bound Method for Stochastic Global Optimization, *Mathematical Programming,* 83:425-450.

31. Pinter, J. (1996) *Global Optimization in Action,* Kluwer, Dordrecht, Netherlands.

32. Robalino, D., Lempert, R. (2000) Carrots and sticks for new technology: Crafting greenhouse gas reduction policies for a heterogeneous and uncertain world, *Integrated Assessment,* **1(1):1–19.**

33. Rogner, H.-H. (1997) An assessment of world hydrocarbon resources, *Annual Review of Energy and the Environment,* **22**:217–262.

34. Strubegger, M., Reitgruber, I. (1995) Statistical Analysis of Investment Costs for Power Generation Technologies, WP-95-109, International Institute for Applied Systems Analysis, Laxenburg, Austria.
35. Wright, S.J. (1996) *Primal -DualInterior-Point Methods,* SIAM, London, UK.

Stochasticity in Electric Energy Systems Planning

A. Ramos, S.Cerisola, Á. Baíllo, and J. M. Latorre

Instituto de Investigación Tecnológica
Universidad Pontificia Comillas
Alberto Aguilera 23
28015 Madrid, Spain

Abstract. Electric energy systems have always been a continuous source of applications of planning under uncertainty. Stochastic parameters that may strongly affect the electric system are demand, natural hydro inflows and fuel prices, among others. A review of some estimation methods used to approximate those parameters is presented. Reliability and stochastic optimisation are widespread techniques used to incorporate random parameters in the decision-making process in electric companies. A unit commitment, a market-based unit commitment, a hydrothermal coordination and a risk management model are typical models that can incorporate uncertainty in the decision framework.

1 Introduction

Uncertainty may be originated, in a broad sense, by the lack of reliable data, measurement errors or parameters representing future information. In electric energy systems planning, uncertainty appears mainly in demand, natural hydro inflows, fuel prices, system availability, electricity prices, competitors' strategies, and regulatory framework. Electric demand has a cyclic pattern, with seasonal, weekly and daily variations along the year. Besides, demand also presents a locational variation depending on the local or regional economic activity. Natural hydro inflows are subject to climate conditions every year and, therefore, also the water flowing into the reservoirs that can be used for electricity production. Fossil fuel prices are subject to geopolitical circumstances. System elements such as power plants and transmission lines are subject to random failures that can affect the capability of the system to supply electricity to final customers. Because of the previous stochastic parameters, electricity prices resulting from market clearing are also subject to stochasticity. Finally, the regulatory frameworks under which many electric energy systems are currently operating are subject to changes to adapt them to new requirements (i.e., emissions market) or to improve their performance by changing some market rules.

Planning[1] and operation decisions of electric systems are certainly complex, with very different time scopes. They can include decades in the case of generation and transmission expansion or just several minutes for the economic dispatch. These decisions are coordinated to achieve the objective of optimal operation of the electric system. This general objective is separated into several others for different time horizons that are implemented in hierarchical decision support tools. In power systems planning the time scope is usually divided into the following levels:

- *Very long term*: for any time ranging from five to fifteen years
- *Long term*: for any time ranging from two to five years
- *Medium term*: any time ranging from one month to two years
- *Short term*: from one week to one month
- *Very short term*: any time below one week

This division is required by the practical infeasibility of finding a model detailed enough to characterize the system. At the same time, the nature of the whole problem is well suited to be functionally decomposed. Longer the time period lower the detail in modelling the system. The purpose of this hierarchical process is to represent adequately the main variables, parameters and characteristics of the electric system affecting each decision level. Besides, it allows managing the complexity of the whole problem. The previously mentioned stochastic parameters can affect the system planning in different time horizons. As previously established, only the relevant stochastic parameters are considered in each time horizon and decision level. For example, stochasticity in demand may affect all the decisions. However, it seems that this influence can be more relevant in the very long term (where expansion decisions are taken) and in the very short term (where unit commitment decisions must be adopted). Uncertainty of natural hydro inflows seems to be relevant in the medium term due to its yearly cycle.

Firstly, in section 2 we present some tables that show the importance of some stochastic parameters. We have used Spain as the case study for presenting real data. In section 3 we present some of the methods that have been used so far to predict future values of stochastic parameters. In section 4 we show some of the mathematical techniques that can be used to deal with uncertainty in electric energy systems incorporated in decision support tools. In section 5 we summarize some of the classical applications and we present how they take into account the uncertainty. Finally, we extract some conclusions and recapitulate the work presented in this chapter.

2   Uncertainty Impact

For proximity and data accessibility, we have chosen to show the impact of the uncertainty of the Spanish electric energy system (see [20]). As a matter

---

[1] Planning is used here for any time horizon for taking decisions apart from the online system operation.

of fact, the energy demand increase for the last five years is shown in the following table. The energy load has increased at an approximate 5 % rate for the last five years (a cumulative 21 %) mainly due to the economic activity. An increment correction is made to include the effect of working vs. non working days and temperature.

Year	Energy	Yearly increment	Corrected yearly increment
	TWh	%	%
2000	195.0	5.8	6.5
2001	205.6	5.4	4.9
2002	211.5	2.8	3.9
2003	225.8	6.8	5.4
2004	235.4	3.5	3.6

Peak load has also increased as shown in the next table. From year 2000 to year 2004 the winter peak load has increased in 4.5 GW (an increment of 13.5 % with respect to winter peak load in year 2000) and the summer peak load in 7.2 MW (an increment of 24.6 % with respect to summer peak load in year 2000) and it is almost the same that the winter peak load. The main reason for this huge increment in summer peak load is the high penetration of air conditioning in new home and hotel developments in Spain. This peak load increment in five years would be equivalent to approximately ten combined cycle gas turbines, which implies two units per year.

Year	Winter peak load GW	Summer peak load GW
2000	33.2	29.4
2001	34.9	31.2
2002	34.3	31.9
2003	37.2	34.5
2004	37.7	36.6

The annual energy coming from natural hydro inflows shows also a great variation along the last years. For example, the hydro energy in year 2003 was 160 % the energy available in year 2002. The percentage of being exceeded corresponds to the value of the cumulative distribution function for that hydro energy.

Year	Hydro energy TWh	Index	Percentage of being exceeded %
2000	26.2	0.90	62
2001	33.0	1.14	27
2002	21.0	0.73	88
2003	33.2	1.15	26
2004	24.6	0.85	68

3   Estimation Methods

In this section we present different techniques used for estimating the evolution of some stochastic parameters along the time, namely demand and natural hydro inflows. With these two parameters, we have tried to show a variety of complementary prediction techniques used in the context of electric energy systems.

3.1   Demand

Load forecasting has always been an important concern for long term expansion decisions, mainly related to yearly peak demand. At this time horizon, the main influence factors are related to the use of electricity by different customers and to the general socioeconomic and demographic parameters. Besides these, weather conditions strongly influence the electric load. In the short term, not only peak is important but also the demand profile and its variation for each day of the week need to be estimated.

Forecasting methods differ depending on the time range they are dealing with, see book [9] for a detailed review. For long term forecasting, *end-use models* and *econometric models* are primarily used. For short term forecasting a large variety of methods from statistical and artificial intelligent techniques are used. Among them, we can mention *regression methods, time series analysis, artificial neural networks, fuzzy logic*, and combinations of them.

End-use models explain the electric demand as a function of the direct use of electricity by different customers (for example, in appliances for domestic users, electric motors or aluminium tons for industrial customers, and air conditioning for commercial customers). So load forecasting is reinterpreted as the estimation of end-user devices and their evolution along the time. Theoretically, this approach is very precise. However, it requires a huge amount of data and can be very sensitive to their quality.

Econometric models use general economic data as factors for explaining electricity consumption. So load forecasting is estimated as a function of economic parameters (such as gross domestic product, customer price index, etc.) obtained by using statistical techniques.

Regression is used to determine the relationship between load consumption and factors such as weather, temperature, day of the week, etc., see [12,23].

Time series methods are based on detecting the intrinsic structure of load data regarding correlation, trend, seasonal variation, etc. ARMA[2] and ARIMA[3] techniques use time and past load as input parameters, see [1].

Artificial neural networks (ANN) are devices able to do nonlinear curve fitting. The outputs of an ANN are nonlinear functions of the inputs. These usually are load, temperature, humidity and weather. Its use in load forecasting has received a lot of attention, see [13] for a recent and exhaustive review of papers.

Fuzzy logic generalizes the classic Boolean logic by associating qualitative ranges to a number value. Therefore, this technique allows the introduction of qualitative data in load forecasting, for example in ANN, see [21].

After the deregulation process that has been carried out by the electricity industry in many countries an important additional factor that may affect load forecasting is price. So sensitivity analysis needs to consider as well demand elasticity in load forecasting.

3.2   Natural Hydro Inflows

Another important source of stochasticity in electric energy systems are natural hydro inflows. Two different techniques are used to include their stochasticity. One is *scenario generation* and the other is *scenario tree generation*.

The first tries to create plausible scenarios for future outcomes of hydro inflows. It usually resorts to time series analysis or other forecasting techniques, see [11]. The second tries to detect the internal structural dependence of the different scenarios previously generated. The scenario tree is then incorporated in multistage decision tools, which are going to be described in the following section. In these models, whose resolution relies on the use of LP, NLP and MIP solvers, uncertainty given by parameters with continuous distributions complicates its resolution because of the necessity of combining simulation techniques with optimisation techniques. For that reason, the choice of an appropriate discrete distribution is crucial for obtaining good results of the associated stochastic optimisation problem.

Among the existing techniques for generating scenario trees, they appear those based on *moment adjustment* [14]. These techniques consist of minimizing the distance between statistical properties of the discrete outcomes given by the scenario tree and those of the underlying distribution. This minimization is carried out through the resolution of a NLP problem. Although this method has been extended to multistage and multivariate distributions [15], the nonlinearity of resulting mathematical problem experiences difficulties

---

[2] AutoRegressive Moving Average.
[3] AutoRegressive Integrated Moving Average.

when a large number of time periods and dimensions in the multivariate distribution needs to be approximated. Another type of methods uses *clustering techniques* to generate the scenario tree [19,16]. This technique adapts iteratively the tree branches to the original data series as a function of its vicinity to a series randomly chosen.

4   Decision Making Methods

In this section we present some of the mathematical methods used to incorporate the uncertainty in the decision support process in electric energy systems. These techniques are *reliability* and *stochastic optimisation*.

4.1   Reliability

It is evident that cost and reliability criteria can be conflicting. A strict reliability criterion may derive in over investment. On the other hand, under investment usually leads to highly unreliable systems. Reliability evaluation in electric energy systems has been for many years an area of research, see the classical reference book from Billinton and Allan [4]. Recently, under deregulated electricity markets it has been a renewed interest in the topic due not only to the recent important blackouts occurred in several systems (for example, in New York, UK and Italy in 2003) but also to new concepts like transmission open access that are being explored. Even more, networks are currently led to operate close to physical limits. The main objective of reliability is to determine some measures or criteria to be used in generation and network capacity or operation planning.

Important aspects to be considered in reliability evaluation are:

1. Load forecast and capability of the system to supply it
2. Possible generator locations for new generators, generation commitment and maintenance scheduling and other unit requirements including fuel availability
3. Possible contingencies in generation or transmission systems and ways to alleviate them

Generation reliability is usually evaluated by analytical methods such as *probabilistic production simulation*, see the seminal papers of Baleriaux [3] and Booth [5] and a comparison of algorithms in [17]. This technique is based on obtaining the cumulative distribution function of the sum of random variables corresponding to load and generation unit failures. Dispatch of generating units is made by iteratively convoluting the random variables. The most common reliability measures obtained by this method are loss of load probability (LOLP) and expected energy not served (EENS). These reliability indexes are frequently used as adequacy criteria for generation expansion and operational planning. For example, a classical planning criterion used for generation expansion has been to have a LOLP lower than 1 day in 10 years.

However, the method only considers the forced outage rate of the units and ignores the frequency and duration of these outages and the operating constraints that play a significant role in short term operations, for example, startup and shutdown time and minimum uptime and downtime. *Monte Carlo simulation* can incorporate some of these characteristics in probabilistic simulation models [22] or in chronological or sequential planning models [10].

Monte Carlo simulation with *variance reduction techniques* (VRT) is also used to evaluate generation and transmission composite reliability, see references [18,6]. Control and antithetic variables are some of the VRT frequently used.

4.2   Stochastic Programming

Within a decision-making framework, many problems can be posed as optimisation problems. This way of modelling considers a set of decision variables, relations among these variables (termed constraints) and an expression of the variables whose value needs to be optimised (the objective function). A problem set in this form is known as a mathematical programming problem. The algebraic expressions that form the constraints and objective function may lead to a LP or NLP problem. Additionally, the nature of the decision variables leads to a continuous problem or to a mixed-integer one. These problems are solved by using a collection of algorithms that are the wide-range subject of research of mathematical programming community. These algorithms include simplex methods, branch & bound methods, methods of feasible directions, etc. From a practical point of view, there exists a wide collection of algorithms already implemented in computer codes available for being used by decision makers. In addition, current algebraic languages give the possibility of modelling a mathematical programming problem and test these algorithms quickly.

The difficulty of the resolution of mathematical programming problems increases when stochasticity is introduced in the problem parameters. The introduction of uncertainty in the context of energy planning is aimed at providing a collection of optimal decisions that have to be taken prior to uncertainty disclosure. This type of stochastic problem is usually denoted as two-stage program and its purpose is to give a solution, which hedges against the uncertain future. This is the most extended way of dealing with uncertainty. There also exist other methods, like those of probabilistic constraints, which produce a solution of a mathematical program such that their constraints are satisfied with some given probability.

Random parameters in stochastic programming (SP) appear as scenarios. The use of scenarios is extended and is a common way of representing stochasticity in multistage problems. These scenarios share part of their stochastic information and create a graph structure, which is denoted in the literature as scenario tree. Contrary to deterministic problems, for which a collection of well-studied algorithms exists, for the moment there is no algorithm that outstands as the leading algorithm to solve stochastic problems. Users and researches are focused on the resolution of the deterministic equivalent problem and in the combination of decomposition techniques to create ad hoc algorithms for specific problems. SP has been widely used as a mathematical programming technique for planning under uncertainty in electric energy systems. Next section describes some examples that deal with uncertainty in different ways depending on the time scope of the model.

5   Characteristic Models

The type of stochastic parameters that enter within energy planning mathematical programming models heavily depends on the considered model. This section reviews classical models, focusing on the presence of stochasticity:

- a unit commitment (UC) model
- a market-based unit-commitment model
- a mid-term hydrothermal coordination model
- a mid-term risk management model

With these models, we try to introduce the treatment given in SP to random parameters like demand, hydro inflows and fuel prices.

Short-term models consider uncertainty in electricity demand. A classical cost-minimization UC model considers uncertainty in the chronological weekly load demand. A market unit-commitment model represents competitors' behaviour by means of their residual demand curve. Uncertainty in competitors' behaviour can be represented as a discrete random variable whose values are the different residual demand curves. In mid-term models, besides uncertainty in demand information, models incorporate uncertainty of hydro inflows and fuel costs. Typically, hydrothermal models use SP to obtain robust decisions for the set of future hydro scenarios. The use of SP is also necessary for risk management models. Finally, stochasticity in fuel costs is employed in one of the presented problems to model future contracts with the purpose of exercising control over minimum benefit scenarios.

The authors have developed the models presented in this section and their references are given in the corresponding sections. These models have been implemented in computer applications and applied to the Spanish electric system.

5.1   Unit Commitment

This problem has to decide the set of generating units that need to be committed as well as their generation levels. In these problems, total variable cost is minimized. Demand appears in classical models as a known parameter and the problem decides the subset of committed units that will provide the required demand. This modelling reflects the traditional regulation framework where an Independent System Operator (ISO) orders to the different companies the amount of energy they had to produce.

The operating cost of thermal units is modelled as a straight line with a fixed operating cost (the intercept) and a variable cost (the slope). This operating cost represents the fuel and operation and maintenance costs.

A weekly model is interpreted as a multiperiod problem where each period comprises a set of hours. A possibility is to consider one period for each hour, summing up 168 periods. The nature of the decision variables turns this optimisation problem into a mixed-integer one. Variables that represent the commitment status of the units are binary and those that represent operating levels are continuous.

The remaining section describes the algebraic model of a weekly UC problem. Consider the following collection of sets, indices, parameters and variables.

Sets

T	Set of periods
I	Set of thermal units

Indexes

t	Index for periods
h	Auxiliar index for periods
i	Index for thermal units

Parameters

D_t	Demand of period t	[MW]
R_t	Spinning reserve coefficient for thermal production in period t	[%]
Dur_t	Duration of period t	[h]
P_i^{max}	Maximum rated capacity of thermal unit i	[MW]
P_i^{min}	Minimum rated capacity of thermal unit i	[MW]
L_i^{up}	Upwards ramp limit of thermal unit i	[MW/h]
L_i^{down}	Downwards ramp limit of thermal unit i	[MW/h]
F_i	Fixed operating cost of thermal unit i	[€/h]
V_i	Variable cost of thermal unit i	[€/MWh]
C_i^{up}	Startup cost of thermal unit i	[€]
C_i^{down}	Shutdown cost of thermal unit i	[€]
τ_i	Minimum uptime of thermal unit i	[h]
κ_i	Minimum downtime of thermal unit i	[h]

Variables

p_{ti}	Operating level of thermal unit i in period t	[MW]
u_{ti}	Commitment status of thermal unit i in period t	{0,1}
s_{ti}^{up}	Startup decision of thermal unit i in period t	{0,1}
s_{ti}^{down}	Shutdown decision of thermal unit i in period t	{0,1}

The UC problem must satisfy the load profile in each load level considered

$$\sum_{i=1}^{I} p_{ti} = D_t \qquad \forall t \tag{1}$$

requiring a spinning reserve operating margin that can be modelled as

$$\sum_{i=1}^{I} (P_i^{max} u_{ti} - p_{ti}) \geq R_t D_t / 100 \qquad \forall t \tag{2}$$

Each thermal unit operating level is bounded between its minimum and maximum rated capacity

$$P_{ti}^{min} u_{ti} \leq p_{ti} \leq P_{ti}^{max} u_{ti} \qquad \forall t, i \tag{3}$$

Variation in a thermal unit power generation is controlled through the ramp constraints

$$L_i^{down} Dur_t \leq p_{ti} - p_{t-1i} \leq L_i^{up} Dur_t \qquad \forall t, i \tag{4}$$

Startup and shutdown decisions are managed with the following constraints

$$u_{ti} - u_{t-1\,i} = s_{ti}^{up} - s_{ti}^{down} \qquad \forall t, i \tag{5}$$

Some advanced UC models include minimum uptime and downtime requirements for switched-on and switched-off thermal units. Committed units are usually required to produce a minimum number of hours before they can stop. Similarly, once they stop, they must also remain offline a minimum number of hours, before they can produce again. These minimum uptime and downtime requirements can be formulated as follows:

$$u_{t+h_t\,i} \geq u_{ti} - u_{t-1\,i} \qquad \forall t, h_t, i \tag{6}$$

$$u_{t+h_t\,i} \leq 1 + u_{ti} - u_{t-1\,i} \qquad \forall t, h_t, i \tag{7}$$

where the set of shifted indexes, controlled by h_t, maybe reduced for those values $h_t \geq 1$ such that

$$\tau_i \leq \sum_{l=0}^{h_t-1} Dur_{t+l} \tag{8}$$

$$\kappa_i \leq \sum_{l=0}^{h_t-1} Dur_{t+l} \tag{9}$$

Given the above variables and constraints, the UC model minimizes the total variable operating cost, given as:

$$\sum_{t=1}^{T}\sum_{i=1}^{I}(Dur_t F_i u_{ti} + Dur_t V_i p_{ti} + C_i^{up} s_{ti}^{up} + C_i^{down} s_{ti}^{down}) \tag{10}$$

Uncertainty in a weekly UC model appears in the randomness of demand load profiles that a generation company faces. For this reason, instead of formulating a single-scenario problem, the company may analyze its decision-making problem by means of a SP problem. Stochasticity in demand profiles can be modelled as a discrete random variable in the form of a scenario tree. A load profile scenario tree is presented in figure 1. It is represented the possible evolution of the demand for a week that begins on Tuesday. It is not considered being uncertain along the very first day. For the second day, Wednesday, two branches appear. These branches branch at the end of the second day producing four scenarios that represent the evolution of the demand profiles for the remaining days of the week.

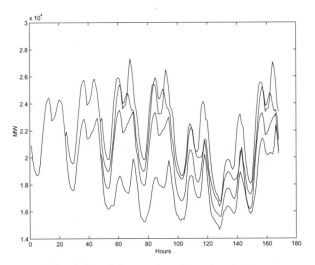

Fig. 1. Load demand profile scenario tree.

5.2   Market-Based Unit Commitment

Classical UC problem changes dramatically if the company operates in an electricity spot market. In this new framework, companies are responsible of their total production, which is no longer decided by an ISO. A day-ahead market is the market that takes place one day before the physical delivery of power production. This market is based on offers submitted by power producers and bids submitted by power purchasers. Offers and bids indicate the price at which producers are willing to sell and purchasers to buy.

In this new context, the objective function of an energy planning problem changes from the traditional cost minimization to a maximization of the company's benefit.

The company's benefit $B(p)$ is defined as the difference between revenues and operating cost $c(p)$. In addition, company's incomes depend on the market price π at which the energy p is sold.

$$B(p) = \pi p - c(p) \tag{11}$$

The energy price is a function of the total amount of energy sold. Similarly, the energy amount that each company is able to sell depends on the final price. Observe that the energy demand (understood as a function of price) needs to be equal to the energy supplied (also understood as a function of price).

$$D(\pi) = \sum_{agents} S^{agents}(\pi) \tag{12}$$

Under the assumption that competitor's behaviour is given by their supply energy functions, the amount of power a single company is able to sell depends on the demand at that price, $D(\pi)$, and the offers of the rest of agents, $S^{rest}(\pi)$

$$R(\pi) = D(\pi) - \sum_{rest} S^{rest}(\pi) \tag{13}$$

expression that gives the residual demand faced by the company, $R(\pi)$. The company's benefit is now given as

$$B(p) = R^{-1}(p)p - c(p) \tag{14}$$

The inverse residual demand function is a staircase function that can be approximated by means of a piecewise linear function. The revenue function is also a non-concave function that can be modelled as a piecewise linear function fig.2). This function is modelled by considering a collection of binary variables to represent the total amount of energy produced as a sum of the quantities of each segment. Price and revenue values can also be modelled in the same way.

Uncertainty is again a relevant ingredient of these new market-based UC models. However, the main source of uncertainty is now the wholesale electricity market, because the decisions made by the rest of agents are not known

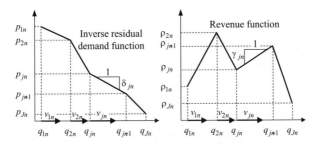

Fig. 2. Piecewise linear residual demand function and revenues' function.

in advance. This uncertainty is implicit into the residual demand function that may be considered a random variable within a SP problem.

When having a completely known residual demand function, the benefit maximization problem is deterministic. This problem determines company's optimal production and price for selling that production. However, if a random residual demand function is given, the benefit maximization problem turns into a SP problem. It should provide an optimal quantity for each one of the residual demand functions involved. This obeys the rules of a supply energy function, although additional conditions about non decreasing values need to be imposed.

The multistage stochastic problem we are about to present considers a realization of uncertainty to be a set of residual demand functions, one function for each period of the problem scope (fig. 3). The reader should note the difference in uncertainty management in this model with respect to that of the weekly UC model and forthcoming models.

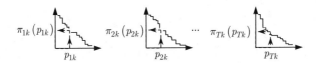

Fig. 3. Single scenario of residual demand functions.

Consider the next collection of sets, indexes, parameters, variables and constraints used in the formulation of a market-based UC problem.

Sets

T	Set of periods
I	Set of thermal units
J	Set of segments to represent the residual demand function
K	Set of scenarios

Indexes

t	Index of periods
i	Index of thermal units
j	Index of segments
k	Index of scenarios

Deterministic parameters

Dur_t	Duration of period t	[h]
P_i^{max}	Maximum rated capacity of thermal unit i	[MW]
P_i^{min}	Minimum rated capacity of thermal unit i	[MW]
L_i^{up}	Upwards ramp limit of thermal unit i	[MW/h]
L_i^{down}	Downwards ramp limit of thermal unit i	[MW/h]
F_i	Fixed operating cost of thermal unit i	[€/h]
V_i	Variable cost of thermal unit i	[€/MWh]
C_i^{up}	Startup cost of thermal unit i	[€]
C_i^{down}	Shutdown cost of thermal unit i	[€]

Stochastic parameters

δ_{tj}^k	Slope of segment j of the residual demand function in period t and scenario k	[€/MW]
γ_{tj}^k	Slope of segment j of the revenue function in period t and scenario k	[€/MW]
π_{tj}^k	Price at segment j of the residual demand function in period t and scenario k	[€]
$\bar{p}_{tj}^k$	Quantity at segment j of the residual demand function in period t and scenario k	[MW]
$\bar{b}_{tj}^k$	Benefit at segment j of the revenue function in period t and scenario k	[€]
$Prob^k$	Probability of scenario k	

The load demand constraints adopt the next expression in this case

$$\sum_{i=1}^{I} p_{ti}^k = p_t^k \qquad \forall t, k \tag{15}$$

where the total amount of energy produced p_t^k in period t and scenario k is modelled by

$$p_t^k = \bar{p}_{t0}^k + \sum_{j=1}^{J-1} p_{tj}^k \tag{16}$$

The total revenue is modelled as a piecewise linear function similarly to the total amount of energy produced.

Variables

v_{tj}^k	Binary variable corresponding to segment j in period t and scenario k	{0,1}
p_t^k	Total production in period t and scenario k	[MW]
p_{tj}^k	Total production of segment j in period t and scenario k	[MW]
p_{ti}^k	Operating level of thermal unit i in period t and scenario k	[MW]
π_t^k	Price in period t and scenario k	[€]
b_t^k	Benefit in period t and scenario k	[€]
u_{ti}^k	Commitment status of thermal unit i in period t and scenario k	{0,1}
$s_{ti}^{up\ k}$	Startup decision of thermal unit i in period t and scenario k	}0,1}
$s_{ti}^{down\ k}$	Shutdown decision of thermal unit i in period t and scenario k	}0,1}
$x_t^{kk'}$	Binary variable related with monotonicity of the supply function in period t and scenarios k and k'	{0,1}

$$b_t^k = \bar{b}_{t0}^k + \sum_{j=1}^{J-1} \gamma_{tj}^k p_{tj}^k \qquad (17)$$

as well as the price obtained when considering the optimal production p_t^k in period t and scenario k.

$$\pi_t^k = \bar{\pi}_{t0}^k + \sum_{j=1}^{J-1} \delta_{tj}^k p_{tj}^k \qquad (18)$$

This piecewise linear modelling requires the next constraints, which force a monotonic use of variables representing segment values.

$$(\bar{p}_{tj}^k - \bar{p}_{tj-1}^k)v_{t+1\ j}^k \le p_{tj}^k \le (\bar{p}_{tj}^k - \bar{p}_{tj-1}^k)v_{tj}^k \qquad (19)$$

$$v_{tj}^k \ge v_{tj+1}^k \qquad j = 1, \ldots, J-1 \qquad (20)$$

Due to uncertainty, limits for thermal units power output are introduced for any of the scenarios considered. Similarly, ramp constraints and startup and shutdown constraints are independently introduced for each scenario k.

$$P_{ti}^{min} u_{ti}^k \le p_{ti}^k \le P_{ti}^{max} u_{ti}^k \qquad \forall t, k \qquad (21)$$

$$L_i^{down} Dur_t \le p_{ti}^k - p_{t-1\ i}^k \le L_i^{up} Dur_t \qquad \forall t, k \qquad (22)$$

$$u_{ti}^k - u_{t-1\,i}^k = s_{ti}^{up\ k} - s_{ti}^{down\ k} \qquad \forall t, k \qquad (23)$$

The former set of constraints is the core of the market-based UC problem with stochasticity in the parameters modelling the residual demand functions. As it has already been commented, the optimal solution provided by this SP problem is a set of quantities and prices that form an offer curve. This curve has to be non decreasing. The following set of constraints is introduced into the model for that reason.

$$p_t^k - p_t^{k'} \geq -x_t^{kk'} M^p \qquad \forall t, k, k' \quad k' > k \qquad (24)$$

$$\pi_t^k - \pi_t^{k'} \geq -x_t^{kk'} M^\pi \qquad \forall t, k, k' \quad k' > k \qquad (25)$$

$$p_t^k - p_t^{k'} \geq -(1 - x_t^{kk'}) M^p \qquad \forall t, k, k' \quad k' > k \qquad (26)$$

$$\pi_t^k - \pi_t^{k'} \geq -(1 - x_t^{kk'}) M^\pi \qquad \forall t, k, k' \quad k' > k \qquad (27)$$

The SP model is completed with the objective function that maximizes the expected benefit.

$$\max \sum_{t=1}^{T} \sum_{k=1}^{K} Prob^k [b_t^k - c(p_t^k)] \qquad (28)$$

where $c(p_t^k)$ indicates the production cost in each period t and scenario k. This cost can be modelled as it has been presented in previous section. The optimal solution for this problem is an offer curve for each period (fig.4). For simplicity in the exposition, a pure thermal generating system has been considered. However, the model has been extended to more complex systems comprising hydro units as well as futures and options [2]. It is necessary to outline that building the offer curve necessarily implies the consideration of stochasticity. This model represents uncertainty in a different way that the weekly UC problem and the next models, where stochasticity is introduced by means of a scenario tree.

5.3   Hydrothermal Coordination

A hydrothermal coordination model considers a generating system with thermal units as well as hydro units, see [8] for further details. Hydro units provide the capability for energy reserve management. In hydrothermal models, a constant coefficient of efficiency for each hydro unit is usually considered and hydro reserves are expressed in terms of energy stored, in MWh. A difference between short-term models and mid-term models appears in the way of

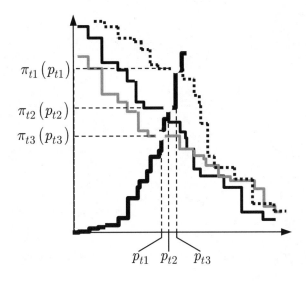

Fig. 4. Stochastic residual demand function and offer curve for a single period.

considering electricity demand. Short-term models usually consider a chronological load profile, while mid-term models use to represent the demand aggregated. Thus, mid-term models usually gather the demand in blocks of peak, shoulder and off-peak hours. Another difference with short-term models appears in the stochastic parameters considered. In short-term models, demand profile (together with units' outage) is the main source of uncertainty. In mid-term models, hydro inflows and fuel costs represent additional sources that must be taken into account when looking for optimal solutions to hedge against uncertainty. In a mid-term model, like the presented in this section, stochasticity enters as a scenario tree. Figure 5 shows a hydro inflows scenario tree. The tree represents an initial inflow value that branches into two possibilities in the second month of the model. The scenario tree branches again in the second and third months producing a final eight-scenario tree.

One of the objectives of a mid-term model is to schedule hydro reserves. A model that minimizes the expected operation cost over the complete time scope can achieve this. Hydro units have a very low cost that is usually neglected. Operating cost is limited to variable costs of thermal units. Reservoir levels are bounded in order to prevent spillage and dramatic scenarios of low reserves.

Thus, hydrothermal models include equations that represent the evolution of the reserves. Let us consider the next collection of sets, indexes, parameters and variables in order to model them.

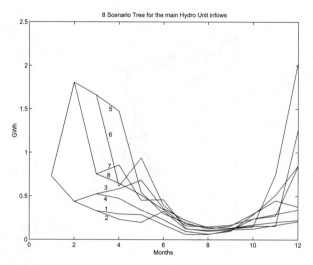

Fig. 5. Scenario tree for hydro inflows to a reservoir.

Sets

T	Set of periods
H	Set of hydro units
K	Set of scenarios

Indexes

t	Index of periods
h	Index of hydro units
k	Index of scenarios

Deterministic parameters

R_j^{max}	Maximum storage capacity of hydro reserve j	[MWh]
R_j^{min}	Minimum storage capacity of hydro reserve j	[MWh]
L_j^{max}	Maximum rated capacity of hydro unit j	[MW]
L_j^{min}	Minimum rated capacity of hydro unit j	[MW]
ρ_j	Pumping efficiency of hydro unit j	[%]

Stochastic parameters

I_j^k	Natural inflows of hydro unit j	
	in period t and scenario k	[MWh]
$Prob^k$	Probability of scenario k	

Variables

r_{tj}^k	Level of hydro reserve j	
	in period t and scenario k	[MWh]
s_{tj}^k	Production level of hydro unit j	
	in period t and scenario k	[MW]
ω_{tj}^k	Pumping level of hydro unit j	
	in period t and scenario k	[MW]

As mentioned, scheduling of hydro reserves can be obtained by introducing some constraints that represent the dynamics of water reserve evolution.

$$r_{tj}^k = r_{t-1,j}^k + I_j^k - Dur_t(s_{tj}^k - \rho_j\omega_{tj}^k)/100 \tag{29}$$

with

$$R_j^{min} \le r_{tj}^k \le R_j^{max} \tag{30}$$

$$L_j^{min} \le s_{tj}^k \le L_j^{max} \tag{31}$$

$$L_j^{min} \le \rho_j\omega_{tj}^k/100 \le L_j^{max} \tag{32}$$

A stochastic mid-term hydrothermal coordination problem gives the possibility of verifying the reserve evolution for the set of hydro scenarios analyzed. The SP problem provides a solution for the first stage that does not anticipate the uncertainty given by natural hydro inflows. An example of this solution is given in the next figure 6. It is depicted the evolution of the hydro reservoir storage level for the hydro unit whose natural inflows are given in figure 5.

5.4   Risk Management Model

Risk is implicit to all activities that take place in energy operation business and planning activities must consider this risk. SP is a suitable tool to carry on with this risk, which appears under different forms depending on the activity considered. As outlined at the beginning of the chapter, short-term operation suffers from the unit failure risk and demand fluctuation. Mid-term operation has to deal with uncertain hydro inflows and fuel prices, see for example [7], and long-term models pay careful attention to different factors, for example demand evolution and regulatory changes.

A risk management model controls the variability of the random variable that represents the operating cost function or the profit function. A variety of methods to measure risk can be introduced into a SP problem. A possibility consists of penalizing those scenarios in which the company cost is greater than a certain reference cost. Similarly, those scenarios whose profits are less than a certain reference profit can be penalized.

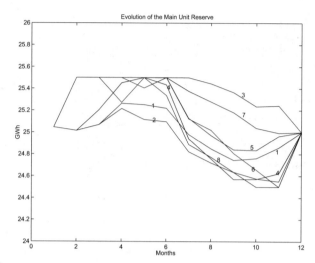

Fig. 6. Reserve evolution of a stochastic mid-term hydrothermal model.

Another alternative consists on introducing a hard limit for the quantile of the distribution function at a given confidence level. This quantile is usually referred to as Value at Risk (VaR) in risk management models. VaR has the additional difficulty, for SP problems, that it requires the use of binary variables for its modelling. Conditional Value at Risk (CVaR) computes the average of scenario profit values that lie under the quantile given by the VaR. CVaR computation does not require the use of binary variables and it can be modelled by the simple use of linear constraints. Figure 7 illustrates the concepts of VaR and CVaR.

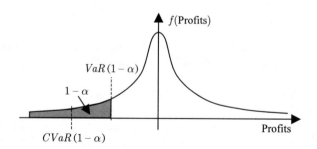

Fig. 7. VaR and CVaR illustration.

A SP model that incorporates risk measures obtains a final solution (cost or profit random variable) with less volatility than the final solution of a model that does not incorporate any measure of risk control. This is observed in the next figure 8. Different distribution functions are depicted for the profit random variable in a mid-term operation planning. It can be observed that the higher the upper limit imposed to the CVaR, the more concentrate the scenarios' profit values. In the following figure 9 the efficient frontier curve is obtained for expected profit and CVaR for the same case.

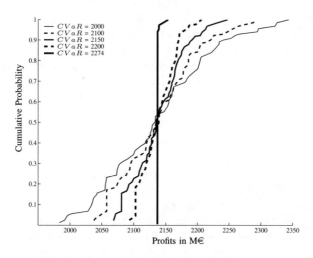

Fig. 8. Comparison of profit's distributions.

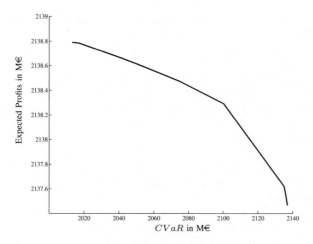

Fig. 9. Efficient frontier.

6 Conclusions

Electric energy systems have been for long time a continuous source of advances and applications of planning under uncertainty and a test bed for many developments to include stochastic parameters. In this chapter, we have presented a review and summary of the impact that the uncertainty may have in electric energy systems. We have presented the methods used to estimate the main stochastic parameters to be considered in power systems, namely demand and hydro inflows. Then, we have examined the two main methodologies that deal with uncertainty. One is reliability computation and the other is stochastic optimisation. Finally, we have presented some characteristic models that include an explicit treatment of parameter uncertainty.

References

1. Amjady, N. "Short-term hourly load forecasting using time-series modelling with peak load estimation capability" *IEEE Transactions on Power Systems* Vol. 16, No. 4, pp. 798-805. Nov 2001.
2. A. Baíllo, M. Ventosa, M. Rivier, A. Ramos "Optimal Offering Strategies for Generation Companies Operating in Electricity Spot Markets" *IEEE Transactions on Power Systems* Vol. 19, No. 2, pp. 745-753. May 2004.
3. Baleriaux, H., Jamoulle, E. et Linard de Guertechin, Fr. "Simulation de l'exploitation d'un parc de machines thermiques de production d'électricité couplé á des stations de pompage" *Revue E* (Edition SRBE), Vol. V, No. 7, pp. 225-245. 1967.
4. Billinton, R. and Allan, R. *Reliability Evaluation of Engineering Systems* Second edition. Plenum Press. New York, 1996.
5. Booth, R.R. "Power System Simulation Model Based on Probability Analysis" *IEEE Transactions on Power Apparatus and Systems* Vol. 91, pp. 62-69. 1972.
6. Borges, C.L.T., Falcao, D.M., Mello, J.C.O. and Melo, A.C.G. "Composite reliability evaluation by sequential Monte Carlo simulation on parallel and distributed processing environments" *IEEE Transactions on Power Systems* Vol. 16, No. 2, pp. 203-209. May 2001.
7. Cabero, J., Baíllo, A., Cerisola, S., Ventosa, M., García Alcalde, A., Perán, F., and Relaño, G. "A Medium-Term Integrated Risk Management Model for a Hydrothermal Generation Company" *IEEE Transactions on Power Apparatus and Systems* Vol. 20, No. 3, pp. 1379-1388. August 2005.
8. Cerisola, S. "Benders decomposition for mixed integer problems. Application to a medium term hydrothermal coordination problem" Ph.D. Thesis. Universidad Pontificia Comillas. April 2004.
9. Chow, J. H., Wu, F. F. and Momoh, J. A. (eds.) *Applied Mathematics for Restructured Electric Power Systems* Springer 2005.
10. CIGRE TF 38.03.13, convened by J. van Hecke "Sequential probabilistic methods (SPM)" *Electra* No. 179. Aug 1998.
11. Dupacova, J. "Scenarios for Multistage Stochastic Programs" *Annals of Operations Research*, 100:25-53, 2000.

12. Haida, T. and Muto, S. "Regression based peak load forecasting using a transformation technique" *IEEE Transactions on Power Systems* Vol. 9, No. 4, pp. 1788-1794. Nov 1994.
13. Hippert, H.S., Pedreira, C.E. and Souza, R.C. "Neural networks for short-term load forecasting: a review and evaluation" *IEEE Transactions on Power Systems* Vol. 16, No. 1, pp. 44-55. Feb 2001.
14. Hoyland, K. and Wallace, S.W. "Generating Scenario Trees for multistage decision problems" *Managment Science* 47:295-307. 2001.
15. Hoyland, K., Kaut, M. and Wallace, S.W. "A heuristic for moment-matching scenario generation" *Computational Optimization and Applications* 24(2-3):169-185. 2003.
16. Latorre, J.M., Cerisola, S., Ramos, A. "Clustering Algorithms for Scenario Tree Generation. Application to Natural Hydro Inflows" *European Journal of Operational Research* (submitted)
    http://www.iit.upcomillas.es/~aramos/papers/Latorre.pdf.
17. Mazumdar, M. and Gaver, D.P. "A comparison of algorithms for computing power generating system reliability indices" *IEEE Transactions on Power Apparatus and Systems* Vol. 103, No. 1, pp. 92-99. Jan 1984.
18. Pereira, M.V.F. and Balu, N.J. "Composite generation/transmission reliability evaluation" *Proceedings of the IEEE* Vol. 80, No. 4, pp. 470-491. Apr 1992.
19. Pflug, G.C. "Scenario Tree Generation for Multiperiod Financial Optimization by Optimal Discretization" *Mathematical Programming* 89:251-271, 2001.
20. Red Eléctrica de España "Annual Report"
    http://www.ree.es/cap07/pdf/infosis/Inf_Sis_Elec_REE_2004_v02.pdf
21. Srinivasan, D., Liew, A. C. and Chang, C. S. "Forecasting daily load curves using a hybrid fuzzy-neural approach" *IEE Proceedings Generation, Transmission and Distribution* Vol. 141, No. 6, pp. 561-567. 1994.
22. Valenzuela, J. and Mazumdar, M. "Monte Carlo computation of power generation production cost under unit commitment constraints" *IEEE Power Engineering Society Winter Meeting*, Vol. 2, pp. 927-930. Jan 2000.
23. Valenzuela, J., Mazumdar, M. Kapoor, A. "Influence of temperature and load forecast uncertainty on estimates of power generation production costs" *IEEE Transactions on Power Systems* Vol. 15, No. 2, pp. 668-674. May 2000.

Stochastic Programming Based PERT Modeling

A. Gouda[1], D. Monhor[2], and T. Szántai[1]

[1] Budapest University of Technology and Economics, Műegyetem rkp. 3,
  H-1111 Budapest, Hungary
[2] Western Hungarian University, Faculty of Geoinformatics, Pirosalma u. 1-3,
  H-8000 Székesfehérvár, Hungary

Abstract. Main drawback of the traditional PERT modeling is that the probabilistic characteristics determined for the project completion time are only valid
when it is supposed that any activity can be started promtly after executing all of
its predecessor activities. This is possible in the case of scheduling computer tasks,
however it is hardly possible in many other cases, like architectural project planning what is one of the the most important applicational areas of PERT modeling.
In the paper a stochastic programming based PERT modeling will be introduced.
This modeling will produce deterministic earliest starting times for the activities of
the project. These deterministic starting times will be attainable with prescribed
probability. So we also get an estimated project completion time what is attainable with the same prescribed probability. Numerical examples will be given for
comparing the traditional and the newly introduced PERT modeling techniques.

1 Introduction

Consider a compact, directed and acyclic network $(\mathcal{N}, \mathcal{A})$ as a representation
of the project. Assume that $\mathcal{N} = \{c_1, \ldots, c_n\}$ is the set of nodes (*events*),
and $\mathcal{A} \subset \mathcal{N} \times \mathcal{N}$ is the set of arcs (*activities*). Without restricting generality,
we may assume that there is exactly one node such that no arc leads into
it and there is exactly one node such that no arc goes out of it. These two
nodes will be called start and terminal nodes, respectively. Let us suppose
that c_1 is the single start node and c_n is the single terminal node,

Each activity has a *duration* (or *length*). The duration (or length) of
a path is the sum of the durations of the arcs contained in the path. Of
particular importance are the paths connecting the start and terminal nodes.
The maximum length of these paths is the shortest time needed to complete
the project and we call it the project completion time. The corresponding
path is the *critical path*.

Suppose that there are m arcs numbered by $1, 2, \ldots, m$. Suppose furthermore that there are p paths, numbered by $1, 2, \ldots, p$, which connect the start
and terminal nodes. The elements of the path–arc incidence matrix $A = (a_{ij})$
are defined as

$$a_{ij} = \begin{cases} 1, & \text{if activity } j \text{ is contained in path } i \\ 0, & \text{otherwise.} \end{cases}$$

We will designate by A_i the ith row of the matrix A $(1 \leq i \leq p)$. Let $\xi = (\xi_1, \ldots, \xi_m)^T$ be the vector of the activity durations. Then the critical path length $R(\xi)$ equals

$$R(\xi) = \max_{1 \leq i \leq p} A_i \xi.$$

Designating by $P_1, \ldots, P_p$ the paths from the origin to the terminal nodes, we may also write

$$R(\xi) = \max_{1 \leq i \leq p} \sum_{j \in P_i} \xi_j.$$

If the durations $\xi_1, \ldots, \xi_m$ are random variables then $R(\xi)$ is a random variable, too. Its probability distribution function (cdf) will be designated by $F(x)$, i.e.,

$$F(x) = P(R(\xi) \leq x). \tag{1}$$

The main drawback of the traditional PERT modeling is that the probabilistic characteristics determined for the project completion time are only valid when it is supposed that any activity can be started promtly after executing all of its predecessor activities. This is possible in the case of scheduling computer tasks, however it is hardly possible in many other cases, like architectural project planning what is the most important applicational area of PERT modeling.

In the paper a stochastic programming based PERT modeling will be introduced. This modeling will produce deterministic earliest starting times for the activities of the project. These deterministic starting times will be attainable with prescribed probability. So we also get an estimated project completion time what is attainable with the same prescribed probability.

The original PERT technique, developed by Malcolm et al. (see [9]), is a technique to approximate the expected duration of the project. Further approximations and bounds to this value are due to D.R. Fulkerson ([6]), C.T. Clingen ([1]), S.E. Elmaghraby ([4]), P. Robillard and M. Trahan ([13], [14]), L.P. Devroye ([2]) and others. Even more important is, from the point of view of applications, to bound or approximate the probability distribution function of the critical path. In connection with this we mention the works by G.B. Kleindorfer ([7]), A.W. Shogan ([15]), A. Nádas ([12]), I. Meilijson and A. Nádas ([10]), B.M. Dodin ([3]), G. Weiss ([21]), D. Monhor ([11]) and S.W. Wallace ([20]). Efficient algorithm for the calculation of the cdf (1) can be found in the paper by J. Long, A. Prékopa and T. Szántai ([8]).

2   The Stochastic Programming Model of PERT

Let us describe the project by the $(\mathcal{N}, \mathcal{A})$ directed graph, which doesn't contain any loop. Here $\mathcal{N}$ is the set of nodes (events) and $\mathcal{A}$ is the set of arcs (activities). Let us designate by $c_j, j = 1, \ldots, n$ the nodes in the set

$\mathcal{N}$, among which c_1 let be the start and c_n let be the terminal node. Let us assign the variable x_j to the node c_j representing the earliest starting time for all activities starting from the node c_j, $j = 1, \ldots, n$. Let us designate by e_i, $i = 1, \ldots, m$ the arcs in the set $\mathcal{A}$ and let us assign the number d_i to the arc e_i as the duration time of the represented activity. If these are deterministic numbers, then the shortest execution time of the whole project represented by the loopless directed graph $(\mathcal{N}, \mathcal{A})$ can be determined by solving the following linear programming problem:

$$x_{f_i} - x_{s_i} \geq d_i, i = 1, \ldots, m$$
$$x_j \geq 0, j = 1, \ldots, n \qquad (2)$$
$$\min(x_n - x_1),$$

where s_i, f_i are the indices of the starting and ending nodes of arc e_i. Obviuosly one can suppose, that $x_1 \equiv 0$ and the problem (2) can be simplified. If the activity duration times $d_i, i = 1, \ldots, m$ are random variables, then let us designate these by $\xi_i, i = 1, \ldots, m$ and let us solve the following jointly probabilistic constrained stochastic programming problem for finding the $x_j, j = 1, \ldots n$ earliest starting times:

$$P(x_{f_i} - x_{s_i} \geq \xi_i, i = 1, \ldots, m) \geq p$$
$$x_j \geq 0, j = 1, \ldots, n \qquad (3)$$
$$\min(x_n - x_1),$$

where p is a prescribed, large enough probability. If the activity starting times determined by the $x_1, \ldots, x_n$ variables according to the optimal solution of the optimization problem (3) are applied then we can garantee at reliability level p that the whole project can be executed without any conflict in the activity starting and executing times.

In the literature of PERT the activity duration times are usually supposed to be independent. In these cases in the stochastic programming problem (3) the joint probability can easiliy be calculated by taking the product of the probabilities calculated from the one dimensional marginal probability distributions. These problems are easy to solve from a numerical point of view.

In the first problem of the next section we will show that the stochastic programming problem (3) can also be solved if we suppose the random activity duration times to be Dirichlet distributed. In the second problem the duration times are supposed to be indepent normally distributed and the model will be solved both by the original PERT optimization technique as it is described in the paper [8] and by the solution of the stochastic programming problem (3). The numerical results will be compared.

3   Numerical Results

Let us regard first the PERT problem given by the loopless, directed graph of Figure 1. The values $d_i, i = 1, \ldots, 15$ denote the duration times of 15 activities and x_j denotes the earliest starting times for all activities starting at event j, $j = 1, \ldots, 8$. The event No. 1 is the start and the event No. 8 is the terminal event and we suppose that $x_1 = 0$. Now, if the activity duration times $d_i, i = 1, \ldots, 15$ are deterministic, then the PERT modell can be regarded as a CPM (*Critical Path Method*) problem and we have to solve the problem (4) according to the linear programming problem (2). The solution component x_8 gives the total execution time of the project and the solution components x_j, $j = 2, \ldots, 7$ give the earliest starting times for the appropriate activities.

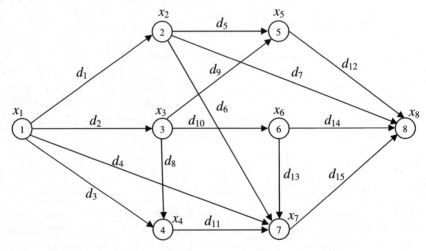

Fig. 1. PERT network

$$
\begin{array}{rcl}
x_2 & & \geq d_1 \\
x_3 & & \geq d_2 \\
x_4 & & \geq d_3 \\
x_7 & & \geq d_4 \\
-x_2 & +x_5 & \geq d_5 \\
-x_2 & +x_7 & \geq d_6 \\
-x_2 & +x_8 & \geq d_7 \\
-x_3 +x_4 & & \geq d_8 \\
-x_3 & +x_5 & \geq d_9 \\
-x_3 & +x_6 & \geq d_{10} \\
-x_4 & +x_7 & \geq d_{11} \\
-x_5 & +x_8 & \geq d_{12} \\
-x_6 +x_7 & & \geq d_{13} \\
-x_6 & +x_8 & \geq d_{14} \\
-x_7 +x_8 & & \geq d_{15} \\
\end{array}
$$

$$x_i \geq 0, i = 2, \ldots, 8$$

$$x_8 \to \min$$

(4)

If however the activity duration times are stochastic, we designate them by $\xi_i, i = 1, \ldots, 15$. In the classical PERT model of Malcolm et al. ([9]) the random activity duration times were assumed to be independent beta distributed on the intervals $[a_i, b_i]$, where the optimistic activity duration times a_i and the pessimistic activity duration times b_i were subjectively determined by experts. In addition the most likely activity duration times M_i were also subjectively determined by experts. Using these values the mean and variance of the random activity duration times were estimated as

$$E(\xi_i) \approx (a_i + 4M_i + b_i)/6,$$

and

$$D^2(\xi_i) \approx (b_i - a_i)^2/36.$$

In this small example we will show the application of the Dirichlet distribution for modeling the joint probability distribution of the random activity duration times. This can be regarded as a natural generalization of the beta distributed activity duration times, as the one dimensional marginal distributions of the Dirichlet distribution are beta distributions.

Let us define the random variables ξ_i as

$$\xi_i = a_i + (b_i - a_i)\eta_i, i = 1, \ldots, 15,$$

where a_i, b_i are the optimistic and the pessimistic estimators of the duration times of the ith activity, and the random variables $\eta_1, \ldots, \eta_{15}$ are Dirichlet distributed with parameters $\vartheta_1 > 0, \ldots, \vartheta_{15} > 0, \vartheta_{16} > 0$. Their joint probability density function is

$$f(x_1, \ldots, x_{15}) = \frac{\Gamma(\vartheta_1 + \ldots + \vartheta_{15} + \vartheta_{16})}{\Gamma(\vartheta_1) \cdots \Gamma(\vartheta_{15}) \Gamma(\vartheta_{16})} x_1^{\vartheta_1 - 1} \cdots x_{15}^{\vartheta_{15} - 1} (1 - x_1 - \ldots - x_{15})^{\vartheta_{16} - 1},$$

if $x_1 \geq 0, \ldots, x_{15} \geq 0$ and $x_1 + \ldots + x_{15} \leq 1$.

Now we have to solve a stochastic programming problem of type (3):

$$P \begin{pmatrix} \frac{1}{(b_1-a_1)}(& -a_1 & +x_2 & & & & & &) \geq \eta_1 \\ \frac{1}{(b_2-a_2)}(& -a_2 & & +x_3 & & & & &) \geq \eta_2 \\ \frac{1}{(b_3-a_3)}(& -a_3 & & & +x_4 & & & &) \geq \eta_3 \\ \frac{1}{(b_4-a_4)}(& -a_4 & & & & & +x_7 & &) \geq \eta_4 \\ \frac{1}{(b_5-a_5)}(& -a_5 & -x_2 & & & +x_5 & & &) \geq \eta_5 \\ \frac{1}{(b_6-a_6)}(& -a_6 & -x_2 & & & & +x_7 & &) \geq \eta_6 \\ \frac{1}{(b_7-a_7)}(& -a_7 & -x_2 & & & & & +x_8) & \geq \eta_7 \\ \frac{1}{(b_8-a_8)}(& -a_8 & & -x_3 & +x_4 & & & &) \geq \eta_8 \\ \frac{1}{(b_9-a_9)}(& -a_9 & & -x_3 & & +x_5 & & &) \geq \eta_9 \\ \frac{1}{(b_{10}-a_{10})}(& -a_{10} & & -x_3 & & & +x_6 & &) \geq \eta_{10} \\ \frac{1}{(b_{11}-a_{11})}(& -a_{11} & & & -x_4 & & & +x_7 &) \geq \eta_{11} \\ \frac{1}{(b_{12}-a_{12})}(& -a_{12} & & & & -x_5 & & +x_8) & \geq \eta_{12} \\ \frac{1}{(b_{13}-a_{13})}(& -a_{13} & & & & & -x_6 & +x_7 &) \geq \eta_{13} \\ \frac{1}{(b_{14}-a_{14})}(& -a_{14} & & & & & -x_6 & +x_8) & \geq \eta_{14} \\ \frac{1}{(b_{15}-a_{15})}(& -a_{15} & & & & & & -x_7 & +x_8) \geq \eta_{15} \end{pmatrix} \geq p \quad (5)$$

$$x_i \geq 0, i = 2, \dots, 8$$
$$x_8 \to \min,$$

where p is the prescribed probability of completing the project for the earliest possible due date. In Table 1 the parameters of the Dirichlet distribution are given. The correlation coefficients between different pairs of activity duration times can be determined from these parameter values. We don't give here their values just remark that all of them are negative and the smallest one equals -0.124409. Table 2 contains the solutions of the linear programming problem (4) for those cases, when the optimistic, the pessimistic and the most likely activity duration times are applied as deterministic values. There are given in the same table the solutions of the stochastic programming problem (5) for three different probability levels: 0.9, 0.95 and 0.99. The parameters of the Dirichlet distribution were taken from the Table 1. In Table 2 the value of the variable x_8 means also the completion time of the project. It can be seen that the deterministic cases do not provide appropriate results. If we work with the optimistic or with the most likely activity duration times then the project will be completed very quickly (less than 200), however if we calculate the reliability level of this solution when the random activity duration times follow the given Dirichlet distribution it will be probably much more less then 0.9. When working with the pessimistic activity duration times then the completion time of the project is too large (300) although the reliability level according to the Dirichlet distribution is probably very high, even more than 0.99. The decision maker can choose from the stochastic versions according to his acceptable reliability level of completing the whole project for the calculated time. We belive this choice will be easier for him than

Table 1. Parameters of the Dirichlet distribution

No.	optimistic estimation	pessimistic estimation	most likely value	θ parameter	expected value	standard deviation
1	45	60	45.043	1.06	45.691	0.642
2	10	40	11.500	2.05	12.674	1.745
3	50	75	50.048	1.04	51.130	1.060
4	10	40	12.957	3.07	14.004	2.083
5	15	45	15.286	1.20	16.565	1.362
6	70	95	70.071	1.06	71.152	1.070
7	40	75	40.133	1.08	41.643	1.511
8	85	95	85.524	2.10	85.913	0.588
9	10	35	10.600	1.05	11.141	1.065
10	45	90	45.011	1.005	46.966	1.878
11	25	45	25.967	2.015	26.752	1.154
12	25	50	25.083	1.07	26.163	1.075
13	30	60	30.071	1.05	31.370	1.278
14	55	75	56.048	2.10	56.826	1.176
15	15	35	15.029	1.03	15.896	0.844
16				1.02		

Table 2. Solutions of the linear and stochastic programming problems

variables	deterministic cases			stochastic cases		
	optimistic	pessimistic	most likely	$p = 0.90$	$p = 0.95$	$p = 0.99$
x_1	0	0	0	0	0	0
x_2	70	130	70.171	79.695	82.784	96.084
x_3	10	40	11.500	15.291	24.443	27.131
x_4	95	135	97.024	104.492	114.908	117.901
x_5	170	255	170.410	177.479	181.694	193.509
x_6	115	175	116.507	126.897	138.993	148.067
x_7	140	225	144.199	162.424	181.097	184.299
x_8	195	300	199.227	226.194	247.067	250.336

the choice between the deterministic versions. The stochastic programming problems were solved by that version of the code PCSP (*Probabilistic Constrained Stochastic Programming*) (see [16]) which can handle Dirichlet and multivariate gamma distributions, too.

As a second numerical example let us consider the network of the Figure 2 what was applied in the paper by J. Long, A. Prékopa and T. Szántai (see [8]).

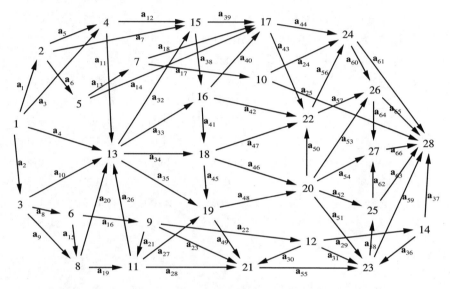

Fig. 2. PERT network from the paper by J. Long, A. Prékopa and T. Szántai (see [8])

This network consists of 66 activities and 28 events, the start event is the first one and the terminal event is the event No. 28. The optimistic and pessimistic estimates of the activity duration times are given in Table 3.

In the PERT network of Figure 2 there exist 1623 paths from the start to the terminal node. Let us designate these paths by $P_j, j = 1, \ldots, 1623$. As the optimistic a_i and the pessimistic b_i estimators of the duration times of the ith activity are lower resp. upper bounds on the ith random activity duration time, one can determine unconditional and conditional lower and upper bounds of the path lengths in the following way:

$$L(P_j) = \sum_{k \in P_j} a_k, \qquad U(P_j) = \sum_{k \in P_j} b_k, \qquad j = 1, 2, \ldots, 1623$$

Table 3. Lower and upper bounds for the duration times of 66 activities (Table 1. in the paper by J. Long, A. Prékopa and T. Szántai (see [8]))

No.	Activity	Lower bound	Upper bound	No.	Activity	Lower bound	Upper bound
1	(1, 2)	24	32	34	(13,18)	47	55
2	(1, 3)	48	56	35	(13,19)	44	52
3	(1, 4)	49	57	36	(14,23)	11	19
4	(1,13)	24	32	37	(14,28)	36	44
5	(2, 4)	21	29	38	(15,16)	39	47
6	(2, 5)	43	51	39	(15,17)	18	26
7	(2,15)	30	38	40	(16,17)	13	21
8	(3, 6)	14	22	41	(16,18)	41	49
9	(3, 8)	28	36	42	(16,22)	42	50
10	(3,13)	29	37	43	(17,22)	38	46
11	(4,13)	36	44	44	(17,24)	27	35
12	(4,15)	19	27	45	(18,19)	26	34
13	(5, 7)	49	57	46	(18,20)	39	47
14	(5,17)	12	20	47	(18,22)	25	33
15	(6, 8)	35	43	48	(19,20)	13	21
16	(6, 9)	28	36	49	(19,21)	16	24
17	(7,10)	15	23	50	(20,22)	29	37
18	(7,17)	26	34	51	(20,23)	42	50
19	(8,11)	33	41	52	(20,25)	33	41
20	(8,13)	46	54	53	(20,26)	43	51
21	(9,11)	41	49	54	(20,27)	44	52
22	(9,12)	47	55	55	(21,23)	22	30
23	(9,21)	42	50	56	(22,24)	46	54
24	(10,24)	40	48	57	(22,26)	19	27
25	(10,28)	37	45	58	(23,25)	33	41
26	(11,13)	27	35	59	(23,28)	39	47
27	(11,19)	26	34	60	(24,26)	15	23
28	(11,21)	31	39	61	(24,28)	48	56
29	(12,14)	38	46	62	(25,27)	27	35
30	(12,21)	48	56	63	(25,28)	26	34
31	(12,23)	29	37	64	(26,27)	29	37
32	(13,15)	32	40	65	(26,28)	22	30
33	(13,16)	20	28	66	(27,28)	20	28

and

$$L(P_j \setminus P_l) = \sum_{k \in P_j \setminus P_l} a_k, \qquad U(P_j \setminus P_l) = \sum_{k \in P_j \setminus P_l} b_k,$$

$$j, l = 1, 2, \ldots, 1623, \quad j \neq l.$$

The first elimination algorithm described in the paper [8] was based on the comparison of the unconditional bounds of the path lengths, while the second elimination algorithm was based on the comparison of the conditional bounds of the path lengths. If, for a given pair of paths P_j and P_l, $U(P_j) \leq L(P_l)$, then P_j is obviously redundant. So the first elimination step can be carried out in $O(p \ln p)$ time. If, for a given pair of paths P_j and P_l, $U(P_j \setminus P_l) \leq L(P_l \setminus P_j)$, then P_j is again obviously redundant. To eliminate all redundant paths this way, we need to perform about $\binom{p'}{2}$ pairwise comparisons if we start with p' paths. Fortunately p', the remaining number of paths after the first elimination algorithm is usually much less than the total number of existing paths from the start to the terminal node.

With the lower and upper bounds on duration times of the activities given in Table 3 the number of the remaining paths after the application of the first path elimination algorithm is 201, while after subsequent application of the second path elimination algorithm only 8 paths will remain as paths ever may become critical. As in these 8 paths only 21 activities are involved, the path–arc incidence matrix reduced to these paths only, has a size of 8 × 21 and it is given in Table 4. This matrix has only 4 linearly independent col-

Table 4. The path–arc incidence matrix of the remained 8 paths (Table 2. in the paper by J. Long, A. Prékopa and T. Szántai (see [8]))

	2	8	15	16	19	21	26	32	38	41	45	46	48	50	51	56	58	60	62	64	66
1	1	1	1	0	1	0	1	1	1	1	1	1	0	1	1	0	1	0	1	0	1	1
2	1	1	1	1	0	1	0	1	1	1	1	1	0	1	1	0	1	0	1	0	1	1
3	1	1	0	1	0	1	1	1	1	1	1	0	1	0	1	0	1	0	1	0	1	1
4	1	1	1	0	1	0	1	1	1	1	0	1	0	1	0	1	0	1	0	1	1
5	1	1	0	1	0	1	1	1	1	1	1	0	1	0	1	0	1	0	1	0	1
6	1	1	1	0	1	0	1	1	1	1	1	0	1	0	1	0	1	0	1	0	1
7	1	1	0	1	0	1	1	1	1	1	0	1	0	0	1	0	1	0	1	0	1
8	1	1	1	0	1	0	1	1	1	1	0	1	0	0	1	0	1	0	1	0	1

umn vectors, so the 8–variate normal probability distribution is restricted to a 4–dimensional subspace, i.e. the distribution is singular. Let us compare the results of the multivariate normal approach published in paper [8] with the results of the new stochastic programming based approach. Suppose the random duration times of activities to be independent and normally distributed with given expected values and variances. The expected values are defined as the arithmetical mean values of their lower and upper bounds given in Table 3. The variances are defined as the squares of the differences between the lower and upper bounds divided by twelve. Now the lengths of the remaining 8 paths have a singular multivariate normal probability distribution, concentrated on a 4–dimensional subspace. The parameters of this multivariate normal probability distribution are given in Table 5.

Table 5. Parameters of the multivariate normal probability distribution of remaining 8 paths

Expected value	508	507	504	503	487	486	483	482
Variance	8.9443	8.9443	8.6410	8.6410	8.6410	8.6410	8.3267	8.3267
Correlation matrix	1.0000	0.8667	0.8971	0.7591	0.7591	0.6211	0.6445	0.5013
	0.8667	1.0000	0.7591	0.8971	0.6211	0.7591	0.5013	0.6445
	0.8971	0.7591	1.0000	0.8571	0.6429	0.5000	0.7412	0.5930
	0.7591	0.8971	0.8571	1.0000	0.5000	0.6429	0.5930	0.7412
	0.7591	0.6211	0.6429	0.5000	1.0000	0.8571	0.8895	0.7412
	0.6211	0.7591	0.5000	0.6429	0.8571	1.0000	0.7412	0.8895
	0.6445	0.5013	0.7412	0.5930	0.8895	0.7412	1.0000	0.8462
	0.5013	0.6445	0.5930	0.7412	0.7412	0.8895	0.8462	1.0000

For comparing the two different approaches first the stochastic programming problem (5) has been solved with different probability levels ranging from 0 to 1. This was done easily by using the AMPL modeling language (see Fourer, R., Gay,D. M. and Kernighan,B. W. [5]) and the LOQO solver (see Vanderbei,R. J. [18] and [19]) as the random duration times of activities were supposed to be independent and normally distributed. The results are given in Table 6. For plotting the graph of the cumulative probability distribution function we needed its values in the interval 530–620 with unit steplength. These values were calculated from the results of Table 6 by linear interpolation.

The results of the multivariate normal approach are given in Table 7.

Table 6. The solutions of the stochastic programming problem for different probability levels

probability level	execution time	probability level	execution time	probability level	execution time
0.01	531.358	0.35	560.794	0.70	577.641
0.05	540.927	0.40	563.087	0.75	580.677
0.10	546.329	0.45	565.351	0.80	584.153
0.15	550.112	0.50	567.624	0.85	588.333
0.20	553.206	0.55	569.945	0.90	593.796
0.25	555.927	0.60	572.353	0.95	602.312
0.30	558.427	0.65	574.897	0.99	619.592

Figure 3 shows the cumulative distribution functions of the project completion times when the two different, multivariate normal and stochastic programming approaches are applied. It can be seen, that the cdf curve produced by the stochastic programming approach runs along significantly higher values than the cdf curve produced by the multivariate normal approach. This means, if we were able to start any activity promptly when all of its predecessor activities are finished in a random instant, then the whole project could be finished in a much shorter time. On the contrary, if we prescribe a deterministic starting time for all of the activities in the project before the starting time of the first activities and guarantee a reliability level to beeing the whole project executable without any conflict, then the project can be finished in a much longer time only. Even so, the decision maker sometimes should accept this longer completion time as he cannot guarantee to start the activities of the project in random time instants.

Table 7. The results of the multivariate normal approach

execution time	probability level	execution time	probability level	execution time	probability level
480	0.0003	500	0.1297	520	0.8831
481	0.0004	501	0.1560	521	0.9042
482	0.0006	502	0.1855	522	0.9223
483	0.0009	503	0.2183	523	0.9377
484	0.0013	504	0.2541	524	0.9506
485	0.0020	505	0.2928	525	0.9613
486	0.0028	506	0.3340	526	0.9701
487	0.0040	507	0.3772	527	0.9770
488	0.0057	508	0.4220	528	0.9826
489	0.0079	509	0.4678	529	0.9869
490	0.0108	510	0.5141	530	0.9904
491	0.0146	511	0.5601	531	0.9931
492	0.0195	512	0.6053	532	0.9949
493	0.0258	513	0.6490	533	0.9964
494	0.0337	514	0.6909	534	0.9975
495	0.0435	515	0.7304	535	0.9983
496	0.0554	516	0.7671	536	0.9989
497	0.0697	517	0.8009	537	0.9993
498	0.0867	518	0.8315	538	0.9996
499	0.1067	519	0.8589	539	0.9997

Finally we remark, that in the introduced new PERT modeling it is also possible to realize some type of on–line control. This can be based on ideas of rolling horizon, i.e., recalculations of the predetermined activity starting times can take place after observing that one or more of the activities cannot be started in the predetermined start time, or in an opposite way, after observing that the predetermined start times became too loose for almost all of the activities.

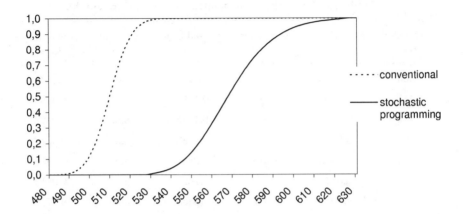

Fig. 3. The cdf's of the project completion times determined by the multivariate normal and stochastic programming approaches.

Acknowledgments. The authors gratefully acknowledge the careful reading and constructive suggestions of two anonymous referees of the paper.

This work was partly supported by the grant No. T047340 of the Hungarian National Grant Office (OTKA).

References

1. Clingen, C.T. (1964): A modification of Fulkerson's PERT algorithm. Operations Research **12**, 629–632
2. Devroye, L.P. (1979): Inequalities for the completion time of stochastic PERT networks. Mathematics of Operations Research **4**, 441–447
3. Dodin, B.M. (1985): Bounding the project completion time distribution in PERT networks. Operations Research **33**, 862–881
4. Elmaghraby, S.E. (1967): On the expected duration PERT type networks. Management Science **13**, 299–306.
5. Fourer, R., Gay, D.M., Kernighan, B.W. (1993): AMPL A Modeling Language for Mathematical Programming. Boyd & Fraser Publishing Company, Danvers Massachusetts
6. Fulkerson, D.R. (1962): Expected critical path lengths in PERT networks. Operations Research **10**, 808–817
7. Kleindorfer, G.B. (1971): Bounding distributions for a stochastic acyclic network. Operations Research **19**, 1586–1601
8. Long, J., Prékopa, A., Szántai, T. (2004): New bounds and approximations for the probability distribution of the length of the critical path. In: Marti,K.,Ermoliev,Yu.,Pflug,G. (Eds.): Dynamic Stochastic Optimization. Lecture Notes in Economics and Mathematical Systems, 532, Proceedings of the IFIP/IIASA/GAMM/–Workshop on "Dynamic Stochastic Optimization", held

at the International Institute for Applied Systems Analysis – IIASA, Laxenburg/Vienna, Austria, March 11–14, 2002, Springer-Verlag, Berlin Heidelberg, 293–320

9. Malcolm, D.G., Roseboom, J.H., Clark, C.E., Fazar, W. (1959): Application of a technique for research and development program evaluation. Operations Research **7**, 646–669

10. Meilijson, I., Nádas, A. (1979): Convex majorization with an application to the length of critical paths. Journal of Applied Probability **16**, 671–677

11. Monhor, D. (1987): An approach to PERT: Application of Dirichlet distribution. Optimization **1**, 113–118.

12. Nádas, A. (1979): Probabilistic PERT. IBM J. Res. Dev. **23**, 339–347

13. Robillard, P., Trahan, M. (1976): Expected completion time in PERT network. Operations Research **24**, 177–182

14. Robillard, P., Trahan, M. (1977): The completion time of PERT network. Operations Research **25**, 15–29

15. Shogan, A.W. (1977): Bounding distributions for a stochastic PERT network. Networks **7**, 359–381

16. Szántai, T. (1988): A computer code for solution of probabilistic constrained stochastic programming problems. In: Ermoliev,Yu.,Wets, R. J-B. (Eds.): Numerical Techniques for Stochastic Programming Problems. Springer Series in Computational Mathematics, Springer Verlag, 229–235

17. Szántai, T. (1997): Probabilistic constrained programming and distributions with given marginals. In: Benes,V.,Stepan,J. (Eds.): Distributions with given marginals and moment problems. Proceedings of the 3rd Conference on "Distributions with Given Marginals and Moment Problems", held at Czech Agricultural University, Prague, Czech Republic, September 2-6, 1996, Kluwer Academic Publishers, Dordrecht Boston London, 205–210

18. Vanderbei, R.J. (1999): LOQO: An interior point code for quadratic programming. Optimization Methods and Software **12**, 451–484

19. Vanderbei, R.J. (1999): LOQO user's manual – version 3.10. Optimization Methods and Software **12**, 485–514

20. Wallace, S.W. (1989): Bounding the expected time-cost curve for a stochastic PERT network from below. Operations Research Letters **8**, 89–94

21. Weiss, G. (1986): Stochastic bounds on distributions of optimal value functions with applications to PERT, network flows and reliability. Operations Research **34**, 595–605

Towards Implementable Nonlinear Stochastic Programming

L. Sakalauskas

Dept. of Operational Research
Institute of Mathematics and Informatics
Akademijos st. 4, Vilnius, 08663 Lithuania
e-mail: sakal@ktl.mii.lt

Abstract. The concept of implementable nonlinear stochastic programming by finite series of Monte-Carlo samples is surveyed addressing the topics related with stochastic differentiation, stopping rules, conditions of convergence, rational setting of the parameters of algorithms, etc. Our approach distinguishes itself by treatment of the accuracy of solution in a statistical manner, testing the hypothese of optimality according to statistical criteria, and estimating confidence intervals of the objective and constraint functions. The rule for adjusting the Monte-Carlo sample size is introduced which ensures the convergence with the linear rate and enables us to solve the stochastic optimization problem using a reasonable number of Monte-Carlo trials. The issues of implementation of the developed approach in optimal decision making, portfolio optimization, engineering are considered, too.

Keywords: stochastic programming, Monte-Carlo method, stopping rules, convergence.

1   Introduction

Optimal decision making in finance, engineering, management, ecology, statistics, etc. is often related with an uncertainty conditioned by variuos causes. If this uncertainty is described in a theoretical probabilistic way, the associated stochastic optimization problem can be formulated as a mathematical programming task with expectations included in the objective function and constraints. The idea of statistical simulation of these expectations is basic in many iterative methods of stochastic programming (SP), whose behaviour is studied theoretically very well, when the number of iterations infinitely increased, i.e., *in principle*, however, in real situations we must make an optimal decision after performing only a finite number of computer aided procedures. In the latter case, development of the concept of *implementable methods* becomes a useful way for practical needs. First monographs, where the methods *of principle* and the *implementable* ones were sequentially distinguished, appeared two-three decades ago (see, i.e., Polak (1971)). Implementable methods for deterministic optimization are best known and investigated at the present time (see Gill *et al.* (1981), Bertsekas (1982), Reklaitis *et al.* (1983), Dennis and Shnabel (1983), etc.).

Following this concept, we treat a method as implementable, if it satisfies the main requirements on finiteness and efficiency, namely, *a simple definition is given for all the parameters of the method, ensuring the finding of optimum with an admissible accuracy in a rational way after a finite number of computer aided procedures.*

The development of stochastic implementable methods meets a lot of numerical problems. Let us mention only the estimation of the objective and constraint functions as well as their gradients, often having rather a complicated analytical form, estimation and control of errors due to randomization of algorithms, stopping rules, etc. Two main issues should be mentioned to overcome obstacles in the creation of methods for practical stochastic optimization:

- the model risk, that appears due to randomization of the numerical search procedure, related with the accuracy of stochastic approximations, stopping rules, etc.;
- the great volume of computations needed to solve the SP problem in a stochastic way.

Development of algorithms of stochastic optimization is grounded frequently by the ideas of stochastic approximation and the Monte-Carlo method. The stochastic approximation theory, focussed on the term of a stochastic quasigradient and certain rules for the step-length regulation, remains a widely theoretically developed approach, because the above-mentioned issues - absence of stopping rules, low convergence, etc. - remain great obstacles for their practical implementation (Robins and Monro (1951), Arrow *et al.* (1958), Vazan (1972), Ermolyev (1976), Mikhalewitch *et al.* (1987), Ermolyev and Wets (1988), Uriasyev (1990), Nurminski (1991), Kushner and Jin (2003), Marti (2005), etc.).

It is well known when applying the Monte-Carlo method to stochastic optimization that an infinite increase in the Monte-Carlo sample size leads to the convergence to a desired solution (see, Rubinstein (1983), Shapiro (1989), etc.). However, the mentioned issues arise again using this fact in numerical implementation: first, it is not always so clear, how to choose the sample size in order to assure the establishment of an optimum with a desired accuracy, second, the numerical iterative optimization, in this case, can be expensive, in particular, for large sample sizes, third, certain problems may occur, as the probabilistic measure of uncertain parameters depends on the optimized ones, etc. A lot of these problems have been considered in (Shapiro and Homem-de-Mello (1998)), including statistical optimality testing and sample size regulation, however, concluding finally that "this requires further theotretical and numerical investigation".

An interesting way to ensure the convergence in stochastic optimization is related to the application of methods with a relative stochastic gradient error. The theoretical scheme of such methods requires the variance of the stochastic gradient to be varied in the optimization procedure so that it remained proportional to the square of the gradient norm (see Polyak (1987)). This approach offers an opportunity to develop implementable algorithms of stochastic optimization. We survey them in this paper using a finite series of Monte-Carlo estimators for algorithms construction (see also Sakalauskas (1992), Sakalauskas (2000), Sakalauskas (2002), Sakalauskas (2004)). The accuracy of the solution and the model risk due to Monte-Carlo randomization are treated in a statistical manner, testing the hypothesis of optimality according to statistical criteria. The rule for adjusting the Monte-Carlo sample size is introduced to ensure the convergence and to find the solution to the SP problem from rational volume of Monte-Carlo trials. Stochastic differentiation techniques are briefly surveyed and a set of Monte-Carlo estimators for stochastic programming is introduced. Some points for practical realization of the developed approach and its implementation in decision making, enginering, finance are considered, too.

2   Stochastic Differentiation and Monte-Carlo Estimators

Let us consider a constrained stochastic optimization problem in general, where expectations are included in the objective function and/or several constraints:

$$
\begin{aligned}
F_0(x) &= E f_0(x, \xi) \to \min, \\
F(x) &= E f(x, \xi) \le 0, \\
\Psi(x) &= E \psi(x, \xi) = 0, \\
\Phi(x) &= 0, \quad x \in R^n_+,
\end{aligned}
\tag{1}
$$

where $F_0\colon R^n \to R$ is the objective scalar function, vector functions $F\colon R^n \to R^m$, $\Psi\colon R^n \to R^l$ describe, respectively, the inequality and equality expected-value constraints as the vector function $\Phi\colon R^n \to R^k$ corresponds to the deterministic ones, $\xi \in \Omega$ is an elementary event in a probability space (Ω, Σ, P_x), the functions $f\colon R^n \times \Omega \to R^m$, $\psi\colon R^n \times \Omega \to R^l$ satisfies certain conditions on integrability and differentiability, the measure P_x is absolutely continuous and parameterized with respect to x in general, i.e., it can be defined by the density function $p\colon R^n \times \Omega \to R_+$, and E is the symbol of mathematical expectation. We restrict ourselves to the interesting theoretical and practical cases, where deterministic constraints should be valid during all the steps of algorithm. Such situations are met, for instance, in the finance portfolio optimization, when the total portfolio weighting should remain the same and strictly equal to one (see Ziemba and Mulvey (1998), etc.). Following this assumption it is useful for further consideration to define the *feasible set*:

$$
W = \left\{ x \,|\, \Phi(x) = 0, x \ge 0 \right\}
\tag{2}
$$

Since mathematical expectations in (1) are computed explicitly only in rare cases, all the more it is complicated to analytically compute the gradients of functions, containing this expression. The Monte-Carlo method is a universal and convenient tool of estimating these expectations and it could be applied to estimate derivatives, too. The procedures of gradient evaluation are often constructed by expressing a gradient as an expectation and afterwards evaluating this expectation by means of statistical simulation (see, Rubinstein (1983), Rubinstein and Shapiro (1993), Kall and Wallace (1994), Prekopa (1995), etc.). Thus, let us consider the expectation

$$F(x) = Ef(x, \omega) \equiv \int_{R^n} f(x, y) \cdot p(x, y) \mathrm{d}\, y, \tag{3}$$

where the function f and the density function p are differentiable with respect to x in the entire space R^n. Let us denote the support of measure P_x as $S(x) = \{y | p(x, y) > 0\}$, $x \in R^n$. Then it is not difficult to see that the vector-column of the gradient of this function could be expressed as

$$\nabla F(x) = E\big(\nabla_x f(x, \omega) + f(x, \omega) \cdot \nabla_x \ln p(x, \omega)\big), \tag{4}$$

(we assume $\nabla_x \ln p(x, y) = 0$, $y \notin S(x)$). We also see from the equation

$$E\nabla_x \ln p(x, \omega) = 0$$

(which is obtained by differentiating the equation $\int_\Omega p(x, y) \mathrm{d}\, y = 1$) that there follow various expressions of the gradient follow. For instance, the formula

$$\nabla F(x) = E\big(\nabla_x f(x, \omega) + (f(x, \omega) - f(x, E\omega)) \cdot \nabla_x \ln p(x, \omega)\big) \tag{5}$$

serves as an example of such an expression.

Thus we see that it is possible to express the expectation and its gradient through a linear operator from the same probability space. Hence, operators (3), (4), and (5) can be estimated by means of the same Monte-Carlo sample. It depends on the task solved, which formula, (4) or (5), is better for use. For instance, expression (4) can provide smaller variances of gradient components than (5), if the variances of ξ components are small. This fact is important when implementing the idea of smoothing in the stochastic approximation, when the smoothing parameter tends to zero (Bartkute and Sakalauskas (2006)).

Let us introduce a set of Monte-Carlo estimators needed for the construction of a stochastic optimization procedure. In solving the problems of kind (1), suppose it is possible to get finite sequences of realizations (trials) of ξ at any point x and after that to compute the values of functions f_0, f, ψ, p as well as of their gradients for these realizations.

Thus, assume the Monte-Carlo sample to be given for some $x \in R^n$:

$$Y = (y^1, y^2, \ldots, y^N), \tag{6}$$

where y^i are independent random vectors identically distributed with the density $p(x, \cdot) \colon \Omega \to R_+$, i.e., copies of ξ. Monte'Carlo estimators of the objective and constraint functions and their sampling variances are as follows:

$$\widetilde{F}_0(x) = \frac{1}{N} \sum_{j=1}^{N} f_0(x, y^j), \tag{7}$$

$$\widetilde{D}_{F_0}^2(x) = \frac{1}{N-1} \sum_{i=1}^{N} \left(f_0(x, y^j) - \widetilde{F}_0(x) \right)^2. \tag{8}$$

Monte-Carlo estimators of the components F, Ψ are obtained in a similar way. We will use a technique of stochastic differentiation developed on the basis of likelihood ratios, which allows the estimation of the objective/constraint functions and the coprresponding gradients, using the same Monte-Carlo sample (see, Rubinstein and Shapiro (1993), Bartkute and Sakalauskas (2006)). Hence, the gradient estimate of the objective function follows by virtue of (4) (or (5)) using the sample (6):

$$\widetilde{\nabla}_x F_0(x) = \frac{1}{N} \sum_{j=1}^{N} g_0^j, \tag{9}$$

where $g_0^j \equiv g_0(x, y^j) = \nabla_x f_0(x, y^j) + f_0(x, y^j) \cdot \nabla_x \ln p(x, y^j)$ is the stochastic gradient, namely, $E g_0^j = \nabla F_0(x)$. The estimators of the components of vector-functions F, Φ and that of their gradients are defined analogously.

Let us introduce the partial Lagrange function of the problem 1):

$$L(x, \lambda, \mu) = F_0(x) + \lambda \cdot F(x) + \mu \cdot \Phi(x), \tag{10}$$

where $\lambda \geq 0$, μ are the vectors of Lagrange multipliers of respective dimension, that may be treated as an expectation of the stochastic Lagrange function

$$l(x, \lambda, \mu, \xi) = f_0(x, \xi) + \lambda \cdot f(x, \xi) + \mu \cdot \varphi(x, \xi).$$

Then the estimator of the gradient of the Lagrange function:

$$\widetilde{\nabla}_x L(x, \lambda, \mu) = \frac{1}{N} \sum_{j=1}^{N} G^j, \tag{11}$$

might be introduced according to (4) (or (5)) as the mean of identically distributed independent vectors

$$G^j = \nabla_x l(x, \lambda, \mu, y^j) + l(x, \lambda, \mu, y^j) \cdot \nabla_x \ln p(x, y^j), \quad i = \overline{1, N}.$$

The sampling covariance matrix

$$A = \frac{1}{N} \sum_{j=1}^{N} (G^j - \widetilde{\nabla}_x L) \cdot (G^j - \widetilde{\nabla}_x L)' \tag{12}$$

will also used later on.

3 Statistical Verification of the Optimality Hypothesis

A possible decision on optimal solution finding should be examined during the optimization process. Let x^+ be the solution of (1). By virtue of the Kuhn-Tucker theorem (see, e.g., Bertsekas (1982)) there exist values. $\lambda^+ \geq 0$, μ^+, such that

$$\left(\nabla L_x(x^+, \lambda^+, \mu^+)\right)_W = 0, \quad \lambda^+ \cdot F(x^+) = 0, \quad \Phi(x^+) = 0, \qquad (13)$$

0where the denotation g_Q denotes the projection of the vector g to some set Q.

Since we estimate the objective/constraint functions and that of their gradients by statistical simulation, we can test only the statistical optimality hypothesis. Thus, the decision on finding optimum could be made on the basis of Monte Carlo estimators of objective/constraint functions and their gradients, if, first, there is no reason to reject the hypothesis on the validity of conditions (13), and second, the objective and the constraint functions are estimated with a permissible accuracy.

Note that the distribution of sampling averages (7), (9), and (11) (as well as that of their projections) can be approximated by the one- and multi-dimensional Gaussian law (see, e.g., (Bhattacharya and Ranga Rao (1976), Box and Wilson (1962), Gotze and Bentkus (1999)). Then it is convenient to test the validity of the first optimality condition in (12) by means of the well-known multidimensional Hotelling T^2-statistics (see, e.g., Aivazian *et al.* (1985), Krishnaiah, and Lee (1980), etc.). Hence, the optimality hypothesis may be accepted for some point x with significance μ, if the following condition is true:

$$(N - n) \cdot (\widetilde{\nabla}_x L)' \cdot A^{-1} \cdot (\widetilde{\nabla}_x L)/n \leq Fish(\mu, n, N - n), \qquad (14)$$

where $Fish(\mu, n, N^t - n)$ is the μ-quantile of the Fisher distribution with $(n, N^t - n)$ degrees of freedom, $\widetilde{\nabla}_x L = \widetilde{\nabla}_x L(x, \lambda)$ is the estimate (11) of the Lagrange function gradient projection to the set (2), and A is the projection of normalizing matrix (12) estimated at the point (x, λ, μ) to the set W and N is the size of sample (6).

Let us consider a numerical experiment for the study of the proposed criteria.

Example 1. Since functions met in practice are typically of a quadratic character with some nonlinear disturbance in a neighbourhood of the optimal point, let us consider a test example

$$F(x) \equiv Ef_0(x + \omega) \to \min,$$

where $f_0(y) = \sum_{i=1}^{n}(a_i y_i^2 + b_i \cdot (1 - \cos(c_i \cdot y_i)))$, $y_i = x_i + w_i$, w_i are random and normally $N(0, d^2)$ distributed, $d = 0.5$, $(a = (8.00, 5.00, 4.30, 9.10, 1.50, 5.00, 4.00, 4.70, 8.40, 10.00)$, $b = (3.70, 1.00, 2.10, 0.50, 0.20, 4.00, 2.00, 2.10, 5.80, 5.00)$, $c = (0.45, 0.50, 0.10, 0.60, 0.35, 0.50, 0.25, 0.15, 0.40, 0.50))$.

Such an example may be treated as a stochastic version of the deterministic task of mathematical programming with the objective function, where the controlled variables are measured under some Gaussian error with the variance d^2. Note that our task is not convex. We examine the distribution of the statistic in criterion (14) by means of statistical simulation, solving the test task, when $p = 0$, $n = 2$. The optimal point is known in this case: $x_+ = 0$. Let the gradient of considered functions be evaluated numerically by the Monte-Carlo estimator following from (9) or (11). Thus, 400 Monte-Carlo samples of size $N = (50, 100, 200, 500, 1000)$ were generated and the T^2-statistics in (14) were computed for each sample. The hypothesis on the difference of empirical distribution of this statistic from the Fisher distribution was tested according to the criteria ω^2 and Ω^2 (see Bolshev and Smirnov (1983)). The value of the first criterion on at the optimal point for $N = 50$ is $\omega^2 = 0.2746$ against the critical value 0.46 ($p = 0.05$), and that of the next one is $\Omega = 1.616$ against the critical value 2.49 ($p = 0.05$). Besides, the hypothesis on the coincidence of empirical distribution of the considered statistics to the Fisher distribution was rejected at the points differing from the optimal one according to the criteria ω^2 and Ω^2 (if $r = |x - x^+| \geq 0.1$). So the distribution of the multidimensional T^2-statistics, in our case, can be approximated rather well by the Fisher distribution even in the case of not very large samples ($N \cong 50$).

Next the dependencies of the stopping probability according to criterion (14) on the distance $r = |x - x^+|$ to the optimal point were studied. These dependencies are presented in Fig. 1 (for confidence $\alpha = 0.95$). So we see that by regulating the sample size we are able to test the optimality hypothesis in a statistical way and to evaluate the objective and contraint functions with a desired accuracy.

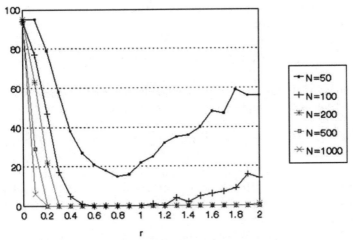

Fig. 1. Stopping probability according to (13), $\mu = 0.95$.

Now we make important remarks as to the control of the model risk due to randomization, because real values of the objective and constraint functions remain unknown and we know only their estimates, produced by statistical simulation. Note, that the upper confidence bound of the constraint function estimate with the appropriate confidence probability has to be used for testing the validity of expected-value inequality constraints, i.e., the second optimality condition in (13). Regarding equality constraints, we have to test wether the appropriate confidence interval includes zero. Besides, the sample size N has to be sufficiently large in order to estimate the confidence intervals of the objective and the constraint functions with a permissible accuracy. In the latter cases it is convenient to use the asymptotic normality again and to approximate the respective confidence bounds by means of the sampling variance (8). Clearly, since the statistical testing of the optimality hypothesis is grounded by the convergence of distribution of sampling estimates to the Gaussian law, additional points could be introduced, considering the rate of convergence to the normal law.

Summarizing we introduce some rules, the violation of at least one of which enables us to reject the optimality hypothesis. Namely, let certain values (x, λ, μ) be given for which Monte-Carlo sample (6) is generated and the corresponding Monte-Carlo estimators are estimated. Then, if:

a) criterion (14) does not contradict the hypothesis that the projection to the feasible set of the Lagrange function gradient is equal to zero (the first condition in (13));

b) the inequality constraint conditions (the second condition in (13)) vanishs with a given probability β:

$$\widetilde{F}_i(x^t) + \eta_\beta \cdot \widetilde{D}_{F_i}(x^t) \leq 0, \quad i = \overline{1, m},$$

c) the equality constraint conditions (third condition in (13)) vanishes with a given probability β

$$\left|\widetilde{\Phi}_i(x^t)\right| \leq \eta_\beta \cdot \widetilde{D}_{\Phi_1}(x^t), \quad i = \overline{1, l},$$

d) the estimated lengths of the confidence interval of the objective function and that of constraints do not exceed the given accuracy ε_i:

$$2\eta_\beta \cdot D_{F_i}/\sqrt{N} \leq \varepsilon_i, \quad i = \overline{0, m}, \quad 2\eta_\beta \cdot D_{\Phi_i}/\sqrt{N} \leq \varepsilon_i, \quad i = \overline{1, l},$$

where η_β is the β-quantile of the standard normal distribution, then there are no reasons to reject the hypothesis on finding optimum. Therefore there are reasons to stop the optimization and make a decision on finding optimum with an admissible accuracy.

If at least one condition a)–d) is not satisfied, then new improved values of (x, λ, μ) should be chosen and other samples generated, etc.

4 Optimization by Monte-Carlo Estimators with Sample Size Regulation

Now we start developing a stochastic optimization procedure by Monte-Carlo estimators. It has been noticed of late that in stochastic optimization only the first order methods are working. Therefore we confine ourselves only to the gradient type methods. Our goal is to show how well-known deterministic approaches might be generalized for the tochastic case. Since the description of the algorithm for full problem (1) would be rather cumbersome, we consider partial cases emphasizing the peculiarities arising in each case. Thus, we focus on unconstrained stochastic optimization, optimization with expected-value constraints and stochastic optimization with deterministic constraints, which should be strictly fullfilled. In the latter case, a nonlinear stochastic optimization with deterministic linear constraints is of interest.

4.1 Unconstrained Stochastic Optimization

Thus, we consider the problem:

$$F(x) = Ef(x, \xi) \to \min, \quad x \in R^n,$$

under the assumptions disscussed in the statement (1). Let an initial point $x^0 \in R^n$ be given, random sample (6) of a certain initial size N^0 be generated at this point, and the corresponding Monte-Carlo estimators (7), (8), (9), (12) be computed. Now, the iterative stochastic procedure of gradient search might be introduced:

$$x^{t+1} = x^t - \rho \cdot \widetilde{G}(x^t), \tag{15}$$

where $\rho > 0$ is a certain step-length multiplier.

Letu s consider the choice of random sample (6) size when this procedure is iterated. Sometimes this sample size is taken to be fixed in all the iterations of the optimization process and choosen sufficiently large to ensure the required accuracy of estimates in all the iterations (see, Antreich and Koblitz (1982), Belyakov *et al.* (1985), Jun Shao (1989), Shapiro (1989), etc.). Very often this guaranteeing size is about 1000–1500 trials or more, and, if the number of optimization steps is large, solution of stochastic optimization problem can obtain substantial computation (Jun Shao (1989)). On the other hand, it is well known that the fixed sample size, although very large, is sufficient only to ensure the convergence to some neighbourhood of the optimal point (see, e.g., Polyak (1987), Sakalauskas (1997)).

Note, that there is no great necessity to compute estimators with a high accuracy on starting the optimization, because then it suffices only to evaluate approximately the direction leading to the optimum. Therefore, one can obtain not so large samples at the beginning of the optimum search and later

on increase the samples size so as to obtain the estimate of the objective function with a desired accuracy only at the time of decision making on finding
the solution to the optimization problem. We pursue this purpose by choosing
the sample size at every next iteration inversely proportional to the square
of the gradient estimator from the current iteration:

$$N^{t+1} \geq \left[\frac{C}{\rho \cdot |\widetilde{G}(x^t)|^2}\right] + 1, \tag{16}$$

where $C > 0$ is a constant, $[\cdot]$ means the integer part of the number. The
following theorems justifies such an approach.

Theorem 1. *Let the function $F: R^n \rightarrow R$, expressed as expectation, be
bounded: $F(x) \geq F^+ > -\infty$, $\forall x \in R^n$, and differentiable, such that the
gradient of this function sattisfies the Lipshitz condition with the constant
$L > 0$, $\forall x \in R^n$.*

*Assume that for any $x \in R^n$ and any number $N \geq 1$ one can obtain sam-
ple (6) of independent identically distributed vectors with the density $p(x, \cdot)$
and compute estimates (7), (9) such, that the variance of the stochastic gra-
dient norm in (9) is uniformly bounded: $E|g(x, \xi) - \nabla F(x)|^2 < K$, $\forall x \in R^n$.*

*Let the initial point $x^0 \in R^n$ and initial sample size N^0 be given andran-
dom sequences $\{x^t\}_{t=0}^\infty$ be defined according to (15), where the sample size
is iteratively changed according to the rule (16), where $C > 0$ is a certain
constant. Then almost surely (a.s.):*

$$\lim_{t \to \infty} |\nabla F(x^t)|^2 = 0,$$

if $0 < \rho \leq \frac{1}{L}$, $C \geq 4K$.

*If in addition, the function $F(x)$ is twice differentiable and $\|\nabla^2 F(x)\| \geq
l > 0$, $\forall x \in R^n$, then the estimate*

$$E\left(|x^t - x^+|^2 + \frac{\rho \cdot K}{N^t}\right) \leq \left(|x^0 - x^+|^2 + \frac{\rho \cdot K}{N^0}\right)\left(1 - \rho\left(l - \frac{K \cdot L^2}{C}\right)\right)^t, \quad t = 0, 1, 2, \ldots,$$

*holds for $0 < \rho \leq \min[\frac{1}{L}, \frac{3}{4 \cdot (1+l)}]$, $C \geq K \cdot \max[4, \frac{L^2}{l}]$, and where x^+ is a
stationary point.*

The proof of the theorem is given by (Sakalauskas (2000)).
    The step length ρ could be determinated experimentally or using the
method of a simple iteration (see, e.g., Kantorovitch and Akilov (1958),
Sakalauskas (1997), Sakalauskas (2000)). The choice of constant C or of the
best metrics for computing the stochastic gradient norm in (16) requires a
separate study. Such a version for regulating the sample size might be proposed for practical application:

$$N^{t+1} = \min\left(\max\left(\left[\frac{n \cdot Fish(\gamma, n, N^t - n)}{\rho \cdot (\widetilde{G}(x^t))' \cdot (A(x^t))^{-1} \cdot (\widetilde{G}(x^t))}\right] + n, N_{\min}\right), N_{\max}\right) \tag{17}$$

where $Fish(\gamma, n, N^t - n)$ is the γ-quantile of the Fisher distribution with $(n, N^t - n)$ degrees of freedom. Minimal and maximal values $N_{\min}$ (usually $\sim$20–50) and $N_{\max}$ (usually $\sim 1000 - -2000$) are introduced to avoid great fluctuations of sample size in iterations. Note that $N_{\max}$ may also be chosen from the conditions on the permissible confidence interval of estimates of the objective function (Sakalauskas (2000)). The choice $C = n \cdot Fish(\gamma, n, N^t - n) \approx \chi_\gamma^2(n)$ and estimation of the gradient norm in a metric induced by the sampling covariance matrix (12) is convenient for interpretation, because in such a case, a random error of the stochastic gradient does not exceed the gradient norm approximately with probability $1 - \gamma$. The rule (17) implies rule (16) and, in its turn, the convergence by virtue of the moment theorem for multidimensional Hotelling T^2-statistics with arbitrarily distributed vectors (see Bentkus and Goetze (1999)).

For stopping of the method (15), (17) it suffices to test the hypothesis of equality of the objective function gradient to zero according to the point a) of the previous section and to verify the admissible length of the confidence interval of this function according to the point d). As follows from the Theorem 1, the method stops a.s. after a finite number of iterations. The numerical study of the method (15), (7) is discussed in the next section, too.

4.2   Nonlinear Optimization with Expected-Value Constraints

Let us consider problem (1) in the absence of deterministic constraints. The stochastic implementation of various versions of constraint optimization procedures distinguishes itself by the same peculiarities. We may analyse these peculiarities in the stochastic programming problem when constructing a stochastic version of the Arrow-Hurvitz-Udzava procedure (Bertsekas (1982)).

Theorem 2. *Let the functions $F_0 \colon R^n \to R$, $F \colon R^n \to R^m$, $\Psi \colon R^n \to R^k$ expressed as expectations, be convex and twice differentiable. Assume that all the eigen-values of the matrix of second derivatives $\nabla_{xx}^2 L(x, \lambda^+, \mu^+)$, $\forall x \in R^n$ of the Lagrange function (10) be uniformly bounded and belonging to some interval $[m, M]$, $m > 0$, and, besides,*

$$\left|\nabla F(x) - \nabla F(x^+)\right| \leq \frac{m}{2 \cdot M} \left|\nabla F(x^+)\right|, \quad \left|\nabla \Psi(x) - \nabla \Psi(x^+)\right| \leq \frac{m}{2 \cdot M} \left|\nabla \Psi(x^+)\right|,$$

$\left|\nabla F(x^+)\right| > 0$, $\forall x \in R^n$, where (x^+, λ^+, μ^+) is the point satisfying the Kuhn-Tucker conditions (13).

In addition, let for any $x \in R^n$ and any number $N \geq 1$ there exists a possibility exists to generate sample (6) and obtain estimates (7) and (11), where the conditions of uniform boundedness on variances and covariances of the estimates introduced are valid: $E(f(x, \xi) - F(x))^2 < d$, $E(\psi(x, \xi) - \Psi(x))^2 < d$, $E|g(x, \xi) - \nabla F(x)|^2 < K$, $\forall x \in R^n$.

Let now the initial point $x^0 \in R^n$, the vectors $\lambda^0 \geq 0$, μ^0, and the initial sample size N^0 be given and the random sequence $\{x^t, \lambda^t, \mu^t\}_{t=0}^{\infty}$ be defined according to:

$$x^{t+1} = x^t - \rho \cdot \widetilde{\nabla}_x L(x^t, \lambda^t),$$
$$\lambda^{t+1} = \left(\lambda^t + \rho \cdot \alpha \cdot \widetilde{F}(x^t)\right)_+,$$
$$\mu^{t+1} = \mu_+^t + \rho \cdot \gamma \cdot \widetilde{\Phi}(x^t),$$

where Monte-Carlo estimators are obtained iteratively varying the sample size N^t according to the rule:

$$N^{t+1} \geq \frac{C}{\rho \cdot |\widetilde{\nabla}_x L(x^t, \lambda^t, \mu^t)|^2},$$

$C > 0$, $\alpha > 0$, $\gamma > 0$ are certain constants.

Then, there exist positive values $\bar{\rho}$, $\bar{C}$ such that

$$E\left(|x^t - x^+|^2 + \frac{|\lambda^t - \lambda^+|^2}{\alpha} + \frac{|\mu^t - \mu^+|^2}{\beta} + \frac{1}{N^t}\right) < B \cdot \beta^t, \quad t = 1, 2, \ldots,$$

for certain values $0 < \beta < 1$, $B > 0$, when $\rho < \bar{\rho}$, $C > \bar{C}$.

The proof is similar to that given by Sakalauskas (2002) in the case of one expected-value constraint and differs only in details. The multipliers ρ, α, γ are chosen typicallly in an experimental way. The sample size regulation rule analogous to (17) might be introduced under similar considerations.

For practical implementation, the modification of the method is useful:

$$x^{t+1} = x^t - \rho \cdot \widetilde{\nabla}_x L(x^t, \lambda^t),$$
$$\lambda^{t+1} = \left(\lambda^t + \rho \cdot \alpha \cdot \left(\widetilde{F}(x^t) + \eta_\beta \cdot D_F(x^t)\right)\right)_+, \tag{18}$$
$$\mu^{t+1} = \mu_+^t + \rho \cdot \gamma \cdot \widetilde{\Phi}(x^t),$$

which assures the validity of the inequality constraint with a given confidence, when only a finite number of iterations is taken.

The method considered is stopped according to the rules a)–d) discussed in the previous section. The numerical study of the algorithm presented is disscussed in the next section, too.

4.3   Stochastic Optimization with Deterministic Linear Constraints

Let us focus on the constrained nonlinear stochastic optimization, when constraints are only deterministic and should be strictly followed during computing. The gradient search approach with a projection to the feasible set would be a chance to create an optimizing sequence, however the problems of "jamming" or "zigzagging" are typical in such a case (Bersekas (1982)). To avoid them, we implement the ε – feasible directions approach. For the sake of simplicity, let us consider a *feasible set* consisting of only linear constraints:

$$W = \{x | Ax = b, \ x \geq 0\}, \tag{19}$$

where $b \in R^m$, A is the $n \times m$-matrix. The set of *feasible directions* is defined as follows:

$$V(x) = \{g \in \Re^n \mid Ag = 0, \forall_{1 \leq i \leq n} (g_j \geq 0, \text{ if } x_j = 0)\}.$$

Assume a certain multiplier $\hat{\rho} > 0$ to be given and the function $\rho_x : V(x) \to \Re_+$ defined:

$$\rho_x(g) = \begin{cases} \min\left\{\hat{\rho}, \min\limits_{\substack{g_j < 0, \\ 1 \leq j \leq n}} \left(-\dfrac{x_j}{g_j}\right)\right\}, & g \neq 0 \\ \hat{\rho}. \end{cases} \tag{20}$$

Thus $x + \rho \cdot g \in W$, as $\rho = \rho_x(g)$ for any $g \in V$, $x \in W$. Now let a certain small value $\hat{\varepsilon} > 0$ be given. We introduce the function $\varepsilon_x : V(x) \to \Re_+$:

$$\varepsilon_x(g) = \hat{\varepsilon} \max_{\substack{1 \leq j \leq n \\ g_j \leq 0}} \left\{ \min\{x_j, -\hat{\rho} \cdot g_j\}\right\}, \quad \forall x \in W, \tag{21}$$

and define the *ε-feasible set*

$$V_{\varepsilon}(x) = \left\{g \mid Ag = 0, \ \forall_{1 \leq i \leq n} (g_j \geq 0, \text{ if } (0 \leq x_j \leq \varepsilon_x(g)))\right\}. \tag{22}$$

Next, let the initial approximation of the solution $x^0 \in W$ and some initial Monte-Carlo sample size N^0 be given. We define the sequence $\{x^t, N^t\}_0^\infty$ in an iterative way by generating Monte-Carlo samples (6) and computing the corresponding Monte-Carlo estimators and setting

$$x^{t+1} = x^t - \rho_t \cdot \widetilde{G}^t, \tag{23}$$

$$N^{t+1} \geq \frac{\hat{\rho} \cdot C}{\rho^t \cdot |\widetilde{G}^t|^2}, \tag{24}$$

where $C > 0$ is a certain constant, $\rho^t = \rho_{x^t}(\hat{G}^t)$, $\widetilde{G}^t$ is an ε-feasible direction at the point x^t (i.e., the projection of gradient estimate (9) to the ε-feasible set (22)). The following theorem provides conditions for the convergence of the method (23), (24).

Theorem 3. *Let the function $F : W \to \Re$ be differentiable, the gradient of this function be Lipshitzian with the constant $L > 0$, $\sup_{x \in W} |\nabla F(x)| < \infty$, $\sup_{x \in W} F(x) < \infty$. Assume the set $W = \{x \in \Re^n | Ax = b, \ x \geq 0\}$ to be bounded and having more than one element, $b \in R^m$, A is the $n \times m$-matrix.*

Let theret be possible to generate samples (6) *for any size* $N > 1$ *and to compute corresponding estimates* (7), (9), *when* $E|f(x, \xi)| < \infty$, $E|g_j| < \infty$, $E|g_j - \nabla F(x)|^p < K$, $\forall x \in W$.

Then, starting from any initial approximation $x^0 \in W$ *and* $N^0 > 1$, *formulae* (22), (23) *define the sequence* $\{x^t, N^t\}_0^\infty$ *so that* $x^t \in W$, *and there exist values* $\bar\rho > 0$, $\varepsilon_0 > 0$, $\bar C > 0$ *such that*

$$\lim_{t\to\infty} \left|\nabla F(x^t)_{V^t}\right|^2 = 0 \quad (a.s.),$$

for $0 < \hat\rho \le \bar\rho$, $0 < \varepsilon \le 0$, $C \ge \bar C$.

The proof of the theorem is given in (Sakalauskas (2004)). Thus, we see that the application of the ε-feasible solution enables us to avoid "jamming" or "zigzagging" as well as the "jumping" arising due to the statistical nature of Monte-Carlo estimators. The linear convergence rate can be proved in principle, too, although the proof becomes rather cumbersome. Note, that, for numerical implementation, the rule similar to (17) is sometimes convenient for practice. The stopping of the method is performed according to points a) and d) of Section 3.

Let us summarize the theoretical results provided in this section. First, we may comment that approaches typical of the gradient search for the first order may be generalized to unconstrained and constrained nonlinear stochastic optimization, using the Monte-Carlo sample regulation proposed. It is important that such an approach makes it possible to ensure the convergence with the linear rate. Interesting comment follows from the linear convergence rate. First, the Monte-Carlo sample size regulation according to rule introduced allows us to develop reasonable, from the computational standpoint, stochastic methods for stochastic optimization. Namely, the method can start from a small initial size $N^0 = 20 - 50$, because there is no need to evaluate Monte-Carlo estimators with a high accuracy at the beginning of optimization, when it suffices only to estimate an approximate direction leading to the optimum. Later the sample size is increased with respect to rule introduced, gaining the values sufficient to evaluate the estimators with an admissible accuracy only at the final stage of optimization, when the gradient becomes small in the neighbourhood of optimum. The numerical experiments and testing corroborate such a conclusion.

On the other hand, as we see, the distance from the optimal solution $|x^t - x^+|^2$ and the sample size N^t have a linear rate of changing, dependent on the constant $0 < \beta < 1$, which is stipulated mostly by the conditionality of the matrix of second derivatives of the Lagrange function and constant C in expressions of the rule. It follows from the proof of the theorem that $\frac{N^t}{N^{t+1}} \approx \beta$ for large t. Then, by virtue of the formula of geometrical progression:

$$\sum_{i=0}^{t} N^i \approx N^t \cdot \frac{Q}{1 - \beta},$$

where Q is a certain constant. On the other hand, note that the stochastic error of estimates at the moment of the stopping decision mostly depends on the sample size (at this moment, say N^t). Hence the ratio of the total number of computations $\sum_{i=0}^{t} N^i$ with the sample size N^t, guaranteeing the permissible accuracy, can be considered as bounded a.s. and not depending on this accuracy. Consequently, if we have a certain resource to compute one value of the objective or constraint function with a permissible accuracy, then in practice the optimization requires only several times more computations. This enables us to create reasonable, from a computational viewpoint, stochastic methods for SP.

5   Numerical Study of Algorithms

We consider here several counterexamples to illustrate theoretical conclusions given by the theorems as well as to explore numerical properties of algorithms.

Example 2. Let us consider the stochastic optimization task with expected-value constraints

$$F(x) \equiv E f_0(x + w) \rightarrow \min,$$
$$P\big(f_1(x + \omega) \le 0\big) - p \ge 0, \tag{25}$$

where $f_1 = \sum_{i=1}^{n}(y_i + 0.5)$, the function $f_0(x)$ is defined and other constants are given in Example 1.

Now, let us consider the results, obtained for this example, by iterating the procedure (18) 40 times and changing the sample size according to the rule (17), where $N^0 = N_{\min} = 50$ and $N_{\max}$ is chosen. The initial data were as follows: $p = (0.0, 0.3, 0.6, 0.9)$, $x^0 = (-1, -1)$, $n = (2, 5, 10)$, $\alpha = 0.1$, $\rho = 20$, $\beta = 0.95$, $\mu = 0.95$, $\gamma = 0.95$, $\varepsilon_0 = 5\%$, $\varepsilon_1 = 0.05$. The optimization was repeated 400 times. Conditions a) – d) were satisfied although once for all the paths of optimization. So a decision could be made on finding the optimum finding with a permissible accuracy for all the paths (the sampling frequency of stopping after t iterations with the confidence intervals is presented in Fig. 2 ($p = 0.6$, $n = 2$)). Mean, minimal, and maximal values of the amount of iterations and that of the total Monte-Carlo trials are presented in Table 1, that are necessary to solve the optimization task ($p = 0.6$, $n = 2$). The amount of iterations and that of total Monte-Carlo trials, needed for stopping, dependent on the dimension n, are presented in Table 2 ($n = 2$). The sampling estimates of the stopping point are $\tilde{x}_{stop} = (0.006 \pm 0.053, -0.003 \pm 0.026)$ for $p = 0$, $n = 2$ (compare with $x_{opt} = (0, 0)$). These results illustrate that the algorithm proposed can be successfully applied when the objective and constraint functions are convex and smooth only in a neighbourhood of the optimum.

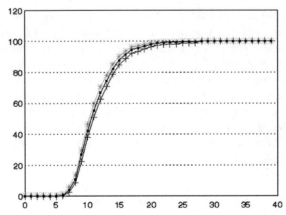

Fig. 2. Frequency of stopping (with confidence interval).

Table 1.

p	Amount of iterations (t)			Total amount of trials $(\sum_t N_t)$		
	Min	Mean	Max	Min	Mean	Max
0.0	6	11.5±0.2	19	1029	2842±90	7835
0.3	4	11.2±0.3	27	1209	4720±231	18712
0.6	7	12.5±0.3	29	1370	4984±216	15600
0.9	10	31.5±1.1	73	1360	13100±629	37631

Table 2.

N	Amount of iterations (t)			Total amount of trials $(\sum_t N_t)$		
	Min	Mean	Max	Min	Mean	Max
2	6	11.5±0.2	19	1029	2842±90	7835
5	6	12.2±0.2	21	1333	3696±104	9021
10	7	13.2±0.2	27	1405	3930±101	8668

The averaged dependencies of the objective function $\widetilde{E}F_0^t$, the constraint $\widetilde{E}F^t$, the Lagrange multiplier $\widetilde{E}\lambda^t$ and the sample size $\widetilde{E}N^t$ by the iteration number t are given in Figs. 3–6 to illustrate the convergence and behaviour of the optimization process ($p = 0.6$, $n = 2$). Also, one path of realization of the optimization process is given to illustrate the stochastic character of this process in these figures.

Example 3. In this example, we consider the two-stage stochastic linear optimization problem. In fact, such problems are solved by means of nonlinear optimization (Shapiro and Homem-de-Mello).

We consider the manpower planning problem (King (1988)), where the employer must decide upon the base level of regular staff at various skill levels.

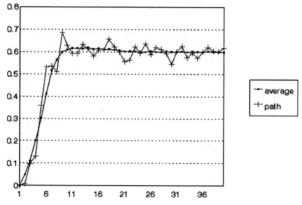

Fig. 3. Change of the objective function.

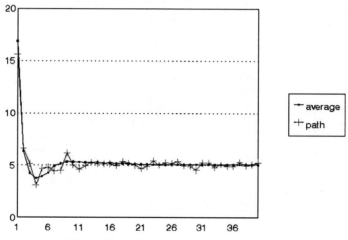

Fig. 4. Change of the constraint function.

The recourse actions available are regular staff overtime or outside temporary help in order to meet unknown demand for service at the minimum cost.

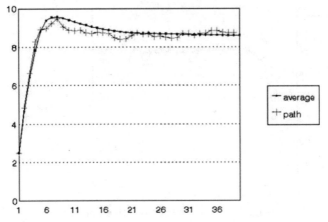

Fig. 5. Change of the Lagrange multiplier.

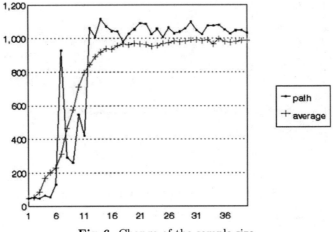

Fig. 6. Change of the sample size.

The problem is as follows:

choose $x = (x_1, x_2, x_3)$ to minimize:

$$F(x, z) = \sum_{j=1}^{3} c_j \cdot x_j + \sum_{t=1}^{12} E \min_{y,z} \left(\sum_{j=1}^{3} (q_j \cdot y_{j,t} + r_j \cdot z_{j,t}) \right)$$

subject to:

$$x_j \geq 0, \quad y_{j,t} \geq 0, \quad z_{j,t} \geq 0,$$

$$\sum_{j=1}^{3}(y_{j,t} + z_{j,t}) \geq w_t - \alpha_t \sum_{j=1}^{3} x_j, \quad t = 1,\ldots,12,$$

$$y_{j,t} \leq 0.2 \cdot a_t x_j, \quad j = 1,2,3, \quad t = 1,\ldots,12,$$

$$\gamma_{j-1}(x_j + y_{j-1,t} + z_{j-1,t}) - (x_j + y_{j-1,t} + z_{j-1,t}) \geq 0,$$
$$j = 2,3, \quad t = 1,\ldots,12,$$

where
    x_j: base level of regular staff at skill level $j = 1,2,3$,
    $y_{j,t}$: amount of overtime help,
    $z_{j,t}$: amount of temporary help,
    c_j: cost of regular staff at skill level $j = 1,2,3$,
    q_j: cost of overtime,
    r_j: cost of temporary help,
    w_t: demand for services at period t,
    α_t: anticipated absentees rate for regular staff at time t,
  γ_{j-1}: ratio of amount of skill level j per amount of $j-1$ required,
the demands ξ^t, $t = \overline{1,12}$, are independent normal: $N(\mu_t, \sigma_t^2)$, where $\mu^t = l \cdot \sigma^t$. Initial data and other details can be found in (King (1988)).

The results of solving this task by the method of unconstrained optimization (16), (18) are given in Table 3 (the confidence interval in 100 USD, the cost of manpower is given in USD).

Table 3.

Manpower amount and cost (in dependence of variation l)			
η	X_1	X_2	X_3	F
0	9222	5533	1106	94,899
1	9222	5533	1106	94,899
10	9376	5616	1046	96,832
30	9452	5672	1036	98,614

Example 4. Let us to consider an application of the developed approach to the optimization of portfolio of the Lithuanian Stock Market with $n = 4$ securities.

We make the analysis of daily returns of the following assets:
ENRG – joint stock company "Lietuvos energija" (energetics),
MAZN – joint stock company "Mazeikiu Nafta" (oil refinery),
ROKS – joint stock company "Rokiskio suris" (dairy products),
RST – joint stock company "Rytu skirstomieji tinklai" (energetics).

Table 4.

	ENRG	MAZN	ROKS	RST	μ_i	σ_i
	Correlations					
ENRG	1	0.0120	0.0010	0.1621	0.5029	0.7439
MAZN	0.0120	1	−0.0310	0.0954	0.4447	0.6414
ROKS	0.0010	−0.031	1	0.0572	0.2609	0.3320
RST	0.1621	0.0954	0.0572	1	0.3327	0.3555

A brief description of data is given in Table 4, where empirical data were fitted by a lognormal model according to the Kolmogorov-Smirnov criterion. The data source is www.nse.lt/nvpb/index_en.php, time period – 2002.01–2003.10.

The portfolio return function is as follows,

$$r(w, \xi) = \sum_{i=1}^{n} w_i \cdot e^{\xi_i},$$

$\xi \succ N(\mu, \Sigma)$, $\mu = (\mu_1, \mu_2, \ldots, \mu_n)$, $\Sigma = [\sigma_{ij}]_1^n$. Portfolio weighting to maximize a probability of portfolio return to exceed the desired threshold R:

$$F(w) = P(r(w, \xi) \geq R) \to \max_{w \in W},$$

was considered with a simple set of constitutional constraints $W = (w | w_i \geq 0, \sum_{i=1}^{n} w_i = 1)$.

This problem was solved as nonlinear stochastic optimization task with strictly valid linear constraints.

Table 5.

t	w_1	w_2	w_3	w_4	Estimate (7) (Confidence)	Hotelling statistics T^2 (Fisher quantile F_σ))	N_t
1	25.0	25.0	25.0	25.0	78.12%(68.92 87.33)	2.04 (2.57)	50
2	39.6	28.1	18.7	13.6	80.83%(73.59 88.08)	2.21 (2.53)	63
3	35.5	42.2	12.4	9.9	78.50%(71.39 85.61)	0.20 (2.51)	72
4	37.3	44.8	11.2	6.7	82.94%(81.14 84.73)	5.55 (2.38)	870
5	40.2	46.1	8.8	4.9	85.12%(82.58 87.67)	1.96 (2.40)	376
6	41.6	48.8	7.3	2.3	83.66%(81.25 86.07)	3.46 (2.39)	459
7	44.3	50.4	5.3	0.0	82.84%(79.92 85.76)	2.58 (2.63)	319
8	49.3	47.7	0.3	0.0	83.14%(80.28 86.00)	0.16 (2.63)	326
9	50.3	49.2	0.5	0.0	84.00%(83.30 84.69)	0.84 (2.61)	5318
10	50.7	49.3	0.0	0.0	84.29%(83.79 84.79)	0.18 (3.00)	9900
						$\Sigma N_t =$	17753

Portfolio weighting according to this objective function by the method developed is shown in Table 5. The gradient of the objective function was expressed in form (3), using the transformation to polar variables described by (Sakalauskas (1998)). The parameters of the method were as follows: $\rho = 2.0$, $\delta = 1\%$, $\gamma = \sigma = \beta = 0.95$, $\varepsilon = 0.7$. We see that, after $t = 10$ iterations and total 17753 Monte-Carlo trials, the probability of the desired portfolio increased from 78.12% (67.92 87.33) to 84.29% (83.79 84.79) (third column), changing the strategies of portfolio sharing with respect to (18) (second column) and choosing the Monte-Carlo sample size with respect to (21) (last column). The total amount of trials $\sum_{i=1}^{t} N_i$ exceeded the Monte-Carlo sample size N_t at the time of the stopping decision only by 1,79 times.

Thus, counterexamples corroborate the theoretical conclusions and show that the proposed approach enables us to solve well conditioned SP problems, where the conditions of convexity and smoothness vanish only in some neighbourhood of the optimal point, with a sufficient permissible accuracy, using the acceptable volume of computations (5–20 iterations and 3000–10000 total Monte-Carlo trials). If we keep in mind that application of the Monte-Carlo procedure usually requires 1000–2000 trials in statistical simulation and estimation of one value of the function, the optimization by our approach can usually require only 3–5 times more computation.

6 Discussion and Conclusions

The stochastic iterative method has been developed to solve the SP problems by a finite sequence of Monte-Carlo sampling estimators. Since the stochastic optimization only of the first order methods are suitable we confined ourselves to the gradient-descent type methods, showing that typical deterministic approaches of unconstrained and unconstrained optimization might be generalised to the stochastic case.

The approach surveyed in this paper is grounded by the stopping procedure and the rule for iterative regulation of Monte-Carlo sample size, taking into account the stochastic model risk. The stopping procedure proposed allows us to test the optimality hypothesis and to evaluate the confidence intervals of the objective and constraint functions in a statistical way. The numerical experiment has shown the acceptability of this procedure, when the Monte-Carlo sample size is $N \geq 50$. The regulation of sample size, when this size is taken inversely proportional to the square of the gradient norm of the Monte-Carlo estimate, makes it possible to solve SP problems rationally, from the computational viewpoint, and guarantees the convergence a.s. at a linear rate. The numerical study and the practical example corroborate theoretical conclusions and show that the procedures developed enable us to solve SP problems with an admissible issible accuracy using the acceptable volume of computations (5–20 iterations and 3000–10000 total Monte-Carlo trials). If we keep in mind that application of the Monte-Carlo procedure usually requires 1000–2000 trials in statistical simulation and estimation of one value of the function, the optimization by our approach can require only 3–5 times more computation.

References

1. Arrow, K.J., Hurwicz, L., and Uzawa, H. eds. (1958). Studies in Linear and Nonlinear Programming. Stanford University Press, Stanford, California.
2. Bartkute V. and Sakalauskas L. (2006). Simultaneous Perturbation Stochastic Approximation for Nonsmooth Functions. European Journal on Operational Research (accepted).
3. Bentkus V. and Goetze F. (1999) Optimal bounds in non-Gaussian limit theorems for U-statistics. Annals of Probability, **27** (1), 454–521.
4. Bertsekas, D.I. (1982). Constrained Optimization and Lagrange Multiplier Methods. Academic Press, Paris-Toronto.
5. Bhattacharya, R.N., and Ranga Rao, R. (1976). Normal Approximation and Asymptotic Expansions. John Wiley, New York, London, Toronto.
6. Ermolyev, Ju.M. (1976). Methods of Stochastic Programming. Nauka, Moscow. 240 pp. (in Russian).
7. Ermolyev Yu., and Wets R. (1988). Numerical Techniques for Stochastic Optimization. Springer-Verlag, Berlin.
8. Ermolyev Yu. and Norkin I.(1995). On nonsmooth problems of stochastic systems optimization. WP-95-96, IIASA, A-2361, Laxenburg, Austria.
9. Jun Shao (1989) Monte-Carlo approximations in Bayessian decision theory. JASA, **84** (407), 727–732.
10. Katkovnik V.J. (1976). Linear Estimators and Problems of Stochastic Optimization. Nauka, Moscow (in Russian).
11. King J. (1988). Stochastic Programming problems: Examples from the Literature. In: Ermolyev Ju and Wets R. (eds.), Numerical Techniques for Stochastic Optimization, Springer-Verlag, Berlin.
12. Kushner H, and Jin G.G. (2003). Stochastic Approximation and Recursive Algorithms and Applications. Springer, NY.
13. Stochastic optimization methods. Springer, NY. Marti K. (2005).
14. Mikhalevitch V.S., Gupal A.M. and Norkin V.I. (1987). Methods of Nonconvex Optimization. Nauka, Moscow (in Russian).
15. Pflug G.Ch. (1988). Step size rules, stopping times and their implementation in stochastic optimization algorithms. Numerical Techniques for Stochastic Optimization, Ermolyev, Ju., and Wets, R. (eds.), Springer-Verlag, Berlin, 353–372.
16. Polyak B.T. (1987). Introduction to Optimization. Translation Series in Mathematics and Engineering. Optimization Software, Inc, Publication Division, New York.
17. Prekopa A. (1980) Logarithmic concave metric and related topics. In: M.A.H. Dempster (ed.) Stochastic Programming. Academic Press, London, 63–82.
18. Rockafellar, R.T., and Wets, R. J.-B. (1994). A Lagrangian finite generation technique for solving linear-quadratic problems in stochastic programming. Mathematical Programming, **28**, 63–93.
19. Rubinstein R. (1983). Smoothed functionals in stochastic optimization. Mathematical Operations Research, **8**, 26–33.

20. Rubinstein R. and Shapiro A. (1993). Discrete Events Systems: Sensitivity Analysis and Stochastic Optimization by the score function method. Wiley, New York, NY.
21. Sakalauskas L. (1997). A centering by the Monte-Carlo method. Stochastic Analysis and Applications, **15** (4), 615–627.
22. Sakalauskas L. (1998). Portfolio management by the Monte-Carlo method. Proceedings of the 23rd Meeting of the European Working Group on Financial Modelling, Crakow, Progress&Business Publ., 179–188.
23. Sakalauskas L. (2000). Nonlinear Optimization by Monte-Carlo estimators. Informatica, **11**(4), 455–468.
24. Sakalauskas L. (2002) Nonlinear stochastic programming by Monte-Carlo estimators. European Journal on Operational Research, **137**, 558–573.
25. Sakalauskas L. (2004). Application of the Monte-Carlo method to nonlinear stochastic optimization with linear constraionts. Informatica, **15**(2), 271–282.
26. Shapiro A. (1989). Asymptotic properties of statistical estimators in stochastic programming. The Annals of Statistics, **17** (2), 841–858.
27. Shapiro A. and Homem-de-Mello T. (1998). A simulation-based approach to two-stage stochastic programming with recourse. Mathematical Programming, **81**, 301–325.
28. Yudin D.B. (1965). Qualitative methods for analysis of complex systems. Izv. AN SSSR, ser. Technicheskaya. Kibernetika, **1**, 3–13 (in Russian).
29. Ziemba W. (1972) Solving nonlinear programming problems with stochastic objective functions. Journal of Financial and Quantitative Analysis, **7**, 1995–2001.

Part V

Policy Issues Under Uncertainty

Endogenous Risks and Learning
in Climate Change Decision Analysis

B. O'Neill, Y. Ermoliev, and T. Ermolieva

Institute for Applied Systems Analysis, Laxenburg, Austria

Abstract. We analyze the effects of risks and learning on climate change decisions. Using a new two-stage, dynamic, climate change stabilization model with random time horizons, we show that the explicit incorporation of ex-post learning and safety constraints induces risk aversion in ex-ante decisions. This risk aversion takes the form in linear models of VaR- and CVaR-type risk measures. We also analyze extensions of the model that account for the possibility of nonlinear costs, limited emissions abatement capacity, and partial learning. We find that in all cases, even in linear models, any conclusion about the effect of learning can be reversed. Namely, learning may lead to either less- or more restrictive ex-ante emission reductions depending on model assumptions regarding costs, the distributions describing uncertainties, and assumptions about what might be learned. We analyze stylized elements of the model in order to identify the key factors driving outcomes and conclude that, unlike in most previous models, the quantiles of probability distributions play a critical role in solutions.

Key words: stochastic nonsmooth optimization, climate stabilization, learning, catastrophic risk.

1 Introduction

Early climate change mitigation analysis tended to frame the climate change problem as a hit-or-miss type of decision making situation, in which a single policy choice is made about appropriate emissions reductions over time. Increasingly, analysis explicitly recognizes that the problem is more accurately described as sequential decision making under uncertainty, with the anticipation that new information will be acquired over time. For example, one reflections of this orientation can be found in discussions of climate change policies [14] that are framed as a choice between acting now or waiting until we know more about the problem [17], [27], [28]. This is a natural framing of the problem given its key characteristics involving uncertainty, irreversibility, and the potential for learning. Emissions of greenhouse gases (GHG) associated with the production and consumption of goods and services lead to long-lived atmospheric concentrations (stocks) of these pollutants that alter Earths climate. On one hand, postponing the reduction of GHG emissions may lead to costly and potentially irreversible climate-related impacts such as reorganization of large-scale ocean circulation patterns or increased frequency of extreme weather-related events. On the other hand, undertaking emission abatements now risks potentially irreversible investments that may turn out to be unnecessary if climate change is less severe than expected.

In economics literature, the importance of learning was first discussed in connections with irreversible investments in 1974 in Arrow and Fisher [2] and Henry [13] without an overall two-stage model being formulated. Arrow and Fisher [2], Henry [13], and Chichilnisky and Heal [3] have concluded that when future damages are uncertain and irreversible, the ability to learn should lead to more active ex-ante emission reductions. On the other hand, irreversibility of capital may lock the economy in a wasteful use of resources. Viscusi and Zeckhauser [25], Dixit and Pindyck [5], Ulph and Ulph [24], and Pindyck [21] showed that the ability to learn in this case should lead to less active ex-ante emission reduction. These competing effects imply that the net effect of learning on ex-ante decisions is an empirical question. Nordhaus [20] and Kolstad [16] examined the effects of learning by using empirically-calibrated integrated assessment models. They concluded that, in fact, learning has insignificant effects on ex-ante abatement policies because the damage losses are not severe enough. A reason for this is that in the most integrated climate and economics models, climate changes are considered as if they occur continuously and as if they can eventually be reversed through ex-post adjustments [28]. These models also use average damages (i.e., they cannot properly capture the effects of abrupt climate change [1] and catastrophic risks [8], [19], [28]). A paper by Fisher and Narain [11] analyzed a two-period model with risk characterized by a parameter introducing high or low climate change damages. Because overall impacts are evaluated by using expected values, the effects of capital irreversibility dominate catastrophic damages in a similar way to other models. Epstein [7] demonstrated that the effects of learning on ex-ante decision depend in general on convexity or concavity of marginal costs, which are very restrictive for climate change policy analysis [24].

In this paper we take a different approach. Instead of using expected damages we explicitly introduce safety constraints by formulating the climate change problem within the two-stage framework of stabilization. We develop a two-stage dynamic STO model with random time horizons and deliberately analyze only stylized linear versions of this model. Our simple analytical analysis allows to avoid otherwise endless number of computational experiments in order to identify effects of various driving forces. In particular, we show that the combination of safety constraints and perspectives of learning in linear models induces potentially strong risk aversion among ex-ante decisions that is characterized by quantile based VaR (Value at Risk) and CVaR (Conditional Value at Risk) risk measures common in the risk literature. As a result we show that, even with a linear net cost function, learning may lead to either less or more restrictive emission reductions, depending on mitigation costs and probability distributions describing key uncertainties.

The paper is organized as follows. Section 2 develops the general model, characterizing climate change risk by the probability of total atmospheric

CO_2 concentrations exceeding a vital random threshold associated with potential ranges of global temperature. It outlines a new type of two-stage dynamic climate-change stabilization STO model with random durations of stages. In general, this model can be solved only numerically and therefore the key factors driving results are difficult to identify. For these reasons, we analyze only stylized aspects of the model; these provide a clearer picture of the various driving forces and show why the ability to learn in the future can lead to either less restrictive or more restrictive ex-ante abatement policies today.

Section 3 presents our basic simplified model. It uses a very simple linear two-stage STO model to illustrate that the results from empirical models can be contradictory because optimal solutions depend on complex nonsmooth interactions among ex-ante and ex-post decisions, costs, and probability distributions. In particular, solutions contain potentially strong risk aversion characterized by quantile-based risk measures that are used for regulating the safety of nuclear plants and insolvency of insurance companies, but also in financial applications, extremal value theory [6], and catastrophic risk management [8].

Section 4 analyzes three extensions of the basic model designed to investigate the consequences of non-linear costs, limitations on stage two reductions (or adaptive capacity), or incomplete learning. Results emphasize again the importance to optimal solutions of quantiles of probability distributions characterizing key uncertainties, and also more strongly and even unconditionally require ex-ante anticipative emissions reduction in addition to ex-post reductions.

A more realistic but still linear dynamic two-stage climate change stabilization STO model is analyzed in Section 6. Similar to Section 3, the explicit incorporation of ex-ante and ex-post decisions induces risk aversions characterized by a dynamic version of a CVaR type risk measure. This may create the misleading impression that a truly risk-based policy analysis has been carried out and, without the explicit introduction of adaptive capacity and additional safety constraints, may provoke a catastrophe. In conclusion, Section 6 emphasizes the importance of the proper models, explicit treatment of uncertainty and risks, more realistic accounting for uncertainty, and robust decisions.

2   Endogenous Climate Change Risk: A General Model

Climate-change integrated assessment models (see e.g., [20]) incorporate economic and geophysical processes that link economic growth with the accumulation of GHG emissions in the atmosphere. The accumulation of CO_2 emissions is the main driving force behind global climate change over multi-decade timescale. The process involves complex interactions between the atmosphere, the terrestrial biosphere, and the oceans. Current integrated assessment models use different carbon-cycle models for computing changes in atmospheric concentrations $M(t)$ resulting from CO_2 emissions $e(t)$ [23]. In general, these models are of the form

$$M(t+1) = f(M(t), e(t), \beta), t = 0, 1, 2, ..., \tag{1}$$

where β is a vector of model parameters. Values $M(t)$ are used in integrated assessment models to compute the increase in the global average temperature as a smooth function of $M(t)$, and damages are typically computed in the form of smooth functions of this temperature increase.

This approach to modeling damages has several weaknesses. For example, a serious underestimation of damages may result from the use of global average temperature as a measure of climate change. Changes in the frequency of extreme weather-related events (e.g., floods, droughts, storms, heat waves) may be more important and may be non-linearly related to changes in the mean temperature. In addition, it has been proposed that beyond particular climate change thresholds, singular catastrophic events may be triggered that will have widespread consequences such as changes in ocean circulation, or disintegration of ice sheets [19].

We therefore develop an alternative approach to modeling climate change damages that treats threshold-type risks explicitly and that also accounts for uncertainty and its potential resolution over time. We first present a brief qualitative summary of the model, and then its mathematical form.

We assume that emissions in the absence of climate policy (specifically, emissions mitigation policy) depend on a wide range of uncertainties in future socio-economic development paths, technological progress, and lifestyles which together can be grouped as scenarios. Furthermore, emissions are also affected by explicit mitigation policies that might be adopted with uncertain costs. Emissions lead to uncertain accumulation of atmospheric concentrations and uncertain climate changes (and therefore damages). We introduce risk by assuming that there is a safety constraint in the form of an atmospheric stabilization target; i.e., a level of greenhouse gas concentration we wish to avoid exceeding with some level of confidence, based on an assumed level of aversion to the risk of incurring serious damages. Introducing stabilization in the form of a chance constraint is consistent with the uncertainty assumed in the problem: as long as uncertainty in the climate system persists, any emissions path will yield at best a range of possible outcomes. Thus, policies can only limit the chance of exceeding any particular target, and the acceptable risk must be defined a priori.

We introduce learning by assuming there are two stages. In the first, decision makers face full uncertainty. After a particular (and uncertain) length of time, some new information about uncertain parameters is revealed, and the second stage begins, at which point emission mitigation policies may be adjusted. The problem is to choose the emissions reduction policies for both stages such that mitigation costs are minimized and the safety constraint is achieved with the desired level of confidence.

This model of concentration-stabilization with learning can be defined mathematically as follows. Let emissions $e(t)$ in (1) depend on scenario uncertainties ω and policy variables x. In this paper, we assume that scenarios are characterized as random variables defined in a probability space Ω, $\omega \in \Omega$, with a probability measure $P(d\omega)$. Thus, for $\Omega = \{1, ..., N\}$, $P(d\omega) = p(s) := P[\omega = s]$, $\sum_{s=1}^{N} p(s) = 1$. Frequently we do not indicate the dependence of random variables on ω if this is clear from the context.

Let us denote by $L(\omega)$ the uncertain target level of CO_2 concentration in the atmosphere. If $M(t, x, \omega)$ denotes the mitigated CO_2 concentration, then the main problem can be formulated as the choice of cost-efficient emission-reduction path that satisfies probabilistic safety constraints on vital but uncertain levels of concentrations

$$P[M(t, x, \omega) \le L(\omega), t = \overline{1, T}] \ge 1 - \gamma, \tag{2}$$

where γ is a risk factor, $0 \le \gamma < 1$, T is a time horizon that may also be uncertain. The violation of these constraints can be regarded as a catastrophic collapse. In the insurance industry, constraints of type (2) regulate risk reserves to prevent insolvency. The typical approach to choosing γ in this industry is not based primarily on evaluating potential damages, but rather on limiting the chance that the insolvency may occur, say, to only once in 800 years, $\gamma = 1/800$. Similarly, the major failure of a nuclear plant is allowed once in 10^7 years, $\gamma = 10^{-7}$. Note that these are expected time horizons and therefore there is the possibility that events may occur at any time.

The abrupt climate change in (2) is modeled by random $L(\omega)$, which is revealed as a shock at random moment $\tau(x, \omega)$, which may also depend on x. Despite a smooth and even linear dependence of function $M(t, x, \omega)$ on x, the left-hand side of (2) is, in general, a nonsmooth and often even a discontinuous risk function [9], [10], [18]). Endogenous catastrophic collapse or the vulnerability of analyzed socio-economic and environmental system is modeled as a violation of constraint (2). In general, the learning may not reveal full information but perhaps only shift ranges of probability distributions. The learning may also not occur at $\tau \le T$, or it may occur very close to T. Because the inertia of the system may not allow constraints (2) to be fulfilled quickly, the probability of a catastrophe conditional on revealed information may drop rapidly below the vital level γ (i.e., constraint (2) emphasizes the importance of proper ex-ante actions).

Learning is reflected by specifying a two-stage dynamic STO framework with random time horizon. At stage 1, the emission-reduction path is defined by ex-ante decisions $x(t)$, $t = 1, 2, ...$, until a random time moment τ when new information is revealed. The new information may also include a new critical time horizon $T(\omega)$ for stage 2 ex-post emission reductions $x(t, \omega)$, $t = \tau + 1, ..., T(\omega)$. A new $T(\omega)$ can be shorter or longer than initial T, what may essentially affect adaptive capacity of the system. In general, the ex-ante policy $x(t)$, $t \le \tau$, includes components for choosing appropriate feasible level of this capacity. The problem is to minimize total emission reduction cost subject to constraints (2).

The resulting model can only be solved numerically. Here, instead of using numerical simulations, we take a different approach. In the following sections we formulate various stylized elements of the model and evaluate them analytically. This allows us to keep the discussion on a simple level, which provides a clear picture of the driving forces, their appropriate treatments and potential outcomes.

3   A Basic Model with Linear Cost Functions

The first stylized version of the general model we analyze aims to simplify it to the maximum extent in order to establish which types of solutions are possible even in the most basic case. Accordingly, we assume there are only two periods; that mitigation costs are certain and linear in reductions; that all other uncertainties can be collapsed into a single variable; and that the learning that takes place before stage two begins completely resolves this uncertainty. The single variable treated as uncertain is total emissions reductions over both periods, and the constraint is also expressed in terms of this variable. A constraint on minimum emissions reduction can be thought of as a concise way to represent several factors (and their uncertainties) that come into play in meeting a target based on environmental outcomes such as atmospheric concentrations, global average temperature levels, or particular impacts. For example, emissions reductions required to meet a target will depend on the target itself (i.e., whether a concentration or temperature target is high or low), on unmitigated reference emissions (because the absolute size of required emissions reductions will depend on the magnitude of uncontrolled emissions in the reference case), and on the system mapping emissions to environmental outcomes (e.g., parameters of the carbon cycle or climate system). Uncertainty in total required emissions reductions can be thought of as reflecting uncertainty in one or more of these different factors. Because all uncertainty is resolved before stage two begins, a chance constraint (as in the general problem) is unnecessary; the constraint can always be met with full certainty.

In Section 3.1 we specify the model mathematically and present its solution, and in Section 3.2 discuss the effects of learning (in comparison to an identical problem with uncertainty but no learning, or with perfect information). In Section 3.3 we provide a discussion of the results.

3.1   Two-Stage Model

A stylized concentration-stabilization problem with learning can be formulated as follows: assume that there are only two time intervals or periods $t = 1, 2$. Define by x_t, $x_t \geq 0$, $t = 1, 2$, a feasible level of emission reduction that can be chosen in period t; C_t, $C_t > 0$, is the known expected abatement cost per unit of emission reduction in period t; $\theta(\omega)$ is the uncertain target value of cumulative emission reductions for two periods. In this problem formulation, $\theta(\omega)$ serves as the safety constraint, $x_1 + x_2 \geq \theta(\omega)$. As discussed above, uncertainty in $\theta(\omega)$ can be thought of as reflecting uncertainty in any one of several different factors.

Assume uncertainty in $\theta(\omega)$ is resolved between periods 1 and 2. The ex-ante decision x_1 is made before the uncertainty in θ is resolved, whereas the ex-post decision x_2 is based on known θ, i.e., x_2 is a function of θ, $x_2(\theta)$. Assume that the ability to fulfill risk constraint $x_1 + x_2 \geq \theta(\omega)$ in period 2 is unbounded (the impacts of this rather unrealistic assumption are analyzed in Section 4. The problem is formulated as the minimization of total expected linear costs

$$C_1 x_1 + C_2 E x_2(\theta) \tag{3}$$

subject to safety constraints

$$x_1 + x_2(\theta) \geq \theta \tag{4}$$

for all θ. This is a classical two-stage STO problem [4], [10], [15], [26].

Clearly, if the ex-ante decision x_1 is irreversible, then the optimal period 2 decision is $x_2^*(\theta) = \max\{0, \theta - x_1\}$, that is, it depends non-smoothly on period 1 decision x_1 (path-dependence) and θ, providing potentially strong cross-period random interactions among decisions. Optimal period 1 decision x_1^* solves the stochastic minimax problem: minimize

$$F(x) = C_1 x + C_2 E \max\{0, \theta - x\}, x \geq 0. \tag{5}$$

Remark 2. Although the initial model (3)-(4) is linear in (x_1, x_2), the introduction of ex-post decision $x_2(\omega)$ induces risk aversion among ex-ante decisions that is defined by implicit non-smooth (in general) function (5). The following *Proposition* summarizes the solution and some important facts about stochastic minimax problem (5). It shows that the induced risk attitudes are characterized by VaR (critical quantile) and CVaR risk measures [22] allowing further to derive main conclusions regarding effects of learning. In extremal value theory [6], CVaR is also known as Mean Shortfall and Mean Excess Loss.

Proposition. Assume that $H(z) = P[\theta \leq z]$ is a continuously differentiable function.
(i) $F(x)$ is a strictly convex continuously differentiable function.
(ii) If $C_1 > C_2$, then $x_1^* = 0$ and $x_2^*(\theta) = \theta$. If $C_1 < C_2$, then the necessary and sufficient condition for optimal x^* reads: x^* is the quantile satisfying equation

$$P[\theta \geq x] = C_1/C_2. \tag{6}$$

(iii) The optimal value $F(x^*)$ has two important representations involving the expected cost uner perfect information, the expected value of perfect information and the CVaR risk measure:

$$F(x^*) = C_1\bar{\theta} + C_1 E[x^* - \theta | \theta \le x^*] = C_2 E\theta I(\theta \ge x^*), \tag{7}$$

where $E[\cdot|\cdot]$ denotes the conditional expectation, the indicator function $I(\theta \ge x) = 1$ if $\theta \ge x$ and $I(\theta \ge x) = 0$ otherwise.

Let us outline the proof.

(i) The convexity of $F(x)$ follows from the convexity of function $\max\{0, \theta - x\}$ which is preserved under the expectation operation. The strict convexity and continuous differentiability of $F(x)$ follow from the continuous differentiability of $H(z)$.

(ii) The minimization of $F(x)$ is a specific case of so-called stochastic minimax problems [9]. From the general results (see, e.g., page 416 of [10]) it follows that $F'(x) = C_1 - C_2 P[\theta \ge x]$. From $C_1 < C_2$, it follows that $F'(0) < 0$, i.e., $x^* > 0$ (assuming $x^* = 0$ we can derive contradiction with assumption $C_1 < C_2$ for small x). As $F(x)$ is a strictly convex function, it follows that (6) is indeed a necessary and sufficient optimality condition.

(iii) The first representation in (7) follows from (6) and the following rearrangements:

$$F(x^*) = C_1 x_1^* + C_2 E \max\{0, \theta - x^*\} = C_1 x^* + C_2 E[\theta - x^*|\theta \ge x^*]P[\theta \ge x^*] = C_1 x^* + C_1(E[\theta - x^*] - E[\theta - x^*|\theta \le x^*]) = C_1\bar{\theta} + C_1(E[x^* - \theta|\theta \le x^*])$$

The second representation in (7) follows from (6) and $E \max\{0, \theta - x^*\} = E\theta I(\theta \ge x^*) - x^* P(\theta \ge x^*)$.

Remark 3. The critical quantile in (6) defines the VaR risk measure; i.e., it indicates the magnitude of emission reduction in stage 1 that, with probability $1 - C_1/C_2$, will be sufficient to meet the safety constraint with no additional reduction required in stage 2. The first term $C_1\bar{\theta}$ in (7) represents the expected cost under perfect information. The second term represents the expected value of perfect information; i.e., the value of learning the true value of θ before stage 1, rather than after stage 1 and before stage 2.

The second equation in (7) defines the CVaR risk measure; i.e., the expected value of abatement costs that will be necessary in stage 2 if emissions reductions in stage 1 are not sufficient to meet the safety constraint. For some distributions it is possible to derive x^* from (6) explicitly. If θ is uniformly distributed on $[a, b]$, then it is easy to see that $x^* = \frac{C_1}{C_2}a + (1 - \frac{C_1}{C_2})b$, i.e., x^* is between optimistic and pessimistic scenarios of required emissions reductions with weights defined by ratio of costs C_1 and C_2.

3.2   Comparative Analysis

The *Proposition* of Section 3.1 allows the comparison of cases of perfect information, full uncertainty, and uncertainty with learning. The optimal condition given by equation (6) defines a quantile of the underlying probability distribution, i.e., it shows the critical dependence of period 1 ex-ante optimal decision on the probability distribution H. Assume that $C_1 < C_2$. In the case of perfect information, i.e., when θ is known at the beginning of the first period, both x_1 and x_2 can be chosen as a function of observable θ. Clearly, the optimal solution is $x_1^* = \theta$, $x_2^* = 0$, i.e., the term $C_1\overline{\theta}$ of the first equation in (7) represents indeed the expected cost under perfect information (assuming $\theta = \overline{\theta}$, i.e., uncertainty is unbiased with respect to the true state of the world). The second term represents the expected value of perfect information because this cost would be eliminated if θ were known before the first-period emission reduction decision had to be made (rather than afterward as in the learning case). In the case of full uncertainty ("without learning"), the optimal decision is $x_1^* = \overline{\theta}$, $x_2^* = 0$, which is also known as the certainty equivalent. The possibility of learning combined with explicit introduction of ex-post decisions specifies optimal period 1 abatements by the quantile satisfying (6). It may exceed the certainty equivalent $\overline{\theta}$ or it may be below this level. As equation (6) shows, this depends on the relative values of costs C_1, C_2, and the probability distribution H. For example, if $C_1/C_2 = 1/2$ and θ has a normal distribution, then optimal ex-ante abatement coincides with the certainty equivalent $x_1^* = \overline{\theta}$. For non-symetric probability distributions, the optimal abatements can be below or above $\overline{\theta}$. An asymmetric probability distribution can be caused, for example, by the interaction of a symmetric probability distribution with an environmental constraint. For example, if the probability density function for $e(\omega)$ is normal, the distribution for $\theta(\omega)$ can still be asymmetric if there is an atmospheric concentration constraint that does not require emissions reductions for all ω.

Remark 4. The certainty equivalent solution $x_1^* = \theta$, $x_2^* = 0$ in the case of full uncertainty (no learning) does not satisfy (4) for all θ, which may lead to a catastrophic collapse of high probability. The only way to fulfill the safety constraint (4) is to choose x_1 from the worst case scenario as $\max_\omega \theta(\omega)$, $\omega \in \Omega$. Clearly, this is an unrealistic and extremely costly solution. This calls for the explicit introduction of safety constraint (2) to provide a trade-off between the cost effectiveness and risk. The optimal solution under full uncertainty is now defined as minimizing $C_1 x_1 + C_2 x_2$ under constraint $x_1 + x_2 \geq z_\gamma$, where z_γ is the minimal z satisfying equation $P[z \geq \theta] = 1 - \gamma$, i.e., as $x_1 = z_\gamma$ and since $C_1 < C_2$, $x_2 = 0$. Clearly, the risk-adjusted solution under full uncertainty $x_1 = x_\gamma$ may be greater or less than $\overline{\theta}$, depending on γ, C_1/C_2, and probability distribution H.

3.3 Discussion

To summarize the key results, we find that even in this basic model, the effect of learning on optimal first stage emissions reductions is ambiguous: it can lead to either more or less emissions reductions than would be undertaken if there were uncertainty but no learning, and the direction and magnitude of this effect depends on the relationship between the assumed marginal costs in stages 1 and 2, and the shape of the probability distribution characterizing the uncertainty in total required emissions reductions.

The nature of the optimal solution with learning is not counter-intuitive. Whether to hedge against the possibility that required reductions will be large (or small) in stage 2 by making larger (or smaller) reductions in stage 1 depends on expectations about the tradeoffs in reduction costs (how much cheaper, or more expensive, will it be to make reductions if they are postponed) and about the likelihood that exceptionally large or small reductions will turn out to be required. For example, if marginal reduction costs are known to be cheaper in stage 2, it is always best to postpone all reductions however large they might be to stage 2. If, however, marginal costs are cheaper in stage 1, then the advantages of these cheaper costs must be balanced against the risk that you will reduce emissions in stage 1 more than turns out to be necessary. This tradeoff takes on a simple functional form in this model: it is optimal in stage 1 to reduce emissions up to the quantile of the uncertainty distribution for total emissions reductions given by the ratio of marginal costs between the two periods. If costs are twice as high in stage 2, it is optimal to reduce in stage 1 up to the median of the uncertainty distribution; if costs are three times as high in stage 2, it is optimal to reduce up to the 33-rd percentile, etc.

Clearly, then, a wide range of solutions are possible depending on the particular ratio of costs and the shape of the uncertainty distribution. In comparison, the solution under uncertainty without learning also depends on the costs and the shape of the uncertainty distribution, but there is no interaction between the two. All reductions are made either in stage 1 or in stage 2, depending on when marginal costs are lower. The amount of reductions made depends on the certainty with which the constraint is desired to be met. If a decision maker wants to be 50 percent sure the constraint is achieved, reductions to the median of the distribution will be made. Thus it is easy to see that no particular relationship between the solution with learning and without need hold. Depending on assumptions, optimal reductions with learning can be smaller, larger, or the same as in the no learning case, even in this simple linear model.

4   Extensions to the Basic Model

Next we examine separately three aspects of the basic concentration-stabilization model with learning presented in Section 3 that can be considered oversimplified. First, we replace the linear cost assumption with nonlinear abatement costs. We show that assuming costs are quadratic in reductions implies that it will always be optimal to make at least some reductions during stage one, a stronger result than occurs in the linear case. We also show that, except in the case of normally distributed uncertainty, the optimal solution will not depend only on the mean value of the uncertainty distribution. Thus deterministic analyses that use only the mean value will be misleading. Second, we assume that there is a limited capacity for making emissions reductions in stage two, an assumption that could be motivated by inertia in technological or socio-economic systems, or by the possibility that stage one (the period before learning occurs) may be long, leaving little time to make reductions in stage two. We show that the assumption of limited adaptive capacity in stage two induces greater optimal emissions reductions in stage one. Third, we assess the implications of incomplete learning, i.e., learning in which uncertainty is not completely resolved before stage two begins. We show that, as in the case of complete learning, the effect of learning on optimal stage one decisions is ambiguous: it can lead to larger, smaller, or the same emissions reductions as would be made if there were no learning. The effect depends on the assumed marginal costs in the two periods, and the nature and likelihood of what might be learned.

4.1 Nonlinear Abatement Costs

Abatement costs are generally modeled as nonlinear functions of emission reductions [12] with a quadratic functional form of a typical assumption (e.g., [20]). To examine the implications of this assumption, we let the cost functions of both periods be of the form $C_i(x) = C_i x^2$ with positive C_1, C_2. Cost function (5) then takes on the form $F(x) = C_1 x^2 + C_2 E(\max\{0, \theta - x\})^2$ and hence $F'(x) = 2C_1 x - 2C_2 E(\theta - x)I[\theta \geq x]$. The first observation we make concerns the necessity of first period reductions. Since $F'(0) = -2C_2 E\theta < 0$ (assuming that $E\theta > 0$) zero reductions in stage one are ruled out independently of the particular values of C_1 and C_2. Compare this results to the case of linear costs in Section 3.1, where $F'(0) = C_1 - C_2 < 0$ if $C_1 < C_2$; i.e., non-zero first period reductions are called for only if costs are less in period 1. With quadratic costs, period 1 reductions are optimal even if $C_1 > C_2$.

Let us illustrate some typical situations that may occur in the case of non-smooth, piece-wise linear functions commonly used in emission-control problems with technology switches. These functions implicitly impose upper or lower bounds on positive ex-ante emission reductions. Assume that $C_2(x) = C_2 x$ and $C_1(x)$ is a piece-wise linear function $C_1(x) = C_1^1 x$ for $0 \leq x \leq a$ and $C_1(x) = C_1^2(x - a) + C_1^1 a$ for $x \geq a$, where $C_1^1 < C_2$ and $C_1^2 > C_2$. It is easy to see that the optimal ex-ante solution has the upper bound $x_1 \leq a$. As $C_1^1 < C_2$ and $C_1^2 > C_2$, the optimal ex-ante decision is defined as follows: let $\bar{x}$ be the solution of equation $P[\theta \geq x] = C_1^1/C_2$. The optimal period 1 decision $x_1^* = a$ if $\bar{x} > a$, and $x_1^* = \bar{x}$ for $\bar{x} \leq a$. Assume that $C_1(x) = C_1 x$, and $C_2(x) = C_2^1 x$ for $0 \leq x \leq a$; $C_2(x) = C_2^2(x - a) + C_2^1 a$ for $x \geq a$, where $C_1 > C_2^1$, $C_1 < C_2^2$. Consider solution $\underline{x}$ of the equation $P[\theta > x] = C_1/C_2^2$. It is easy to see that the optimal period 1 decision $x_1^* = \underline{x}$ for $\underline{x} \geq a$ and $x_1^* = 0$ for $\underline{x} < a$, i.e., it has the lower bound $x_1 \geq a$.

4.2   Limited Adaptive Capacity

We next introduce the possibility that the capacity for emissions abatement in stage two may be limited. There are several possible motivations for this assumption. First, in cases with learning, Learning may occur slowly and the second period may begin late, leaving little time for reductions to be made. Second, inertia in technological and socio-economic systems may limit feasible reductions over a given time period [12]. The path-dependencies (inertia) of the technological and socio-economic systems producing greenhouse gases are critical for dealing with abrupt changes. Without inertia, the switching from one emission path to another would be instantaneous. In reality, energy production systems cannot be changed overnight. As a result, the possibilities for emissions reductions will not be bounded.

Limited adaptive capacity can be modeled most simply by constraints $x_2 \leq \beta$ with positive random β which becomes known from learning at stage 2. Without the safety constraint of type (2), the optimal stage 2 decision $x_2 = \min[\beta, \max\{0, \theta - x_1\}]$ cannot in general satisfy safety constraints (4) for all θ. As a consequence, the probability of a catastrophe can be rather high, calling for explicit introduction of type (2) safety constraint $P[x_1 + x_2 \geq \theta] = 1 - \gamma$. Since $x_2 \leq \beta$, this requires ex-ante emission reduction commitments $x_1 \geq x_\gamma$, where x_γ is minimal non-negative x satisfying equation $P[x \geq \theta - \beta] = 1 - \gamma$. Therefore, in order to prevent a catastrophic collapse with sufficient confidence, there must be minimal ex-ante emission reductions sufficient to satisfy the safety constraint in stage 2. Hence, stage one emission abatement is in general larger when limited adaptive capacity is explicitly assumed than when possible future emissions reductions are assumed to be unbounded. This can be evaluated properly by analyzing the STO model with safety constraints (2).

4.3   Incomplete Learning and Safety Constraint

Next we replace the assumption that uncertainty is completely resolved before stage two with the much more realistic assumption that learning is only partial. As an example, we consider the case in which learning affects the prior distribution, $H(z)$, by shifting the range of uncertainty. We first present the quantitative analysis, then discuss the results in qualitative terms.

Let us assume that $H(z) = P[\theta \leq z]$ is a mixture $H(z) = E_\xi H(\xi, z) = \int H(y, z)dG(y)$ of distribution $H(\xi, z)$ with unknown ξ characterized by a probability distribution $G(y) = P[\xi \leq y]$. The learning reveals only ξ at the beginning of period 2. For example, $H(z)$ can be a mixture of distributions $H(\xi, z)$ with probability mass concentrated in different subregions from the support of $H(z)$; in reality, these distributions could reflect differing views on the damages that would be associated with particular emissions pathways. (Note that if the support of $H(\xi, z)$ is a singleton, then the learning of ξ reveals the true value of θ. For the sake of illustration, let $H(z)$ be a mixture of two distributions $H_0(z)$ and $H_1(z)$, that is, $\xi H_0(z) + (1 - \xi)H_1(z)$, where $\xi = 0$ with probability p and $\xi = 1$ with probability $1 - p$, that is, $H(z) = pH_0(z) + (1 - p)H_1(z)$. Since only ξ is observed, the period 2 decision $x_2(\xi)$ can not fulfill constraints (4), and the safety constraint has to be written as in (2):

$$P[x_1 + x_2(\xi) \geq \theta(\xi)] \geq 1 - \gamma, \xi = 0, 1, \qquad (8)$$

where $\theta(\xi)$ has the posterior probability distribution $H_\xi(z)$. For a given ξ and γ, let us define $z_\gamma(\xi)$ as the minimal z, satisfying equation $P[z \geq \theta(\xi)] = 1 - \gamma$. Then equations (8) are equivalent to the equations $x_1 + x_2(\xi) \geq z_\gamma(\xi)$, which are similar to (4). The optimal period 2 decision $x_2(\xi) = max\{0, z_\gamma(\xi) - x_1\}$, and optimal x_1 has to minimize $F(x) = C_1 x + C_2[p \max\{0, z_\gamma(0) - x\} + (1 - p)\max\{0, z_\gamma(1) - x\}]$.

Function $F(x)$ does not have continuous derivatives. Therefore, the optimality condition cannot be derived from the Proposition of Section 3.1. $F(x)$ is a piece-wise continuous linear function which can be characterized as the following. Assume, for example, that $z_\gamma(0) < z_\gamma(1)$, then for $0 \leq x < z_\gamma(0)$, $F(x) = C_1 x + C_2[p(z_\gamma(0) - x) + (1 - p)(z_\gamma(1) - x)] = (C_1 - C_2)x + C_2\bar{z}_\gamma(\xi)$. For $z_\gamma(0) \leq x < z_\gamma(1)$, $F(x) = C_1 x + C_2(1 - p)(z_\gamma(1) - x) = (C_1 - C_2(1 - p))x + C_2(1 - p)z_\gamma(1)$, and for $x \geq z_\gamma(1)$, $F(x) = C_1 x$.

The optimal ex-ante solution hedges against the different contingencies. It is characterized as follows: $x = 0$, if $C_1 > C_2$. Otherwise, $x = z_\gamma(0)$, if $C_1 - C_2(1 - p) > 0$, and $x = z_\gamma(1)$, if $C_1 - C_2(1 - p) < 0$.

This solution can be understood in more qualitative terms as follows. If marginal costs are lower in stage two, then it is best to make all reductions in stage two after learning has taken place. If marginal costs are lower in stage one, then, in general it pays to make reductions in period 1 that are as large as possible. However, as was the case in the basic model in Section 3, this benefit of period 1 reductions must be weighed against the risk of making more reductions than turn out to be required. After learning takes place at the end of period 1, the optimal solution is to make reductions such that the total reduction is either $z_\gamma(0)$ or $z_\gamma(1)$. Thus the minimal first period reduction is $z_\gamma(0)$. If first period costs are very low, or the chance that $\xi = 1$ is very high, then it is optimal to make the larger first period reduction $z_\gamma(1)$, accepting the chance that $\xi = 0$ and that reductions $z_\gamma(1) - z_\gamma(0)$ will have been unnecessary.

Let us compare this ex-ante period 1 optimal "with-learning" solution to the optimal "without-learning" solution $x_1^* = z_\gamma$, $x_2^* = 0$ derived from minimization of (3) under safety constraint $P[x_1 + x_2 \geq \theta] \geq 1 - \gamma$, i.e., $x_1 + x_2 \geq z_\gamma$, where z_γ is the minimal z satisfying constraint $P[z \geq \theta] = 1 - \gamma$. Due to the monotonicity of $P[x \geq \theta]$ w.r.t. x, we can derive inequalities among these decisions by comparing $P[x \geq \theta]$ for $x = z_\gamma$ with $x = z_\gamma(0)$, $z_\gamma(1)$.

Assume that $H_0(z)$, $H_1(z)$ have continuous derivatives, the support of distribution $H_0(z)$ is interval $[a_0, b_0]$, and the support of $H_1(z)$ is interval $[a_1, b_1]$, where $a_1 > b_0$. If $C_1 - C_2(1 - p) < 0$, then the optimal "with-learning" period 1 solution $x = z_\gamma(1)$ from $[a_1, b_1]$. If $x \in [a_1, b_1]$, then $P[x \geq \theta] = p + (1-p)H_1(x)$. Since $H_1(z_\gamma(1)) = 1 - \gamma$, then $P[x \geq \theta] = p + (1-p)(1-\gamma) = 1 - \gamma + \gamma p$ for $x = z_\gamma(1)$. As $\gamma p > 0$, then the optimal "without-learning" decision $x = z_\gamma$ satisfying $P[x \geq \theta] = 1 - \gamma$ is less demanding (smaller) than $x = z_\gamma(1)$, i.e., learning increases the optimal ex-ante emission reductions. This conclusion is reversed in the case $C_1 - C_2(1-p) > 0$. Indeed, let $x \in [a_0, b_0]$. Then $P[x \geq \theta] = pH_0(x)$, $H_0(z_\gamma(0)) = 1 - \gamma$ and for $x = z_\gamma(0)$, $P[x \geq \theta] = p(1-\gamma)$ (i.e., the optimal "without-learning" decision $x = z_\gamma$ is greater than the optimal "with-learning" decision $x = z_\gamma(0)$).

Therefore we find that in the case of incomplete learning, as in the case of complete learning we assumed in Section 3, the effect of learning is ambiguous: it can lead to either larger, smaller, or the same optimal emissions reductions in stage one as would occur under uncertainty without learning. The particular nature of the effect is determined by the marginal costs of reductions in the two periods, and the assumed likelihood of what will be learned at the start of stage two in the example presented here, the likelihood that one of two competing uncertainty distributions will turn out to be supported by the new information received. The size of the effect is determined by the shape of the distributions themselves, and the certainty with which it is desired to achieve the safety constraint.

5   A Dynamic Stabilization Problem

In this section we extend the two-period model presented in Section 3 to multiple periods. In this more general form, the problem becomes similar to catastrophic-risk-management problems discussed in [8]. As discussed in Section 3, the solution of the two-stage model had strong connections with CVaR-type risk measures. Here we show that the dynamic multi-period model also has strong connections with dynamic versions of CVaR risk measures. However, we caution that this resemblance may create the impression of a truly risk-based policy analysis when in fact, without the explicit introduction of additional safety constraints, the solution could provoke a catastrophic collapse.

Assume that CO_2 emission paths are characterized by exogenous scenarios as in Section 3. Let us consider $R_t = \sum_{k=0}^{t} x_k$, where decision variables $x_k \geq 0$, $k = 0, 1, ..., t$, $t \leq T$. We can think of x_k as a feasible level of CO_2 emission reduction at the beginning of period k. At time $t = 0, 1, ...,$ the target value on total emission reduction R_t in period t is given as a random variable ρ_t. It is assumed that exact value of ρ_t is revealed at a random time $t = \tau$. Since τ is uncertain, the decision path $x = (x_0, x_1, ..., x_T)$ for the whole time horizon has to be chosen ex-ante in period $t = 0$ to "hit" the target ρ_t, $R_\tau \geq \rho_\tau$, at $t = \tau$ in a sense specified further by (10). At random $t = \tau$, the decision path can be revised for the ramaining available time. Similar to the model of Section 3.1, consider a stream of linear random costs $v(x) =$

$\sum_{t=0}^{T}[c_t x_t + d_t \max\{0, \rho_t - R_t\} I(\tau = t)]$, where $c_t > 0$, $d_t > 0$, $t = 0, 1, ..., T$ are known ex-ante and ex-post abatement costs. In this model we assume a rather strong irreversibility of investments: the multiplier $I(\tau = t)$ affects only the ex-post costs. The analysis of the case when it affects also ex-ante costs is similar. The expected value of $v(x)$ can be written as

$$V(x) = \sum_{t=0}^{T}[c_t x_t + \rho_t d_t \max\{0, \rho_t - \sum_{k=0}^{t} x_k\}], \tag{9}$$

where $p_t = P[\tau = t]$.

Let us consider a path x^* minimizing $V(x)$ subject to $x_t \geq 0$, $t = 0, 1, ..., T$. Assume that $V(x)$ is a continuously differentiable function (e.g., a component of random vector $\rho = (\rho_0, \rho_1, ..., \rho_T)$ has a continuous density function). Also, assume for now that there exist positive optimal solution $x^* = (x_0^*, x_1^*, ..., x_T^*)$, $x_t^* > 0$, $t = 0, 1, ..., T$. Then from the optimality condition for stochastic minimax problems similar to Section 3, it follows that for $x = x^*$,

$$V_{x_t} = c_t - \sum_{k=t}^{T} p_k d_k P[\sum_{s=0}^{k} x_s \leq \rho_k] = 0, t = 0, 1, ..., T,$$

From this sequentially for $t = T, T - 1, ..., 0$, it follows that

$$P[\sum_{k=0}^{T} x_k \leq \rho_T] = c_T/p_T d_T, P[\sum_{k=0}^{t} x_k \leq \rho_t] = (c_t - c_{t+1})/p_t d_t, t = 0, 1, ..., T-1. \tag{10}$$

Since $E \max\{0, \rho_t - R_t\} = E\rho_t I(\rho_t \geq R_t) - R_t P[\rho_t \geq R_t]$, then from (10) it follows that $V(x^*) = E d_\tau \rho_\tau I(\rho_\tau \geq R_\tau)$, or

$$V(x^*) = p_0 d_0 E[\rho_0 I(\rho_0 \geq R_0^*)] + p_1 d_1 E[\rho_1 I(\rho_1 \geq R_1^*)] + ... \\ + p_T d_T E[\rho_T I(\rho_T \geq R_T^*)],$$

which can be viewed as a dynamic CVaR risk measure. Equations (10) can be used for analysing desirable dynamic risk profiles, say, time independent risk profiles with a given risk factor γ as in (2): $1 - \gamma = c_T/p_T d_T = (c_t - c_{t+1})/p_t d_t$, $t = 0, 1, ..., T - 1$, which can be achieved by decisions affecting parameters c_t, d_t, p_t.

Remark 6. Equations (10) are derived from the existence of the positive optimal solution x^*. It is easy to see that the existence of this solution follows from $c_T/p_T d_T < 1$, $0 \leq (c_t - c_{t+1})/p_t d_t < 1$, $t = 0, ..., T - 1$, and the monotonicity of quantiles β_t, $\beta_0 < \beta_1 < ... < \beta_T$ defined by equations

$$P[\beta_T \leq \rho_T] = c_T/p_T d_T, P[\beta_t \leq \rho_t] = (c_t - c_{t+1})/p_t d_t, t = 0, 1, ..., T - 1.$$

Indeed, the first requirement guarantees that $x_0^* > 0$, $\sum_{k=0}^{t} x_0^* > 0$, $t = 1, 2, ..., T$. From the second requirement follows that $x_0^* + x_1^* > x_0^*$, i.e., $x_1^* > 0$, and so on.

If probability p_t rapidly decreases to 0, e.g., if p_t is associated with a rare catastrophic event, then from (10) it follows that ex-ante abatements are positive for a relatively short initial interval defined by inequality $(c_t - c_{t+1})/p_t d_t < 1$. This misleading conclusion is due to a strong assumption of unlimited capacity for emission reductions, which is a standard assumption of climatic-economic integrated assessment models (see discussions in [12] and [28]). Similar to conclusions of Section 4.2, this requires an adequate treatment of risks by additional safety constraints (2) to prevent catastrophes.

6   Concluding Remarks

This paper analyzes the effects of risks and learning on climate change decisions using a two-stage, dynamic model that assumes a concentration-stabilization constraint. It shows that learning can lead either to larger or smaller first period emissions reductions, compared to the optimal reduction under uncertainty without learning, and that this effect can either be large or small. The direction and magnitude of the learning effect is determined by a number of interacting factors. For example, in a simple linear model with deterministic mitigation costs but uncertainty in total required emissions reductions, the learning effect depends on how mitigation costs evolve over time, the shape of the uncertainty distribution in required emissions reductions, the confidence with which the safety constraint (i.e., stabilization level) is desired to be met, and, in the case of incomplete learning, the probability distribution describing the anticipated learning possibilities. Introducing a more realistic nonlinear cost function with increasing marginal costs induces a higher level of first period emissions reductions compared to the linear case. We also analyze the case of limited capacity to make reductions in period 2, motivated by either uncertain timing of learning or uncertain inertia in socio-economic systems, and show how this consideration can induce a minimum level of first period reductions. Finally, framing the problem in dynamic terms as a multi-period problem with an uncertain time path of required cumulative emissions reductions shows that the problem has strong connections with dynamic versions of CVaR risk measures. This may create the misleading impression that risks are being properly managed, and unless additional safety constraints are introduced, could provoke a catastrophe. Given the multiple influences on the learning effect, we conclude that drawing practical conclusions on the likely effect of learning on climate change decisions is an empirical question requiring analysis with models capable of adequately representing endogenous risks, abrupt changes, realistic learning rates, inertia, and path dependent costs.

Acknowledgment. The authors are grateful to the participants of the IFIP/IIASA/GAMM workshop on Coping with Uncertainty, December 13-16, 2004, IIASA, and to the anonymous referees for critical suggestions that led to improvements of this paper.

References

1. Alley, R.B., Marotzke, J., Nordhaus, W.D., Overpeck, J.T., Peteet, D.M., Pielke Jr., R.A., Pierrehumbert, R.T., Rhines, P.B., Stocker, T.F., Talley, L.D., Wallace, J.M. Abrupt Climate Change. Science. **299**.

2. Arrow, K.J., Fisher, A.C. (1974) Environmental Preservation, Uncertainty, and Irreversibility. Quarterly Journal of Economics. **88**, 312-319.

3. Chichilnisky, G., Heal, G. (1993) Global Environmental Risks. Journal of Economic Perspectives. **7(4)**, 65-86.
   Analysis.

4. Dantzig, G., Madansky, A. (1961) On the Solution of Two-stage Linear Programs under Uncertainty. Proc. Fourth Berkeley Symposium on Mathematical Statistics and Probability. **1**, 165-176, Univ. California Press, Berkley. .

5. Dixit, A.K., Pindyck, R.S. (1994) Investments under Uncertainty. Princeton University Press.

6. Embrechts, P., Klueppelberg, C., Mikosch, T. (2000) Modeling Extremal Events for Insurance and Finance. Applications of Mathematics, Stochastic Modeling and Applied Probability. Springer Verlag, Heidelberg.

7. Epstein, L.G. (1980) Decision Making and the Temporal Resolution of Uncertainty. International Economic Review, **21(2)**, 269-282.

8. Ermoliev, Y., Ermolieva, T.Y., MacDonald, G., and Norkin, V. (2000) Stochastic Optimization of Insurance Portfolios for Managing Exposure to Catastrophic Risks. Annals of Operations Research. **99**, 207-225.

9. Ermoliev, Y.M., Norkin, V.I. (1997) On Nonsmooth and Discontinuous Problems of Stochastic Systems Optimization. European Journal of European Research. **101**, 230-244.

10. Ermoliev, Y., Wets, R. (Eds., 1988) Numerical Techniques for Stochastic Optimization. Computational Mathematics, Springer Verlag, Berlin.

11. Fisher, A.C., Narain, U. (2003) Global Warming, Endogenous Risk, and Irreversibility. Environmental and Resource Economics. **25**, 395-416.

12. Ha-Duong, M., Hourcade, J.-C., Grubb, M. (1997) The Influence of Inertia and Uncertainty upon Optimal CO_2 Policies. Nature. **390**, 270-274.

13. Henry, C. (1974) Investment Decisions under Uncertainty: The Irreversibility Effect. American Economic Review. **64/6**, 1006-1012.

14. IPCC (2001) Climate Change 2001: The Scientific Basis. Technical Report. Intergovernmental Panel on Climate Change.

15. Kall, P., Wallace, S.W. (1994) Stochastic Programming. J. Wiley, Chichaster.

16. Kolstad, C.D. (1996) Learning and Stock Effects in Environmental Regulations: The Case of Greenhouse Gas Emissions. Journal of Environmental Economics and Management. **31**, 1-18.

17. Manne, A.S., Richels, R.G. (1992) Buying Greenhouse Insurance: The Economic Costs of Carbon Dioxide Emission Limits. Cambridge, Mass., MIT Press.

18. Marti, K. (2005) Stochastic Optimization Methods. Springer, Berlin, Heidelberg.

19. O'Neill, B., Oppenheimer, M. (2002) Dangerous Climate Impacts and the Kyoto Protocol. Science. **296**, 1971-1972.

20. Nordhaus, W.D. (1994) Managing the Global Commons: The Economics of Climate Change. Cambridge, Mass.: MIT Press.

21. Pindyck, R.S. (1999) Irreversibilities and the Timing of Environmental Policy. Working Paper 99-005, Center for Energy and Environmental Policy Research, Massachusetts Institute of Technology, Cambridge, MA.
22. Rockafellar, T., Uryasev, S. (2000) Optimization of Conditional Value-at-Risk. The Journal of Risk. **2/3**, 21-41.
23. Schultz, P.A., Kasting, J.F. (1997) Optimal Reduction in CO_2 Emissions. Energy Policy. **25/5**, 491-500.
24. Ulph, A., Ulph, D. (1997) Global Warming, Irreversibility and Learning. Economic Journal. **107/442**, 636-650.
25. Viscusi, W.K., Zeckhauser, R. (1976) Environmental Policy Choice under Uncertainty. Journal of Environmental and Economic Management. **3**, 97-112. Kiev (In Ukrainian).
26. Yastremskij, A. (1983): Stochastic Models of Mathematical Economics. Vischa Shkola, Kiev (In Russian).
27. Webster, M. (2002) The Curious Role of "Learning" in Climate Policy: Should We Wait for More Data? The Energy Journal. **23/2**, 97-119.
28. Wright, E.L., Erickson, J.D. (2003) Incorporating Catastrophes into Integrated Assessment: Science, Impacts, and Adaptation. Climate Change. **57**, 265-286.

Pricing Related Projects

S. D. Flåm[1],* and H. I. Gassmann[2]**

[1] Department of Economics, University of Bergen, 5007 Norway;
sjur.flaam@econ.uib.no.
[2] School of Business Administration, Dalhousie University, Halifax, Nova Scotia
Canada B3J 3J5; Horand.Gassmann@dal.ca.

Abstract. This paper deals with project evaluation from a portfolio perspective. The chief motivation stems from pricing bundles of related projects, all affected by uncertainty, when markets are imperfect or absent.

Novelties come by construing single projects as "players" of a transferable-utility, stochastic, cooperative game. Stochastic programming then provides state-dependent Lagrange multipliers associated to coupling constraints. Granted concave payoff functions, these multipliers not only emulate market clearing and formation of contingent, Arrow-Debreu prices; they also generate core solutions and project evaluations.

1 Introduction

This paper considers evaluation of several, interdependent, uncertain projects, not all properly priced by markets. Examples include public investment in diverse sorts of infrastructure. Chief concerns are with how any single project will perform alongside others. In particular, what is it worth if added to an already existing portfolio?

There are four novelties contained in this paper. First, it casts investment choice as a *cooperative production game*, featuring single projects as individual "players". Second, it argues that any *core* point incorporates project values. Third, it applies *stochastic optimization,* coupled with *Lagrangian duality,* to advocate a computable, explicit evaluation — in the core. Fourth, with a view towards lacking, fictitious or potential markets — and with an eye to *fundamental welfare theorems* — it suggests that optimal dual variables substitute for market-clearing *Arrow-Debreu contingent prices.*

Admittedly, optimal dual variables, alias *Lagrange multipliers* or *shadow prices,* have long served for project appraisal [14] and cost-benefit analysis [9,19]. Often though, those prices are presumed deterministic — whence not

* Thanks are due Finansmarkedsfondet, Norges Bank and Ruhrgas for financial support – and University of Manchester UK for great hospitality. This paper was first drafted at the economics department there. The research was also supported by the Norwegian Research Council under its RENERGI program.
** Research supported in part by a grant from the Natural Sciences and Engieering Research Council of Canada. We thank Bjørn Sandvik and the referee for valuable comments.

always in step with stochastic optimization. To wit, when some constraint appears uncertain ex ante, so does the associated shadow price as well. Further, by mentioning prices, one implicitly alludes to markets. But frequently, the latter might be imperfect or lacking. If so, reasons are strong to synthesize or simulate perfect counterparts. Most likely these resemble competitive asset markets. There though, no asset is priced in isolation from others. So, by way of analogy, each project had better be seen as member of a surrounding ensemble.

Cases in point comprise public investment in diverse energy sources or transportation modes. These are coupled via capacities, resources or tasks.[1] Such coupling suggests the use of cooperative game theory — and especially of the solution concept called the *core*, expressly concerned with synergies, efficiency, and incentive compatibility. Further, in our setting, when core solutions are generated by prices, players may plan as though a perfect exchange market were in smooth operation.

So, what this paper offers is a blend of two chief components, drawn from different strands of literature but rarely seen together. One concerns the theory of *market* or *production games*, emphasizing the constructive and efficient notion of *core solutions* [24]. The other component comprises optimization techniques, designed to deal with stepwise resolution of uncertainty [3]. Central in both strands are *Lagrangian duality* and *shadow prices* [17]. In fact, standard optimization procedures typically produce state- and time-dependent prices that clear spot markets. As already mentioned, those markets could be fictitious or real.

The paper stresses the said features. In doing so it aims at reaching several kinds of readers. Included are managers who, while concerned with numbers and computation, keep an eye on applicable, handy theory. Also addressed are economists who look at how randomness and simple computation connects to markets and coalition games. In particular, the material below may interest financial analyst and actuaries who derive bread and butter from pricing papers and policies, but are less used to endogenous "fundamentals". And last, but not least, we address mathematicians or computer scientists, not all well informed about how closely some of their main constructs fit to the operation and clearing of markets.

To reach so diverse a readership the paper is organized correspondingly. Sections 2 and 3 may suffice to illustrate some key points of linear programming — in particular, its Lagrangian duality theory. Section 4 is oriented towards cooperative game theory. It reviews the concept of the *core*, omitting any mention of time and contingencies. Section 5 remedies that omission by fleshing out investment planning as a two-stage, stochastic production game. Time and again core allocations will emerge via dual optimal solutions

---

[1] To disregard these links may sometimes appear reasonable — and especially so when initial price estimates only serve as proxies. But quite often such practise merits more criticism than justification.

to pooled programs. Core points produced in that manner correspond exactly with Arrow-Debreu contingent prices. Section 6 concludes by mentioning important extensions and caveats.

For simplicity we restrict attention to investments that ex ante can be formalized as one- or two-stage linear programs affected by uncertainty. Extensions and proofs are found in [7].[2]

The following notations apply. Whenever referring to an individual *project*, its index i serves as subscript. The *time* horizon stretches merely one period forwards, from the present, denoted 0, to the next period, labelled 1. When necessary, the appropriate time occurs as superscript. All variables construed as *prices* are written with a star. Thus, in a standard (dot) inner product $x^* \cdot x$, the first vector (with a star) denotes a *price* regime while the second (without a star) stands for a *quantity* vector. Admittedly, such notation is non-standard, but it saves letters and facilitates duality. Since technologies act as linear operators here below, we write all vectors as rows, and in coordinate-free form.

Uncertainty revolves around which *state of the world* $s \in S$ will happen next period. Ex ante, s is predictable only up to probability $P(s) > 0$; ex post *one* s materializes and becomes perfectly known. Since computation and modelling are our chief concerns, we do not hesitate in assuming S finite. Let E denote the expectation operator generated by P. When some decision, say x_i^1, concerning project i during period 1, depends on s, we should for correctness write $x_i^1(s)$. For simplicity, however, most often we omit mention of s since no confusion is likely to arise.

2 Pooling of Single-Stage Linear Problems Subject to Uncertainty

Throughout consider a fixed and finite set I of projects (or investment opportunities). For motivation, and as prelude to greater generality, suppose project $i \in I$ is modelled as a single-stage, one-shot linear program. Specifically, if standing alone, it amounts to the planning problem

$$\max_{x_i} \; x_i^* \cdot x_i \text{ subject to } \mathcal{A}_i x_i \leq y_i, \text{ and } x_i \geq 0. \tag{1}$$

Here x_i is construed as an activity (decision or design) vector in some finite-dimensional space $\mathbb{X}_i$ ordered (coordinatewise) by $\geq$. That activity pattern generates revenue $x_i^* \cdot x_i$ when evaluated in terms of a prescribed (known) price vector x_i^* and a (dot) inner product. Further, y_i denotes a given endowment bundle, owned by project i, and codified as a vector in another, ordered finite-dimensional space $\mathbb{Y}$. Finally, the "technology" $\mathcal{A}_i$ represents a linear operator mapping $\mathbb{X}_i$ into $\mathbb{Y}$.

---

[2] Extensions to multi-stage and nonlinear instances cause no conceptual difficulties, but these will, of course, tax the algorithmic and computational effort.

Problem (1) might come up ex post, after s has been unveiled. In that case, for correctness, we should emphasize that data $[x_i^*, \mathcal{A}_i, y_i]$ — whence any optimal solution x_i — may depend on s. However, to simplify notation, we often suppress mention of the state.

In any case, whether regarded ex ante or ex post, project i may produce uneven revenues from various activities, or face technological bottlenecks, or suffer from resource scarcity. Then, why not let various projects share synergies with each other? In extremis, why not bring all objectives, technologies and resources together to form the following grand planning problem?

$$\max \sum_{i \in I} x_i^* \cdot x_i \text{ subject to } \sum_{i \in I} \mathcal{A}_i x_i \leq \sum_{i \in I} y_i, \text{ and all } x_i \geq 0. \quad (2)$$

Note that individual resources y_i are presumed perfectly divisible and transferable. Note also that (2) pools objectives, resources and technologies. The advantages of doing so are evident: Inefficient or excessively endowed projects will furnish resources, while other, more efficient projects can undertake production. This simple idea immediately raises the question: *How can potential advantages of cooperation and coordination be secured and split equitably?* The following result tells how. For the statement let $\mathcal{A}_i$ have transpose $\mathcal{A}_i^*$. The latter maps $\mathbb{Y}$-prices y^* into $\mathbb{X}_i$-prices $\mathcal{A}_i^* y^*$.

Proposition 2.1 (Ex post pooling and sharing) *Fix any ex post state $s \in S$ together with realized data $[x_i^*(s), \mathcal{A}_i(s), y_i(s)]$. Suppose $y^*(s)$ solves the associated dual problem*

$$\min y^*(s) \cdot \sum_{i \in I} y_i(s) \text{ s. t. } x_i^*(s) \leq \mathcal{A}_i^*(s) y^*(s) \text{ for all } i \text{ with } y^*(s) \geq 0. \quad (3)$$

Then, no party or coalition $C \subseteq I$ of projects can, in the realized state s, do better than accepting value $v_C(s) := \sum_{i \in C} v_i(s)$ where

$$v_i(s) := y^*(s) \cdot y_i(s). \quad \square$$

As customary, problem (3) provides a minimal evaluation $y^*(s)$ of unit "production factors" so that *no* "output price" $x_i^*(s)$ strictly exceeds the corresponding imputation $\mathcal{A}_i^*(s) y^*(s)$. Now, instead of evaluating project i ex post, separately for each realized state s, what is it worth ex ante? We address this question right away:

Proposition 2.2 (Ex ante evaluation of random-yield, linear projects) *Let $s \mapsto y^*(s)$ be the profile of shadow prices mentioned in Proposition 2.1. Then, ex ante, and in expectation, project i commands a value*

$$\bar{v}_i := E v_i = E[y^* \cdot y_i] = \sum_{s \in S} y^*(s) \cdot y_i(s) P(s).$$

No cooperation within any coalition $C \subseteq I$ of projects can together generate expected value $> \bar{v}_C := \sum_{i \in C} \bar{v}_i$. □

Typically, $E[y^* \cdot y_i] \neq E[y^*] \cdot E[y_i]$, a phenomenon well known in finance concerned with covariances and beta pricing [13]. In particular, as brought out below in Proposition 4.3, since the mapping from aggregate endowment $y(s) := \sum_{i \in I} y_i(s)$ to possible price $y^*(s)$ is monotone decreasing, a *law of demand* holds, namely

$$[y^*(s) - \bar{y}^*(s)] \cdot [y(s) - \bar{y}(s)] \leq 0 \text{ for every } s,$$

hence $E[(y^* - \bar{y}^*) \cdot (y - \bar{y})] \leq 0$. As a result, any project i well endowed precisely when the others are suffering from scarcity, gets a premium above and beyond the deterministic counterpart $E[y^*] \cdot E[y_i]$. Put differently: *anti-correlated* or *negatively associated* projects receives some mark-up [15]. They provide insurance to other projects. This is illustrated next:

Example 2.1: Joint energy production. Consider generation of hydro-power, using two plants $i \in I = \{1, 2\}$. Plant 1 draws all water from short term precipitation and is best furnished in chilly, wet years. Conversely, plant 2, which merely taps melting water under a glacier, is best situated in dry, warm years. Since years are presumed either dry or wet, $S := \{dry, wet\}$. Plant i has contracted an obligation to deliver the amount of electricity e_i, independent of $s \in S$. Potential shortfall χ_{short} in production must be covered through purchases in the spot market (at a premium rate). Overproduction $\chi_{surplus}$, if any, is dumped on the grid.

Each producer worships maximization of own revenue, given effective capacity and revealed demand. Thus, in state s, if operating alone, producer i would solve problem (1). Specifically, let $x_i^* := [p, p_{short}, p_{surplus}] \in \mathbb{R}^3$ and $x_i := [\chi, \chi_{short}, \chi_{surplus}] \in \mathbb{R}_+^3$, to get the following problem instance

$$\left.\begin{array}{lllll} \max & p\chi & +p_{short}\chi_{short} & +p_{surplus}\chi_{surplus} & \\ \text{s.t.} & \chi & & & \leq cap_i(s) \\ & \chi & +\chi_{short} & -\chi_{surplus} & = e_i \\ & \chi & , \chi_{short} & , \chi_{surplus} & \geq 0. \end{array}\right\} \quad (4)$$

Here $cap_i(s)$ denotes the effective capacity of plant i in state s. Thus $y_i(s) := [cap_i(s), e_i]$. Choose

$$[p, p_{short}, p_{surplus}] = [15, -25, 10], \quad e_1 = e_2 = 12.5, \quad \text{and}$$

$$\begin{array}{ll} cap_1 = [cap_1(dry), cap_1(wet)] = & [10, 20], \\ cap_2 = [cap_2(dry), cap_2(wet)] = & [20, 10]. \end{array}$$

The state-dependent, go-alone shadow price vectors are $y_1^*(wet) = y_2^*(dry) = [35, -20]$ and $y_1^*(dry) = y_2^*(wet) = [45, -35]$, the subscript always referring to the plant. These price vectors reflect the diametral opposite positions of the two plants. Similarly, the optimal values are $v_1(wet) = v_2(dry) = 450$ and $v_1(dry) = v_2(wet) = 12.5$.

The pooled problem is just like (4), replacing there $cap_i(s)$ by $cap_1(s) + cap_2(s)$ and e_i by $e_1 + e_2$. Most importantly, after that replacement, no uncertainty prevails in the aggregate; years come out equal. Consequently, the associated dual vector $y^* = [35, -20]$ becomes constant, and the overall optimal objective value stabilizes at 550. If dry and wet years occur with equal probability, then ex ante each producer gets 206.25 if going alone. Upon pooling programs, however, each receives the greater value 275.

3   Pricing of Linear Investment Projects

Section 2 had no direct bearing on investment decisions. Explained next is how easily the perspective above can be enlarged to accommodate such decisions. It turns out that formats (1) and (2) still apply with minor modifications.

As stated, for simplicity, consider merely two time periods, denoted 0 and 1. Correspondingly, in that restricted but still dynamic setting, the variable $x_i = (x_i^0, x_i^1)$ has two components. The up-front time-0 component x_i^0 denotes the ex ante immediate decision, committed *before* s becomes known. It is followed by a time-1 decision x_i^1, implemented ex post, *after* s is revealed. Problem (1) assumes a form that mirrors the presence of these two stages:

$$\max \; x_i^{0*} \cdot x_i^0 + E(x_i^{1*} \cdot x_i^1) \quad \text{s.t.} \quad \begin{cases} A_i^{00} x_i^0 & \leq y_i^0 \text{ with } x_i^0 \geq 0. \\ A_i^{10} x_i^0 + A_i^{11} x_i^1 \leq y_i^1 \text{ with } x_i^1 \geq 0. \end{cases} \tag{5}$$

We hesitate to burden the reader with notation. But, begging a little indulgence, if one agrees to write vectors $x_i^* := (x_i^{0*}, x_i^{1*})$, $y_i := (y_i^0, y_i^1)$, and defines the inner product

$$\langle x_i^*, x_i \rangle := x_i^{0*} \cdot x_i^0 + E(x_i^{1*} \cdot x_i^1), \quad \text{and the matrix } \mathcal{A}_i := \begin{bmatrix} A_i^{00} & 0 \\ A_i^{10} & A_i^{11} \end{bmatrix},$$

problem (5) fits the mold

$$\max \; \langle x_i^*, x_i \rangle \quad \text{subject to} \quad \mathcal{A}_i x_i \leq y_i, \text{ and } x_i \geq 0, \tag{6}$$

quite in line with (1). The only crucial point to remember is that $s \mapsto x_i^1(s)$ is a contingent rule whereas x_i^0 is not. Just as before, programs (6) pool naturally into

$$\max \; \sum_{i \in I} \langle x_i^*, x_i \rangle \quad \text{subject to} \quad \sum_{i \in I} \mathcal{A}_i x_i \leq \sum_{i \in I} y_i, \text{ and all } x_i \geq 0. \tag{7}$$

Our chief interest is again with dual optimal solutions, or shadow prices, associated with problem (7). Any such price $y^* = (y^{0*}, y^{1*})$ consists of two components. The stage-0 component y^{0*} is deterministic. It prices endowments already on hand. In contrast, the stage-1 component y^{1*} is a *contingent price regime* $s \mapsto y^{1*}(s)$, serving to evaluate resources in various states s. The overall effect ex ante of the composite price $y^* = (y^{0*}, y^{1*})$ on any endowment $y = (y^0, y^1)$ is

$$\langle y^*, y \rangle := y^{0*} \cdot y^0 + E(y^{1*} \cdot y^1) = y^{0*} \cdot y^0 + \sum_{s \in S} y^{1*}(s) \cdot y^1(s) P(s).$$

Proposition 3.1 (Ex ante pooling) *Suppose* $y^* = (y^{0*}, y^{1*})$ *is a shadow price for problem* (7), *meaning that it solves the corresponding dual problem*

$$\min \left\langle y^*, \sum_{i \in I} y_i \right\rangle \text{ subject to } x_i^* \le \mathcal{A}_i^* y^* \text{ for all } i, \text{ and } y^* \ge 0.$$

Then, ex ante no party or subset $C \subset I$ *can do better, upon acting alone, than by accepting value* $v_C := \sum_{i \in C} v_i$ *where*

$$v_i := \langle y^*, y_i \rangle = y^{0*} \cdot y^0 + \sum_{s \in S} y^{1*}(s) \cdot y^1(s) P(s). \quad \Box \qquad (8)$$

Example 3.1: Investing in joint electricity production. Continuing the previous example, let us assume that producer i must choose his capacity $x_i^0 \in [0, k_i]$ at stage 0, before the water abundance becomes known. That choice is followed by the production decision at stage 1. His stand-alone problem (1) can be rewritten in deterministic equivalent form as

$$\max \quad x_i^{0*} \cdot x_i^0 + \sum_{s \in S} x_i^{1*}(s) \cdot x_i^1(s) P(s)$$

subject to $x_i^1(s) = [\chi, \chi_{short}, \chi_{surplus}](s)$ satisfying the constraints in (4), and $0 \le \chi(s) \le x_i^0 \le k_i$ for all s.

In addition to the specifications of Example 2.1, let us posit $x_1^{0*} = x_2^{0*} = -2.5$ and $k_1 = k_2 = 25$, with $P(\cdot)$ uniform, to have go-alone capacities $x_1^0 = x_2^0 = 20$ and expected optimal values $\bar{v}_1 = \bar{v}_2 \approx 78$. However, the pooled problem, and formula (8), assigns to each project the significantly larger ex ante value 100.

4   Project Portfolios and Core Solutions

It is time now to review the preceding development in terms of theory. This section offers some theoretical statements and discussion but no proofs. Focus is on linear programs — stated in static, deterministic form — and on pooling of these. Mention of their dynamic and stochastic nature must wait until the next section. In the interim our sole purpose is to clarify how cooperation relates to fictitious or real markets.

As before, there is a finite, non-empty set I of economic projects or investment opportunities. Some might already be realized; others are still at the planning stage. Project $i \in I$, if realized and evaluated in isolation from the others, yields net present value

$$\mathcal{V}_i = \sup\{\langle x_i^*, x_i \rangle : \mathcal{A}_i x_i \leq y_i \text{ and } x_i \geq 0\}. \tag{P_i}$$

Here x_i is the decision variable. It resides in a finite-dimensional Euclidean space $\mathbb{X}_i$, equipped with inner product $\langle \cdot, \cdot \rangle$ and vector order $\geq$. There is another, ordered space $\mathbb{Y}$, of similar kind, that contains the specified vector y_i. Finally, $x_i^* \in \mathbb{X}_i$ is a prescribed cost vector, and the linear operator $\mathcal{A}_i$ maps $\mathbb{X}_i$ into $\mathbb{Y}$.

The notation in (P_i) is admittedly somewhat uncommon but chosen to facilitate a subsequent dual perspective in which the symbol $*$ invariably is attached to prices and to operators on these. For interpretation of problem (P_i) one may regard project i as obliged to contend with resource bundle y_i and linear technology $\mathcal{A}_i$. The decision variable x_i then represents the activity, design, or input pattern of project i.

Problem (P_i) is not meant to model Robinson Crusoe's insular project planning. Indeed, the set I is *not* a singleton; it comprises more than one project. Important in that regard is the similarity between individual problems (P_i). In particular, note that all y_i belong to the same ordered vector space $\mathbb{Y}$. This crucial feature leads us to ask: *why not regard all projects as parts of one integrated enterprise?*

For the sake of that argument, suppose projects in $C \subseteq I$ formed a concerted business. By pooling resources and sharing technologies, that integrated endeavor could achieve pooled value

$$\mathcal{V}_C := \mathbb{V}_C(y_C) := \sup\left\{\sum_{i \in C}\langle x_i^*, x_i\rangle : \sum_{i \in C}\mathcal{A}_i x_i \leq y_C \ \forall x_i \geq 0\right\}. \tag{P_C}$$

Here $y_C := \sum_{i \in C} y_i$ is the total resource endowment held by concern C. Plainly, the resulting *superadditivity*

$$\mathcal{V}_{C \cup C'} \geq \mathcal{V}_C + \mathcal{V}_{C'} \text{ for all disjoint coalitions } C, C' \subseteq I,$$

reflects on the advantages of cooperation and overall planning. Let $\mathcal{A}_i^*$ denote the transpose operator of $\mathcal{A}_i$. Associated with the *primal problem* (P_C) is a *dual problem* (P_C^*) with optimal value

$$\inf\{\langle y^*, y_C\rangle : x_i^* \leq \mathcal{A}_i^* y^* \text{ for all } i \in C, \text{ and } y^* \geq 0\}. \tag{P_C^*}$$

While (P_C) deals with *activity planning*, problem (P_C^*) concerns *proper pricing* of resources y_C. Now suppose that the dual problem (P_I^*) for the *grand coalition* $C = I$ admits a solution y^* such that $\langle y^*, y_I\rangle = \mathcal{V}_I$. For brevity, call any such y^* an *equilibrium* or *shadow price*. As it turns out, a scheme for *profit sharing* in which project i receives value

$$v_i := \langle y^*, y_i \rangle, \tag{12}$$

proves both efficient and incentive compatible:

Proposition 4.1. (Linear production games and core solutions) *Consider a finite family I of projects where an integrated concern $C \subseteq I$, if going alone, would obtain economic value $\mathcal{V}_C$ as defined in (P_C). Then, any shadow price y^* generates a profit sharing (12) that satisfies*

$$\begin{cases} \text{(I)} & \textit{Pareto optimality: } \sum_{i \in I} v_i = \mathcal{V}_I \text{ and} \\ \text{(II)} & \textit{stability:} \qquad \sum_{i \in C} v_i \geq \mathcal{V}_C \text{ for all } C \subset I. \ \square \end{cases} \tag{13}$$

A vector $(v_i) \in \mathbb{R}^I$ that solves system (13), is said to reside in the *core* [4,23]. The simplicity of formula (12) is telling. Project $i \in I$ is paid $\langle y^*, y_i \rangle$ for making its technology $\mathcal{A}_i$ and resource endowment y_i available in a larger context. Then, according to (I), the total, most efficient value $\mathcal{V}_I$ is achieved and fully split. Also, by (II) no smaller consortium $C \subset I$ of projects could do better by going it alone. Reflecting on the latter feature, the inequalities in (II) might be referred to as *participation constraints*.

For further interpretation of Proposition 4.1 imagine that projects get the possibility, but have no obligation, to exchange resources (i.e., bundles in $\mathbb{Y}$) at fixed a price regime y^*. Plainly, when offered this more flexible setting, free of direct externalities, no party can fare worse:

Proposition 4.2. (Exchange market, duality, complementarity, and Walras' law)
• *Price-based, perfectly competitive exchange of resources can harm no coalition $C \subseteq I$ of projects. Indeed, for every feasible profile $(x_i)_{i \in C}$ and price y^* in problems (P_C) and (P_C^*) respectively, it holds that*

$$\sum_{i \in C} \langle x_i^*, x_i \rangle \leq \left\langle y^*, \sum_{i \in C} \mathcal{A}_i x_i \right\rangle \leq \langle y^*, y_C \rangle.$$

Weak duality *thus obtains in that $\sup(P_C) \leq \inf(P_C^*)$. So, setting $v_i := \langle y^*, y_i \rangle$, as recommended in (12), the participation constraint $\mathcal{V}_C \leq \sum_{i \in C} v_i$ always holds.*
• *Suppose $\sup(P_I)$ is attained. Then a (P_I^*)-feasible y^* is an equilibrium price if and only if there exists a (P_I)-feasible profile (x_i) with*

$$\sum_{i \in I} \langle x_i^*, x_i \rangle \geq \left\langle y^*, \sum_{i \in I} \mathcal{A}_i x_i \right\rangle \geq \left\langle y^*, \sum_{i \in I} y_i \right\rangle.$$

Any such pair (x_i) *and* y^* *optimally solve problems* (P_I) *and* (P_I^*), *respectively, and* **strong duality** *obtains in that* $\max(P_I) = \min(P_I^*)$.

• *Strong duality is attained precisely when the* **complementarity condition**

$$0 \le y^* \perp \sum_{i \in I}(\mathcal{A}_i x_i - y_i) \le 0$$

is satisfied. The latter amounts to **Walras' law** *by saying that pooled excess demand* $\sum_{i \in I}(\mathcal{A}_i x_i - y_i) \le 0$ *should have no value under prices* $y^* \ge 0$. $\square$

To appreciate Propositions 4.1-2, note that *excess demand*

$$\mathcal{E} := \sum_{i \in I}(\mathcal{A}_i x_i - y_i)$$

prevails in the exchange market. It may well happen that $\mathcal{E} = 0$, in which case the said market literally balances, meaning total demand $\sum_{i \in I} \mathcal{A}_i x_i$ equals total supply $\sum_{i \in I} y_i$. But insisting on $\mathcal{E} = 0$ might be overly stringent. What matters is rather to offset supply with excess demand $\mathcal{E} \le 0$ worth $\langle y^*, \mathcal{E} \rangle = 0$. Thus, a shadow price, first singled out to "balance" a transferable-utility cooperative game [4,23], also balances the resource market.

Elementary economics — or simple heuristics — indicate the way to grasp shadow prices. Imagine (temporarily) there being only *one* resource. Further, also for the sake of argument, suppose the grand coalition I obtains *differentiable* payoff $\mathbb{V}_I(y_I)$, defined in (P_C) when $C = I$. While $\mathbb{V}_I'(y_I) < y^*$, that coalition would gain by buying less in the resource market. Similarly, as long as $\mathbb{V}_I'(y_I) > y^*$, its resource demand had better be increased. So, apparently, an equilibrium price y^* prevails if and only if $\mathbb{V}_I'(y_I) = y^*$.

The preceding argument collapses, of course, when and where $\mathbb{V}_I(\cdot)$ cannot be differentiated in a classical sense. To set things right, reconsider problem (P_I) with standard Lagrangian

$$L_I(\mathbf{x}, y^*) := \sum_{i \in I} \{ \langle x_i^*, x_i \rangle + \langle y^*, y_i - \mathcal{A}_i x_i \rangle \}.$$

Assume the optimal value $\mathbb{V}_I(y_I)$ is finite. Recall that $\mathbb{V}_I(\cdot)$ is declared *superdifferentiable* at y_I if and only if it admits at least one *supergradient* $y^* \ge 0$ there. Then, one writes $y^* \in \partial \mathbb{V}_I(y_I)$, which means that

$$\mathbb{V}_I(y) \le \mathbb{V}_I(y_I) + \langle y^*, y - y_I \rangle \text{ for all } y \in \mathbb{Y}.$$

Proposition 4.3. (Shadow price as marginal payoff and saddle point) *Suppose* $\mathbb{V}_I(y_I) = \sup(P_I)$ *is finite. Then,* $y^* \in \partial \mathbb{V}_I(y_I)$ *if and only if* $y^* \ge 0$ *and* $\sup_{\mathbf{x}} L_I(\mathbf{x}, y^*) = \sup(P_I)$. $\square$

Proposition 4.3 invoked a generalized derivative — namely, the superdifferential of convex analysis — to extend the neoclassical optimality condition, stating that *the imputed value of marginal resources used in perfectly competitive projects should equal a common shadow price.*

5   Stochastic Production Games

Here we apply Proposition 4.1 in a setting wide enough to accommodate a large class of investment problems. There are still two decision stages denoted 0 and 1. At the outset uncertainty prevails as to which $s \in S$ will materialize next.

Corresponding to the two stages, decision x_i has an immediate, up-front component $x_i^0 \geq 0$ and a next-stage, recourse strategy $s \mapsto x_i^1(s) \geq 0$. Problem (P_i) now assumes the form:

$$\left.\begin{array}{l} \text{maximize} \ \langle x_i^*, x_i \rangle = \langle x_i^{0*}, x_i^0 \rangle + E \langle x_i^{1*}, x_i^1 \rangle \ \text{subject to} \\ x_i^0 \geq 0, \quad \mathcal{A}_i^{00} x_i^0 \leq y_i^0, \ \text{and} \\ x_i^1(s) \geq 0, \mathcal{A}_i^{10}(s)x_i^0 + \mathcal{A}_i^{11}(s)x_i^1(s) \leq y_i^1(s) \ \text{for all} \ s \end{array}\right\} \quad \text{(2-stage } P_i)$$

These problems pool into the following overall concern:

$$\left.\begin{array}{l} \text{maximize} \ \sum_{i \in I} \langle x_i^*, x_i \rangle \ \text{subject to} \\ x_i^0 \geq 0, \quad \sum_{i \in I} \mathcal{A}_i^{00} x_i^0 \leq y_I^0, \ \text{and} \\ x_i^1(s) \geq 0, \sum_{i \in I} \left\{ \mathcal{A}_i^{10}(s)x_i^0 + \mathcal{A}_i^{11}(s)x_i^1(s) \right\} \leq y_I^1(s) \ \forall s. \end{array}\right\} \quad \text{(2-stage } P_I)$$

Proposition 5.1. (Portfolio pricing of linear projects) *Suppose problem* (2-stageP_I) *admits a shadow price* $y^* = \left[y^{0*}, y^{1*}(\cdot) \right] \geq 0$. *Then*

$$v_i = \langle y^{0*}, y_i^0 \rangle + E \langle y^{1*}, y_i^1 \rangle$$

is the ex ante, cooperative value of project i. *A planner contemplating to add project* i *to an existing portfolio* $I \backslash i$, *should regard* v_i *instead of the "smaller" optimal value of problem* (2-stage P_i). $\square$

6   Concluding Remarks

Investment theory has increasingly emphasized the importance of uncertainty over returns, the irreversibility of particular installments, and the opportunities to wait and see.[3] Said theory deals, however, mostly with *one* project or *one* firm — whence merely with *one* decision maker — at a time. In contrast, many planning problems — and notably those regarding environmental management, technological reliability, public welfare and health — revolve around *several* projects, typically not traded in markets. And most importantly: the enterprises at hand might "belong" to separate agents or authorities. In short, there could be as many or more owners than projects. Then, *how can potential gains from concerted efforts be secured and shared?*

In addressing that question, this paper has emphasized three things. First, it gives priority to the concept of the core — central, natural, and most applied in cooperative game theory [6–8,18,11,22]. Second, it points to explicit,

---

[3] References include [1], [2] and [5].

computable solutions, defined in terms of *contingent prices* [13,15]. Third, in formalizing investment planning as multi-stage optimization under gradual resolution of uncertainty, it indicates how powerful techniques of stochastic programming can be made to bear on analysis and evaluation [20].

One sees then how some projects merit marked-up values because they insure or stabilize the pooled output. Broadly, these are projects that produce in "counter-cyclical" fashion; they swing up precisely when others turn down. In a different jargon: they provide *recourse*. Such projects can appear of no value while alone, but qualify well together with others.

For ease of exposition, the main focus of this note has been on linear instances, deterministic versions of which have been studied in [10,12,18,21]. Extensions to nonlinear, stochastic settings are found in [6,7,22].

Some important caveats remain though. For one: if payoff functions or technologies exhibit increasing returns to scale, our dual approach, couched in terms of Lagrangians, is fraught with fundamental problems. Core solutions may then be unattainable via linear pricing — or quite simply, be unavailable. Likewise, when some activities or items are indivisible, it may well happen that shadow pricing gives incorrect evaluations. If so, those evaluations are all overestimates.

References

1. A. B. Abel, A. K. Dixit, J. C. Eberly and R. S. Pindyck, Options, the value of capital, and investment, *The Quarterly Journal of Economics* **11** (3), 753-777 (1996).
2. K. J. Arrow and A. C. Fisher, Environmental preservation, uncertainty, and irreversibility, *The Quarterly Journal of Economics* **88**, 312-319 (1974).
3. J. R. Birge and F. V. Louveaux, *Introduction to Stochastic Programming*, Springer, Berlin (1997).
4. O. N. Bondareva, Some applications of linear programming methods to the theory of cooperative games, *Problemy Kibernetiki* **10**, 119-139 (1963).
5. A. Dixit and R. S. Pindyck, *Investment under Uncertainty*, Second Printing, Princeton University Press (1996).
6. P. Dubey and L. S. Shapley, Totally balanced games arising from controlled programming problems, *Mathematical Programming* **29**, 245-276 (1984).
7. I. V. Evstigneev and S. D. Flram, Sharing nonconvex cost, *Journal of Global Optimization* **20** (3-4), 257-271 (2001).
8. S. D. Flram, G. Owen and M. Saboya, Large production games and approximate core solutions, Typescript (2005), to appear.
9. E. M. Gramlich, *Benefit-Cost Analysis of Governmental Programs*, Prentice Hall, New Jersey (1981).
10. D. Granot, A generalized linear production model: a unifying model, *Mathematical Programming* **43**, 212-222 (1986).
11. W. Hildenbrand, Cores, in: J. Eatwell, M. Milgate and P. Newman (eds.) *The New Palgrave: A dictionary of economics*, Palgrave MacMillan, New York (1998).

12. E. Kalai and E. Zemel, Generalized network problems yielding totally balanced games, *Operations Research* **30** (5), 998-1008 (1982).

13. S. LeRoy and J. Werner, *Principles of Financial Economics*, Cambridge University Press (2001).

14. E. H. Londero, *Shadow Prices for Project Appraisal*, Edward Elgar, Cheltenham UK (2003).

15. M. Magill and M. Quinzii, *Theory of Incomplete Markets*, MIT Press (1996).

16. A. Mas-Colell, M. D. Whinston and J. R. Green, *Microeconomic Theory*, Oxford University Press, Oxford (1995).

17. P. K. Newman, Duality, in: J. Eatwell, M. Milgate and P. Newman (eds.) *The New Palgrave: A dictionary of economics*, Palgrave MacMillan, New York (1998).

18. G. Owen, On the core of linear production games, *Mathematical Programming* **9**, 358-370 (1975).

19. D. W. Pearce and C. A. Nash, *The Social Appraisal of Projects: A Text in Cost-Benefit Analysis*, J. Wiley & Sons, New York (1981).

20. A. Ruszczyński and A. Shapiro (eds.), *Stochastic Programming, Handbook in Operations Research and Management Science* **10**, Elsevier (2003).

21. D. Samet and E. Zemel, On the core and dual set of linear programming games, *Mathematics of Operations Research* **9** (2), 309-316 (1994).

22. M. Sandsmark, Production games under uncertainty, *Computational Economics* **14** (3), 237-253 (1999).

23. L. S. Shapley, On balanced sets and cores, *Naval Research Logistics Quarterly* **14**, 453-461 (1967).

24. L. S. Shapley and M. Shubik, On market games, *Journal of Economic Theory* **1**, 9-25 (1969).

Precaution: The Willingness to Accept Costs to Avert Uncertain Danger

C. Weiss

Edmund A. Walsh School of Foreign Service Georgetown University
37th and O Sts., N.W. Washington DC 20057 USA

1 Introduction

Decision makers coping with environmental threats of unknown size and probability need to understand both the science underlying the threat and the uncertainty connected with this science. They also need to understand the level of disagreement among experts - and among the public at large - regarding both the science and its associated uncertainty. Only in this way can they properly assess the political, financial or social costs that they are willing to incur in order to avoid or mitigate this threat, whether it be due to a natural hazard – climate change, earthquake, or destruction of stratospheric oxygen, for example – or to a proposed human intervention or innovation, such as ocean dumping or genetically modified crops. In order to support such decision-making, technical experts need to be able to communicate the level of technical uncertainty they associate with a given threat in a reasonably precise and understandable form.

In this paper, we review and assess two proposed scales, or standard vocabularies, for expressing the subjective estimate of the likelihood that a given assertion is true, given the strength of the underlying evidence. Assertions of this form may range from speculative hypotheses to well–established theories. We then use these scales as the basis for an analytic framework within which to express differing levels of precaution, or risk avoidance, and to distinguish

Charles Weiss is Distinguished Professor and Chair of Science, Technology and International Affairs at the Edmund A. Walsh School of Foreign Service at Georgetown University. A Harvard trained biochemical physicist, he was the first Science and Technology Adviser to the World Bank, and served in that capacity from 1971-86. He taught and helped launch the program in science and technology policy at the Woodrow Wilson School at Princeton University, and has been Visiting Professor at the University of Pennsylvania, Visiting Scholar at the University of California (Berkeley), Course Director at the Foreign Service Institute of the US Department of State, and Professorial Lecturer at the Nitze School of Advanced International Studies at Johns Hopkins University. He is the author of numerous publications on international science and technology policy. He thanks Katherine Couturier for excellent assistance in preparing the figures.

disagreements over precaution from disagreements over science or over scientific uncertainty.

This framework enables us to put the Precautionary Principle in a larger perspective as a clear and valuable statement of one side of the debate over the policy implications of scientific uncertainty. We then argue that a second principle, a proposed Principle of Innovation and Adaptive Management, is needed to complement the Precautionary Principle and to prevent an unnecessarily narrow view of precaution from stifling desirable innovation. This new principle is consistent with the emerging view of precaution in recent literature.

2   Scales of Subjective Uncertainty

The best known scale of subjective technical uncertainty is the quantitative, seven-step scale used by the Inter-Governmental Panel on Climate Change (IPCC), which assigns adjectival phrases, ranging from "virtually certain" to "very unlikely," to seven ranges of Bayesian probability[1]. As a complement, the author has proposed a qualitative, 12-step "legal" scale derived from the standards of proof used in various branches of US law, and has correlated this scale with the IPCC scale and with the informal measures used by various scientists in assessing the likelihood that a given hypothesis will turn out to be true[2].

The IPCC and the "legal" scales are both Bayesian scales that express someone's subjective estimate of uncertainty, as opposed to frequentist probabilities based on experience. These Bayesian scales are scales of uncertainty, not of risk. That is, they apply to situations in which the outcome of events cannot be foreseen because of underlying ignorance, rather than to situations in which the probability of an event can be defined from empirically based statistics. The percentages do not refer the statistical probability of an event based on past experience ("The chances of a flood of this magnitude happening in any given year is one in a thousand"), but rather to the subjective probability that an assertion is true ("I'd give 2:1 odds that half of today's incidence of coronary artery disease will turn out to have been due to a bacterial infection."). The two scales are summarized in Table1[3]. Either may

---

[1] Inter-Governmental Panel on Climate Change (2001) Report of Working Group I. Available at http://www.ipcc.ch/pub/spm22-01.pdf. All URLs are accurate as of 24 May 2005.

[2] Charles Weiss, "Expressing Scientific Uncertainty," *Law, Probability and Risk*, 2, 25-46 (2003).

[3] The "legal" scale of uncertainty set forth in table 1 has been modified from the scale as set forth in the reference of footnote 2 in two ways: (1) a new step, that of "reasonable indication," has been introduced so as to provide "mirror symmetry" around the step corresponding to 50 % probability. (2) The top and bottom step are now expressed as "insufficient to support even a hunch" and "virtually

be used to represent the subjective views of individuals, or alternatively, to represent the distribution of views among experts, stakeholders, members of the general public, or other populations. They can also be used to separate assertions of scientific fact from estimates of the uncertainty associated with those assertions. For example, one may imagine that two scientists or two advocates may agree on the most likely conclusions that may be drawn from a given set of scientific data, but may disagree in the degree of uncertainty associated with that interpretation.

Both the IPCC and the legal scales are summary scales that can take into account the "pedigree" of the model underlying the conclusions, a property defined as a "systematic, multi-criteria evaluation of the different phases of the production of the knowledge base".[4] Both scales can also be used to summarize subjective views regarding the three dimensions of uncertainty identified by Walker et all and to the uncertainty in the choice of frequentist frequency distributions used in various branches of engineering work in order to deal with unpredictable variables, such as the mechanical stresses on structures or vehicles[5].

Each of the two scales has its advantages and disadvantages. The IPCC scale allows the probabilities associated with individual links in a chain of evidence to be multiplied in order to give the overall probability of the entire chain. On the other hand, the legal scale may be more suited to the needs of people or groups who are uncomfortable with numbers and hence are unlikely to carry out these multiplications, simple though they may be. For such people, the advantage of the legal scale is that it is expressed in words that are reasonably familiar and are "anchored" in situations with which they can identify.

Both, the IPCC and the legal scales make negative assertions (such as that a project poses no danger) subject to standards of proof complementary to the standards applicable to the corresponding positive assertion (in this case, that the project is in fact dangerous). As a quantitative scale, the IPCC scale has a significant advantage in dealing with such negative statements, since the subjective probability that an assertion is not true is simply one minus the subjective probability that it is true. $(P(A) = \left(1 - P(-A)\right))$. This relation is not as straightforward in the legal scale. Nevertheless, as is

---

certain," replacing "impossible" and "certain," neither of which belongs in a scale of uncertainty because they do not admit of any uncertainty. The correlations shown in Table 1 are the author's.

[4] Joroen P. Vandersluijs, "Uncertainty and Precaution in Environmental Management: Insights from the UPEM Conference," Environmental Modeling and Software (IEMSS-2004 special issue); Silvio Funtowitz and Jeremy Ravetz, *Uncertainty and Quality in Science for Policy* (Dordricht: Kluwer, 1990).

[5] W.E. Walker et al., "Defining Uncertainty: A Conceptual Basis for Uncertainty Management in Model-Based Decision Support," *Integrated Assessment* 4, 5-17 (2003).

shown in Table 2, the treatment of negative assertions is remarkably good, especially considering that the various legal standards of proof have evolved for quite different purposes and were never intended to be complementary. For example, if a person thinks that there is evidence "beyond a reasonable doubt" (the step in the legal scale that corresponds to 99 % probability) that a particular assertion is true, (s)he still could reasonably be supposed to hold a "fanciful conjecture" that it is not true, this being the step corresponding to < 1 % subjective probability[6]. On the other hand, it would be unreasonable for her to believe that there were "reasonable grounds for suspicion" (1-10 % probability, the next higher step in the scale) that it is false.

3   The Precautionary Principle and the Willingness to Incur Costs

The scales of uncertainty discussed in the previous section allow us to set forth a framework within which to discuss the willingness to incur costs to avert or mitigate an uncertain threat. The major principle addressing such threats in international law is the Precautionary Principle, which states that action to protect the environment from the danger of severe and irreversible damage does not need to wait for rigorous scientific proof. This principle was initially enunciated in opposition to the doctrine that no intervention was necessary or justified until and unless evidence for the danger in question reached the level of rigorous scientific proof[7].

In the international legal literature, the Precautionary Principle appears in two forms. In its "weak" formulation, the Precautionary Principle asserts that the absence of rigorous proof of danger does not justify inaction. The Rio Declaration, for example, states that

---

[6] The terms "inarticulable hunch" and "fanciful conjecture," which are here equated, have different legal origins. "Inarticulable hunch" is used here as a standard insufficient to justify a "*Terry* stop," a "minimally intrusive" pat-down for weapons permitted when a police officer has a "reasonable, articulable suspicion" that a crime is afoot. (*Terry v. Ohio*, 392 U.S. 1, 88 S.Ct. (1968)). "Fanciful conjecture" is used to define a concern that does not rise to the level of a "reasonable doubt" that would justify a verdict of not guilty in a criminal trial. "A reasonable doubt is an actual and substantial doubt arising from the evidence, from the facts or circumstances shown by the evidence, or from the lack of evidence on the part of the state, as distinguished from a doubt arising from mere possibility, from bare imagination, or from fanciful conjecture." (*Victor v. Nebraska; Sandoval v. California* 1994 511 U.S. 1, 114 S.Ct.1239)

[7] For references on the history and legal status of the Precautionary Principle, see Charles Weiss, "Scientific Uncertainty and Science-Based Precaution," *International Environmental Agreements: Politics, Law and Economics* 3, 137-166 (2003), footnote 1.

"Where there are threats of serious or irreversible damage, lack of full
scientific certainty shall not be used as a reason for postponing
cost-effective measures to prevent environmental degradation." [8]

In this statement, the fear of danger forces the consideration of precautionary
intervention but does not require such intervention actually to take place.

In its "strong" formulation, by contrast, the Precautionary Principle de-
clares that the absence of rigorous proof does require precautionary action to
be taken, and that the burden of proof lies with the proponent of an action to
show that it does not pose a danger of environmental harm. The Wingspread
Statement on the Precautionary Principle, for example, states that

"When an activity raises threats of harm to human health or the
environment, precautionary measures *should be taken* [author's italics]
even if cause and effect relationships are not fully established scientifically.
[The] proponent of the activity, rather than the public, should bear the
burden of proof." [9]

In a previous paper, the author pointed out two logical flaws in both of
these formulations of the Precautionary Principle, which substantially reduce
its value as a practical guide to action[10]. The first of these flaws is that
neither formulation deals with the issue of "willingness to pay", i.e., the
costs – economic, political and social – that would be justified by the need to
address a threat whose magnitude is agreed but whose reality is uncertain.

Either the IPCC or the "legal" scale can be used as part of an analytical
framework to clarify the issues involved. We may represent this utility, or
willingness to incur costs, as a function of three variables: $U = f(r, c, p)$,
where

$U =$ A utility function that gives the cost that a person, group,
       organization or government is willing to incur in order to
       avert the threat in question;

$r =$ The anticipated damage, i.e., the net cost if the threat were
       to come to pass;[11]

---

[8] Rio Declaration on Environment and Development (1992), In Edith Brown Weiss
et. al., International Environment Law: Basic Instruments and References, 1992-
1999 (Ardsley NY. Transnational Press, 1999).

[9] The Wingspread Declaration is the declaration of a meeting of non-
governmental experts held in Wingspread WI in January 1998. See
http://www.sehn.org/state.html.

[10] Charles Weiss, *op. cit.*

[11] This anticipated damage is assumed to be agreed in any particular situation. In
principle, these costs may be technical, political, social or even cultural and may
or may not be quantifiable. In most situations, these costs cannot be reduced to
a singel number or even a single dimension, so that these curves are unavoidably
oversimplifications. In purely economic terms, thr net cost might be expressed,
for example, by the discounted value of the sum of cost of repair of the damages,

$c =$ The subjective probability (or other measure of certainty) that the threat is real;[12] and

p is a variable that measures the attitude of a person, group, organization or government toward precaution, or (equivalently) the willingness to accept risk or level of risk aversion.

Figure 1 shows utility functions[13] $U(c)$ at three levels of anticipated threat (r) for each of three attitudes toward precaution, making nine curves in all. The curves are sketches of theoretical archetypes and are not derived from data. In an earlier work, we classified different levels of precaution into five archetypes: environmental absolutist, cautious environmentalist, environmental centrist, technological optimist, and scientific absolutist.[14] For simplicity of presentation, only three of these five are represented in Figure 1. These may be conceived as discrete values of p. If the threat in question is connected to a human intervention, such as a project or a technological innovation, the value of p expresses the balance in a given individual between the avoidance of risk ("Look before you leap") and the welcoming of innovation ("Nothing ventured, nothing gained"),

In Figure 1, each level of precaution is represented by a curve of a different type: dotted, dashed, or solid line. Within each level of precaution, the curves labeled a, b and c represent, respectively, the willingness $U(c)$ to incur costs at different levels of certainty for each of three levels of threat (r): (a) utter catastrophe (an asteroid striking the Earth, for example), (b) serious and irreversible damage (pollution of the bottom of a pristine lake or aquifer, for example), and (c) serious but reversible damage (pollution of the surface water of a major river, for example). By comparing curves a, b, and c of the same type, the reader may trace the change in $U(c)$ at a given level of precaution as the level of threat increases. Similarly, for each level of anticipated threat, Figure 1 shows three curves, each of a different type and representing a different level of precaution.

---

the losses incurred while the damage is being repaired, and the losses that cannot be repaired, or the opportunity costs of the innovation that has been thwarted. A fuller representation would take into account the distribution of these costs among different groups. Our previous treatment (reference, footnote 11) considered only a single representative value of anticipated cost.

[12] A fuller treatment would take into account the fact that a given individual may entertain a subjective probability distribution of the level of risk, independent of his level of confidence in that probability distribution. For example, he may have a "reasonable belief" that there is a 70 % chance of major damage, a 10 % chance of minor damage, and a 20 % probability that there will be no damage at all.

[13] See David M. Kreps, *A Course in Microeconomic Theory* (Princeton NJ: Princeton University Press, 1990).

[14] Charles Weiss, *op. cit.*. We have changed the shape of the function U in Figures 1-3 from that shown in the figures in this reference.

We use the Bayesian probabilities of the IPCC scale in Figure 1 so as to make the units of the x-axis linear. (Our earlier work used the legal scale of uncertainty, resulting in a non-linear scale for the x-axis[15]). The scale of the y-axis represents different levels of intervention by verbal descriptions rather than numbers, since many of the costs are unquantifiable.[16] If the y scale could be made to be well defined and linear in dollars or other appropriate unit, the slope at any point on the curve would represent the price in that unit that a person or group described by a given level of precaution would be willing to pay for a marginal increase in the subjective feeling of certainty that the threatened danger will not take place.[17]

For the archetypal "environmental absolutist" - say, a radical eco-activist - curves 3a-c of Figure 1 depict a demand for action that begins at very low probability and indeed is relatively independent of the degree of certainty of the threat. The "environmental centrist" requires somewhat more certainty before insisting on action. The "scientific absolutist", on the other hand, insists on a much higher degree of certainty before being willing to take strong measures, and is willing to accept significant costs for unlikely threats only if they are truly catastrophic in scope.

The curves of Figure 2 represent the levels of intervention for which the group or individual is willing to pay in order to avert a threat of serious and irreversible danger, at three different levels of precaution (p). The curves are parametrized by the certainty (c) connected with the risk in question, and constitute utility curves for increasingly strict attitudes toward precaution. In essence, the rise of each curve from left to right depicts the extra cost of increased levels of precaution at each level of certainty. If the cost curve could be made quantitative, linear and continuous, the slope of the curve would represent the price to be paid for a marginal increase in precaution at the given level of anticipated cost.

At all but the very highest levels of certainty, the "scientific absolutist" – who nowadays is typically a representative of a polluting industry – is depicted in Figures 2 and 3 as very reluctant to accept additional costs so much so that the utility function representing his or her preferences for certainty (dotted line in Figure 1) is actually concave upward, indicating a welcoming attitude toward risk. Only (s)he would not take comprehensive measures to deal with the threat of serious and irreversible damage at any but the very highest levels of certainty (the top two curves in Figure 2).

---

[15] Charles Weiss, *op.cit.*

[16] These verbal formulations are derived from a study of the response of the international community to several global environmental threats. See Charles Weiss, *op. cit.*

[17] If the person making the judgment were utterly indifferent to risk, the anticipated cost were expressed in dollars, and the y-axis were linear, the curve in Figure 1 for $r = r_0$, where r_0 is the anticipated cost, would be a straight line with a slope of $r_0/100$.

The nine curves of Figure 3 show the levels of intervention that the individual or group is willing to pay for as a function of the seriousness of the anticipated threat, at each of three levels of uncertainty and three levels of precaution. As might be expected, the curves converge at the highest levels of certainty of utter catastrophe, but diverge considerably at lower levels of anticipated cost and higher levels of certainty, with environmental absolutists insisting on relatively vigorous action even at low levels of certainty and anticipated cost, while scientific absolutists insist on a much higher level of certainty, even in the face of higher catastrophe.

Figures 1-3 have not been vetted by users, and are intended to provide a framework for posing questions regarding uncertainty and precaution rather than precise decision support. One need not accept the specific, more or less arbitrary shape of the curves to use them as a framework for clarifying the distinctions among assertions of scientific fact, the uncertainty associated with those assertions, and the question of what to do about the threat they describe. The usefulness of this representation in actual communication and decision support needs to be tested with focus groups and actual decision situations.

4   A Proposed Principle of Innovation and Adaptive Management

The second major limitation of the Precautionary Principle as a practical guide to decisions regarding technological innovation is that it implicitly conceives policy decisions as choices between a risky innovation or intervention, on the one hand, and a presumably safe status quo, on the other. In practice, the Principle is typically employed as a means of discouraging proposals for intervention or innovation. In contrast, many if not most actual policy decisions involve a choice between (or among) risky alternatives.

In retrospect, this narrow view of precaution stems in large part from past experience with "technological lock-in", in which decision makers may never have been given the opportunity to consider the merits of a possibly harmful intervention, or at best are faced with "go/no-go" choices concerning proposals for technological interventions that have acquired considerable momentum and would require major efforts to stop, even though they may never have been properly analyzed or justified.[18]

Drawing on this experience, more recent interpretations of the Precautionary Principle have emphasized the need to frame issues broadly enough to encourage consideration of a variety of alternative strategies, so as to allow time to gather more information before making a commitment to a particular approach, and to reduce the prospect that the decision makers will be

---

[18] Poul Harramoes et al, eds., *The Precautionary Principle in the 20th Century: Late Lessons from Early Warnings* (London: Earthscan Publications, 2002).

confronted with an unpleasant choice down the road. In this broader view, blocking an innovation or intervention is only one of a number of possible measures by which the Precautionary Principle may be implemented.[19]

This literature on the Precautionary Principle thus seems to be edging toward a position consistent with the approach of adaptive management, which is defined as "a systematic process for continually improving management policies and practices by learning from the outcomes of operational programs".[20] Such adaptive management would be discouraged by a strict application of the Precautionary Principle in its present form, which has the effect of rationalizing opposition to paths whose dangers are not well understood from the beginning. To be sure, adaptive management has its limitations. It is appropriate to those situations that do not require immediate and definitive results, in which increased research may be expected to yield information that will create opportunities for improved management in the reasonably near future, and in which unanticipated consequences of the proposed innovation are unlikely to be catastrophic or irreversible. It also requires that the necessary research and monitoring actually take place and that the results be fed into the management regime. Even so, it is applicable to the majority of long-term environmental issues.

In order to create space in the legal regime for this more flexible approach, we propose a new Principle of Innovation and Adaptive Management: "that research and development on a new technology that promises major benefits not be unreasonably blocked until the detailed implications of this technology are well understood, as long as adequate provisions are made for research and monitoring, and for incorporating their results into management strategy".[21]

The new Principle balances and complements the Precautionary Principle, and ensures that the opportunities that could be created by innovation are not lost because of an over-zealous application of precaution. It is in-

---

[19] Joel Tickner, ed., *Precaution, Environmental Science, and Preventive Public Policy* (Washington: Island Press, 2003); M. Kaiser, "Multi-Stakeholder Application of the Precautionary Principle: The Importance of Transparent Values," paper presented at the UPEM Conference in Copenhagen, 7-9 June 2004; and Poul Harramoes, *op. cit.*

[20] Pamela A. Wright, "Monitoring for Forest Management Unit Scale Sustainability: The Local Unit Criteria and Indicators Development Test (The LUCID Test)" US Department of Agriculture and US Forest Service, Inventory and Monitoring Institute Report ♮4, 2002, available at http://www.fs.fed.us/institute/lucid/final_report. *Ibid.* See also National Research Council, *Our Common Journey* (Washington; Natianal Academy Press).

[21] We have here modified the formulation given in our earlier paper, Charles Weiss, "Scientific Uncertainty and Science-Based Precaution," *op. cit.*, by adding the clause beginning "as long as." We are also changing the name of the proposed Principle, which in our earlier paper we called the "reasonableness principle," by replacing the value-laden term "reasonableness," which might imply that any criticism of the proposed Principle would be unreasonable.

tended to frame the debate on the wisdom of a given innovation between advocates of precaution who emphasize risks and dangers, on the one hand, and advocates of innovation who emphasize opportunities, on the other. It is deliberately couched as a general principle that mirrors the generality of the Precautionary Principle. Both points of view are legitimate and deserve to be heard. Indeed, we would argue that adaptive management, properly implemented, is an essential element of precaution.

5   Conclusion: A Framework for Balanced Precaution

We have presented a logical framework to guide decisions involving scientific uncertainties, based on the following elements[22]:

- Scales and standard vocabularies of subjective (Bayesian) uncertainty that facilitate a clear statement of the degree of uncertainty surrounding a scientific assertion;
- A framework for distinguishing between the state of scientific knowledge, on the one hand, and the uncertainty surrounding that knowledge, on the other;
- A framework for distinguishing both of these from one's attitude toward risk and precaution: i.e., one's willingness to accept costs to avert uncertain danger as a function of its size and uncertainty;
- A logical framework for decisions regarding the costs of averting uncertain dangers
- A legal Principle of Innovation and Adaptive Management that facilitates the balanced consideration of risks and opportunities in decisions between alternatives whose scientific underpinnings are uncertain.

The various elements of this framework should be useful in distinguishing the different issues involved in public discussions of these frequently complex and difficult issues. Specifically, the scales of uncertainty should be useful in presenting scientific information to policy makers and the general public, and in explaining the distinction between scientific assertions and their associated uncertainty. One may imagine, for example, experts on opposite sides of a policy dispute acknowledging that they agree on the likelihood that a given danger is real and of approximately of a given cost and magnitude, but disagree on the costs they would accept to forestall or mitigate that danger. Conversely, they might disagree about the level of uncertainty sorrounding the science underlying the danger but agree on what they would recommend if they were to come to agree that the danger was real. The utility functions for precaution and subjective uncertainty should be a useful framework for

---

[22] Both the "legal" scale of uncertainty and the Principle of Innovation and Adaptive Management are slightly modified from the author's earlier work cited in footnotes 2 and 7.

distinguishing differing attitudes toward precaution from differences of opinion regarding science and the uncertainty connected with science and help avoid the more extreme versions of "sizbd science", on the one hand, and excessive precaution, on the other. The Principle of Innovation and Adaptive Management complements the Precautionary Principle by framing the debate between advocates concerned with the risks of innovation, and those more inclined to value its benefits. This in turn facilitates consideration of adaptive management when this is appropriate to a particular situation.

The scale and framework were originally developed for practitioners of science policy, especially those concerned with international environmental issues. The overall framework should have broad application in law, intelligence, the history of science, and other fields as well. The Bayesian scales of uncertainty may also facilitate subjective estimates of the validity of the assumptions and proxies that are inevitable in the models used in some branches of engineering. It would be desirable to subject the scale, the framework and the proposed principle to practical test.

Table 1. Scales of Certainty

Level	Bayesian Probability	IPCC Scale	Informal Scientific	Scale based on Legal Standards of Proof	Legal Situation where Standard of Proof Applies
11	100%	(Not on scale)	Firmly established, has stood the test of time	Virtually certain	Exceeds criminal standard
10	99%	"Virtual certain"	Rigorously proven	"Beyond a reasonable doubt"	Criminal conviction
9	90 - 99 %	"Very likely"	Substantially proven	"Clear and convincing evidence"	Quasi-penal civil actions, such as termination of parental rights
8	80 - 90 %		Very probable	"Clear showing"	Granting temporary injunction
7	67 - 80 %	Likely	Probable	"Substantial and credible evidence"	Referring evidence for impeachment
6	50 - 67 %	Medium likelihood	"If I must choose, this seems more probable than not"	"Preponderance of the evidence"	Most civil cases
5	33 - 50 %		Evidence is fairly strong but not preponderant	"Clear indication"	Proposed criterion for nighttime X-Ray or body cavity searches
4	20 - 33 %		Increasing evidence worth a major effort to verify	"Probable cause", "Reasonable belief"	Field arrest, Search incident to arrest, Search warrant; Arraignment or indictment
3	10 - 20 %	Unlikely	Plausible, backed by some evidence	"Reasonable indication"	Initiate FBI investigation or trade inquiry
2	1 - 10 %		Suggestive	"Reasonable, articulable grounds for suspicion"	Stop and frisk for weapons
1	< 1 %	Very unlikely	Unlikely	"No reasonable grounds for suspicion," "Inchoate hunch", "Fanciful conjecture"	Does not justify stop and frisk
0	0 %	(Not on scale)	Violates well established laws	Insufficient even to support a hunch or conjecture	Action taken could not have resulted in the crime being charged

Table 2. Legal Scale of Uncertainty for Positive and Negative Statements

Bayesian Probability for A	Uncertainty for Positive Statements: If A is Thought to be True to this Standard, ...,	Corresponding Uncertainty for Negative Statements: ...then (-A) Could Still be Thought True to this Standard	Bayesian Probability for (-A)
100 %	Virtually certain	No grounds even for a hunch	0 %
99 %	"Beyond a reasonable doubt"	"No reasonable grounds for suspicion", "Inchoate hunch", "Fanciful conjecture"	< 1 %
90 - 99 %	"Clear and convincing evidence"	"Reasonable, articulable grounds for suspicion"	1 - 10 %
80 - 90 %	"Clear showing"	"Reasonable indication"	10 - 20 %
67 - 80 %	"Substantial and credible evidence"	"Probable cause"; "Reasonable belief"	20 - 33 %
50 - 67 %	"Preponderance of the evidence"	"Clear indication"	33 - 50 %
33 - 50 %	"Clear indication"	"Preponderance of the evidence"	50 - 67 %
20 - 33 %	"Probable cause"; "Reasonable belief"	"Substantial and credible evidence"	67 - 80 %
10 - 20 %	"Reasonable indication"	"Clear showing"	80 - 90 %
1 - 10 %	"Reasonable, articulable grounds for suspicion"	"Clear and convincing evidence"	90 - 99 %
< 1 %	"No reasonable grounds for suspicion", "Inchoate hunch", "Fanciful conjecture"	"Beyond a reasonable doubt"	99 %
0 %	No grounds even for a hunch	Virtually certain	100 %
	If A is Thought to be True to this Standard ...	... Then (-A) Cannot Reasonably be Thought to be True to the Standard Listed Next Below.	

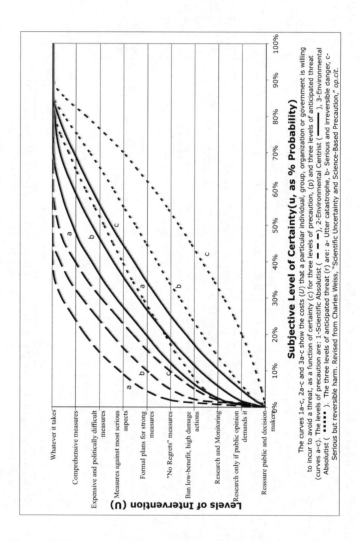

Fig. 1. Willingness to Accept Cost to Avert an Uncertain Threat

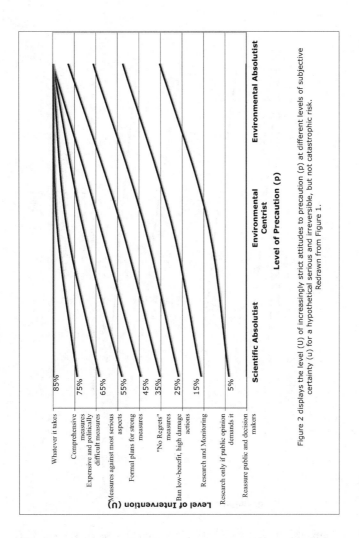

Fig. 2. The Cost of Increasing Precaution as a Function of Increasing Certainty of a Threat of Serious and Irreversible Damage
Curves 1-11 display the cost (U) of increasingly strict attitudes to precaution (p) at different levels of subjective certainty (c) for a hypothetical serious and irreversible, but not catastrophic risk. The levels of certainty and the cost axis are those of Figure 1, curves 1b-3b.

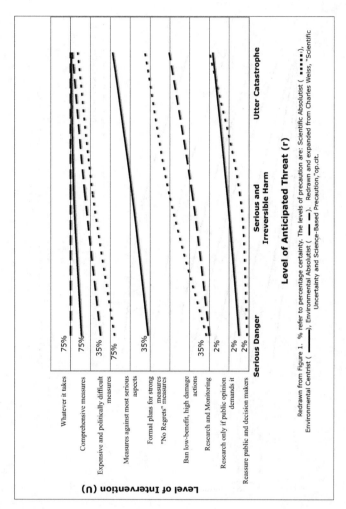

Fig. 3. Willingness to Accept Costs as a Function of Increasing Certainty and Anticipated Threat

Lecture Notes in Economics and Mathematical Systems

For information about Vols. 1–483
please contact your bookseller or Springer-Verlag

Vol. 495: I. Konnov, Combined Relaxation Methods for Variational Inequalities. XI, 181 pages. 2001.

Vol. 496: P. Weiß, Unemployment in Open Economies. XII, 226 pages. 2001.

Vol. 497: J. Inkmann, Conditional Moment Estimation of Nonlinear Equation Systems. VIII, 214 pages. 2001.

Vol. 498: M. Reutter, A Macroeconomic Model of West German Unemployment. X, 125 pages. 2001.

Vol. 499: A. Casajus, Focal Points in Framed Games. XI, 131 pages. 2001.

Vol. 500: F. Nardini, Technical Progress and Economic Growth. XVII, 191 pages. 2001.

Vol. 501: M. Fleischmann, Quantitative Models for Reverse Logistics. XI, 181 pages. 2001.

Vol. 502: N. Hadjisavvas, J. E. Martínez-Legaz, J.-P. Penot (Eds.), Generalized Convexity and Generalized Monotonicity. IX, 410 pages. 2001.

Vol. 503: A. Kirman, J.-B. Zimmermann (Eds.), Economics with Heterogenous Interacting Agents. VII, 343 pages. 2001.

Vol. 504: P.-Y. Moix (Ed.), The Measurement of Market Risk. XI, 272 pages. 2001.

Vol. 505: S. Voß, J. R. Daduna (Eds.), Computer-Aided Scheduling of Public Transport. XI, 466 pages. 2001.

Vol. 506: B. P. Kellerhals, Financial Pricing Models in Con-tinuous Time and Kalman Filtering. XIV, 247 pages. 2001.

Vol. 507: M. Koksalan, S. Zionts, Multiple Criteria Decision Making in the New Millenium. XII, 481 pages. 2001.

Vol. 508: K. Neumann, C. Schwindt, J. Zimmermann, Project Scheduling with Time Windows and Scarce Resources. XI, 335 pages. 2002.

Vol. 509: D. Hornung, Investment, R&D, and Long-Run Growth. XVI, 194 pages. 2002.

Vol. 510: A. S. Tangian, Constructing and Applying Objective Functions. XII, 582 pages. 2002.

Vol. 511: M. Külpmann, Stock Market Overreaction and Fundamental Valuation. IX, 198 pages. 2002.

Vol. 512: W.-B. Zhang, An Economic Theory of Cities.XI, 220 pages. 2002.

Vol. 513: K. Marti, Stochastic Optimization Techniques. VIII, 364 pages. 2002.

Vol. 514: S. Wang, Y. Xia, Portfolio and Asset Pricing. XII, 200 pages. 2002.

Vol. 515: G. Heisig, Planning Stability in Material Requirements Planning System. XII, 264 pages. 2002.

Vol. 516: B. Schmid, Pricing Credit Linked Financial Instruments. X, 246 pages. 2002.

Vol. 517: H. I. Meinhardt, Cooperative Decision Making in Common Pool Situations. VIII, 205 pages. 2002.

Vol. 518: S. Napel, Bilateral Bargaining. VIII, 188 pages. 2002.

Vol. 519: A. Klose, G. Speranza, L. N. Van Wassenhove (Eds.), Quantitative Approaches to Distribution Logistics and Supply Chain Management. XIII, 421 pages. 2002.

Vol. 520: B. Glaser, Efficiency versus Sustainability in Dynamic Decision Making. IX, 252 pages. 2002.

Vol. 521: R. Cowan, N. Jonard (Eds.), Heterogenous Agents, Interactions and Economic Performance. XIV, 339 pages. 2003.

Vol. 522: C. Neff, Corporate Finance, Innovation, and Strategic Competition. IX, 218 pages. 2003.

Vol. 523: W.-B. Zhang, A Theory of Interregional Dynamics. XI, 231 pages. 2003.

Vol. 524: M. Frölich, Programme Evaluation and Treatment Choise. VIII, 191 pages. 2003.

Vol. 525: S. Spinler, Capacity Reservation for Capital-Intensive Technologies. XVI, 139 pages. 2003.

Vol. 526: C. F. Daganzo, A Theory of Supply Chains. VIII, 123 pages. 2003.

Vol. 527: C. E. Metz, Information Dissemination in Currency Crises. XI, 231 pages. 2003.

Vol. 528: R. Stolletz, Performance Analysis and Optimization of Inbound Call Centers. X, 219 pages. 2003.

Vol. 529: W. Krabs, S. W. Pickl, Analysis, Controllability and Optimization of Time-Discrete Systems and Dynamical Games. XII, 187 pages. 2003.

Vol. 530: R. Wapler, Unemployment, Market Structure and Growth. XXVII, 207 pages. 2003.

Vol. 531: M. Gallegati, A. Kirman, M. Marsili (Eds.), The Complex Dynamics of Economic Interaction. XV, 402 pages, 2004.

Vol. 532: K. Marti, Y. Ermoliev, G. Pflug (Eds.), Dynamic Stochastic Optimization. VIII, 336 pages. 2004.

Vol. 533: G. Dudek, Collaborative Planning in Supply Chains. X, 234 pages. 2004.

Vol. 534: M. Runkel, Environmental and Resource Policy for Consumer Durables. X, 197 pages. 2004.

Vol. 535: X. Gandibleux, M. Sevaux, K. Sörensen, V. T'kindt (Eds.), Metaheuristics for Multiobjective Optimisation. IX, 249 pages. 2004.

Vol. 536: R. Brüggemann, Model Reduction Methods for Vector Autoregressive Processes. X, 218 pages. 2004.

Vol. 537: A. Esser, Pricing in (In)Complete Markets. XI, 122 pages, 2004.

Vol. 538: S. Kokot, The Econometrics of Sequential Trade Models. XI, 193 pages. 2004.

Vol. 539: N. Hautsch, Modelling Irregularly Spaced Financial Data. XII, 291 pages. 2004.

Vol. 540: H. Kraft, Optimal Portfolios with Stochastic Interest Rates and Defaultable Assets. X, 173 pages. 2004.

Vol. 541: G.-y. Chen, X. Huang, X. Yang, Vector Optimization. X, 306 pages. 2005.

Vol. 542: J. Lingens, Union Wage Bargaining and Economic Growth. XIII, 199 pages. 2004.

Vol. 543: C. Benkert, Default Risk in Bond and Credit Derivatives Markets. IX, 135 pages. 2004.

Vol. 544: B. Fleischmann, A. Klose, Distribution Logistics. X, 284 pages. 2004.

Vol. 545: R. Hafner, Stochastic Implied Volatility. XI, 229 pages. 2004.

Vol. 546: D. Quadt, Lot-Sizing and Scheduling for Flexible Flow Lines. XVIII, 227 pages. 2004.

Vol. 547: M. Wildi, Signal Extraction. XI, 279 pages. 2005.

Vol. 548: D. Kuhn, Generalized Bounds for Convex Multistage Stochastic Programs. XI, 190 pages. 2005.

Vol. 549: G. N. Krieg, Kanban-Controlled Manufacturing Systems. IX, 236 pages. 2005.

Vol. 550: T. Lux, S. Reitz, E. Samanidou, Nonlinear Dynamics and Heterogeneous Interacting Agents. XIII, 327 pages. 2005.

Vol. 551: J. Leskow, M. Puchet Anyul, L. F. Punzo, New Tools of Economic Dynamics. XIX, 392 pages. 2005.

Vol. 552: C. Suerie, Time Continuity in Discrete Time Models. XVIII, 229 pages. 2005.

Vol. 553: B. Mönch, Strategic Trading in Illiquid Markets. XIII, 116 pages. 2005.

Vol. 554: R. Foellmi, Consumption Structure and Macroeconomics. IX, 152 pages. 2005.

Vol. 555: J. Wenzelburger, Learning in Economic Systems with Expectations Feedback (planned) 2005.

Vol. 556: R. Branzei, D. Dimitrov, S. Tijs, Models in Cooperative Game Theory. VIII, 135 pages. 2005.

Vol. 557: S. Barbaro, Equity and Efficiency Considerations of Public Higer Education. XII, 128 pages. 2005.

Vol. 558: M. Faliva, M. G. Zoia, Topics in Dynamic Model Analysis. X, 144 pages. 2005.

Vol. 559: M. Schulmerich, Real Options Valuation. XVI, 357 pages. 2005.

Vol. 560: A. von Schemde, Index and Stability in Bimatrix Games. X, 151 pages. 2005.

Vol. 561: H. Bobzin, Principles of Network Economics. XX, 390 pages. 2006.

Vol. 562: T. Langenberg, Standardization and Expectations. IX, 132 pages. 2006.

Vol. 563: A. Seeger (Ed.), Recent Advances in Optimization. XI, 455 pages. 2006.

Vol. 564: P. Mathieu, B. Beaufils, O. Brandouy (Eds.), Artificial Economics. XIII, 237 pages. 2005.

Vol. 565: W. Lemke, Term Structure Modeling and Estimation in a State Space Framework. IX, 224 pages. 2006.

Vol. 566: M. Genser, A Structural Framework for the Pricing of Corporate Securities. XIX, 176 pages. 2006.

Vol. 567: A. Namatame, T. Kaizouji, Y. Aruga (Eds.), The Complex Networks of Economic Interactions. XI, 343 pages. 2006.

Vol. 568: M. Caliendo, Microeconometric Evaluation of Labour Market Policies. XVII, 258 pages. 2006.

Vol. 569: L. Neubecker, Strategic Competition in Oligopolies with Fluctuating Demand. IX, 233 pages. 2006.

Vol. 570: J. Woo, The Political Economy of Fiscal Policy. X, 169 pages. 2006.

Vol. 571: T. Herwig, Market-Conform Valuation of Options. VIII, 104 pages. 2006.

Vol. 572: M. F. Jäkel, Pensionomics. XII, 316 pages. 2006

Vol. 573: J. Emami Namini, International Trade and Multinational Activity, X, 159 pages, 2006.

Vol. 574: R. Kleber, Dynamic Inventory Management in Reverse Logisnes, XII, 181 pages, 2006.

Vol. 575: R. Hellermann, Capacity Options for Revenue Management, XV, 199 pages, 2006.

Vol. 576: J. Zajac, Economics Dynamics, Information and Equilibnum, X, 284 pages, 2006.

Vol. 577: K. Rudolph, Bargaining Power Effects in Financial Contracting, XVIII, 330 pages, 2006.

Vol. 578: J. Kühn, Optimal Risk-Return Trade-Offs of Commercial Banks, IX, 149 pages, 2006.

Vol. 579: D. Sondermann, Introduction to Stochastic Calculus for Finance, X, 136 pages, 2006.

Vol. 580: S. Seifert, Posted Price Offers in Internet Auction Markets, IX, 186 pages, 2006.

Vol. 581: K. Marti; Y. Ermoliev; M. Makowsk; G. Pflug (Eds.), Coping with Uncertainty, XIII, 330 pages, 2006 (planned).

Vol. 582: J. Andritzky, Sovereign Default Risks Valuation: Implications of Debt Crises and Bond Restructurings. VIII, 251 pages, 2006 (planned).

Vol. 583: I.V. Konnov, D.T. Luc, A.M. Rubinov (Eds.), Generalized Convexity and Related Topics, IX, 469 pages, 2006 (planned).

Vol. 584: C. Bruun, Adances in Artificial Economics: The Economy as a Complex Dynamic System. XVI, 296 pages, 2006.

Vol. 585: R. Pope, J. Leitner, U. Leopold-Wildburger, The Knowledge Ahead Approach to Risk, XVI, 218 pages, 2007 (planned).

Vol. 586: B. Lebreton, Strategic Closed-Loop Supply Chain Management. X, 150 pages, 2007 (planned).